MAGNETOELECTRIC INTERACTION PHENOMENA IN CRYSTALS

MAGNETOELECTRIC INTERACTION PHENOMENA IN CRYSTALS

Edited by

ARTHUR J. FREEMAN
Northwestern University
Evanston, Illinois,
U.S.A.

and

HANS SCHMID
Battelle Geneva Research Center,
Carouge, Geneva,
Switzerland

GORDON AND BREACH SCIENCE PUBLISHERS
London New York Paris

Editorial office for the United States of America
Gordon and Breach, Science Publishers, Inc.
One Park Avenue
New York, N.Y. 10016

Editorial office for France
Gordon & Breach
7-9 rue Emile Dubois
Paris 75014

The papers published in this book are the Proceedings of a Symposium held at Battelle Seattle Research Center, Seattle, Washington U.S.A., May 21–24, 1973

The symposium was sponsored by Battelle Memorial Institute, Columbus, Ohio. In accord with that part of the charge of its founder, Gordon Battelle, to assist in the further education of men, it is the commitment of Battelle to encourage the distribution of information. This is done in part by supporting conferences and meetings and by encouraging the publication of reports and proceedings.

The papers are also available in the *International Journal of Magnetism* in individual issues of Volumes 4, 5 and 6.

Library of Congress catalog card number 74 15759. ISBN 0 677 15985 4. Printed in Great Britain.

Preface

This book contains the invited lectures, the contributed seminar papers, and selected discussion contributions of the Symposium on Magnetoelectric Interaction Phenomena in Crystals, held at the Battelle Seattle Research Center, Seattle, Washington, U.S.A., 21–24 May 1973.

The magnetoelectric effect in crystals—i.e. the induction of a magnetization by means of an electric field and the induction of a polarization by means of a magnetic field—was already presumed to exist by Pierre Curie in the past century. It has been searched for in vain by various workers early in this century. On the basis of symmetry considerations—taking time reversal symmetry into account—Landau and Lifshitz (1956) came, however, to the conclusion that the phenomenon "can in principle exist." Shortly after that statement, Dzialoshinsky pointed out that the symmetry of antiferromagnetic Cr_2O_3 would be a good candidate. This was rapidly followed by the experimental demonstration of the phenomenon by Astrov (1960) and Folen, Rado and Stalder (1961). Since then many other magnetoelectric materials were discovered. Stimulated by the announcement by Smolensky and Joffe (1958) of the synthesis of the first antiferromagnetic ferroelectric perovskites, and in an attempt to find interesting magnetoelectric properties, many workers tried to find antiferromagnetically and ferromagnetically ordered ferroelectrics and antiferroelectrics. By these efforts a large variety of magnetoelectrically interesting compounds have been discovered, part of which are remarkable because of interactions between spontaneous polarization (or antipolarization) and spontaneous magnetization (or antimagnetization).

It was therefore considered as timely to bring together the various specialists in the field, to assess achievements and particularly the problems remaining to be solved, thereby to stimulate the advancement of this polyvalent field of solid state physics and, last but not least, to cultivate invaluable personal contacts between workers of diverse background.

It is a pleasure for me to express my warm thanks to all contributors to and participants in this symposium, all of whom have worked hard and done their best to come close to the objectives set out.

Although the magnetoelectric effect has not yet made inroads in commercial applications, it has already become an important new tool for solving scientific problems; its helpfulness is now definitely established for clarifying magnetic symmetries as a complementary tool for neutron diffraction, for precise determination of antiferromagnetic phase transitions, study of the critical behaviour of the magnetoelectric effect close to phase transitions, the study of spin flop and metamagnetic transitions, and the production and study of single domains of antiferromagnetic phases. It became clear, during this symposium, that magnetoelectric materials were not rare by far. Hence important developments in the future can be foreseen, both in basic understanding as well as in scientific and technological applications.

Through the kind and helpful cooperation of the *International Journal of Magnetism* it has been possible to refine the papers by cycling them via the standard refereeing procedure. This has caused some time delay, of course, but it is hoped that it will be largely compensated by the advantages gained. Special thanks are due to Professor A. J. Freeman and Mrs. G. Penovich, of the *International Journal of Magnetism*, for their perservering help, and to Gordon and Breach Science Publishers for their conscientious work.

Thanks are also due to Drs. V. J. Folen and G. T. Rado for their help in polishing up the discussion contributions.

Grateful acknowledgment is extended to Mr. L. M. Bonnefond, the local coordinator of the symposium at Battelle Seattle Research Center, whose never fading helpfulness was so essential for this event and the Proceedings to materialize, and who unfailingly did everything—together with Julie Swor and Virginia Ritchey as secretaries—for the participants to feel so comfortable throughout the meeting.

At Battelle-Geneva it was Dr. E. Ascher's many suggestions and stimulation in discouraging moments which proved so essential during the preparation of the symposium, and Mrs. L. Amiguet and Mrs. R. Gygi, who had to cope with all the lengthy correspondence.

To all goes my deep-felt gratitude.

HANS SCHMID
Geneva

Table of Contents

† Invited lectures.

PART IV APPLICATIONS AND MISCELLANEOUS

Part I

Microscopic Mechanisms and Theories

PRESENT STATUS OF THE THEORY OF MAGNETOELECTRIC EFFECTS†

GEORGE T. RADO

Naval Research Laboratory, Washington, D.C. 20375

The present understanding of the physical origins of magnetoelectric (ME) effects in crystals is assessed. Thus the emphasis is on atomic ME mechanisms, on their temperature dependences and on comparisons of their predictions with experiments. All these mechanisms involve changes due to an applied electric field of some parameter in a spin Hamiltonian. A detailed example is given of an atomic ME mechanism applicable to materials containing iron group transition metal ions having an orbital singlet ground state and of an atomic ME mechanism applicable to materials containing rare earth ions. The applicability of ME effects to determinations of the critical behavior of the sublattice magnetization is described. Also presented is a discussion of the thermodynamic theory of ME effects in antiferromagnets and in ferromagnets, and a thermodynamic treatment of ME effects arising from the interplay of piezomagnetism and piezoelectricity. Remarks on the history of ME effects and on some additional topics are also included.

I INTRODUCTION

The main purpose of this paper is to assess the present understanding of the physical origins of magnetoelectric (ME) effects in crystals. Thus its emphasis is on atomic ME mechanisms, on their temperature dependences and on comparisons of their predictions with experiments. The term "ME" denotes, unless specifically qualified, the linear magnetoelectric effect only, i.e., the appearance of an induced magnetization which is proportional to an applied electric field and the appearance of an induced electric polarization which is proportional to an applied magnetic field.

After some remarks on the history of ME effects (Section II), we introduce the various ME tensors by briefly discussing linear and non-linear ME effects from the point of view of thermodynamics and magnetic symmetry (Section III). While the form of these tensors can be predicted on the basis of the magnetic point group (i.e., directional symmetry) of a given crystal, the measured values of the tensor components must be explained by means of atomic theories. Also to be noted is that thermodynamic treatments of ME effects in saturated ferromagnets differ from those in antiferromagnets.

Turning then to the known atomic mechanisms of ME effects (Sections IV and V), we point out that they all involve changes due to an electric field of some parameter in a spin Hamiltonian and that they apply to two types of materials, to be referred to as type 1 and type 2, respectively. In type 1 materials (Section IV) the magnetic ions are iron group transition metal ions whose ground state in the total crystalline electric field is an orbital singlet whereas in type 2 materials (Section V) the magnetic ions are rare-earth ions. We present and discuss an example of an atomic ME mechanism applicable to each of these types of materials. The applicability of ME effects to determinations of the critical behavior of the sublattice magnetization is also described.

A phenomenological rather than an atomic mechanism of ME effects is the interplay of piezomagnetism and piezoelectricity (Section VI). We show by means of a thermodynamic treatment that this mechanism must exist in any piezoelectric ferromagnet if it is unclamped and magnetically saturated. Brief comments on various additional topics (Section VII) are also presented.

II. HISTORICAL REMARKS

The crucial event in the early history of ME effects is probably the work of Landau and Lifshitz[1] who pointed out that the time-reversal transformation is not an independent symmetry element of a

† Presented at the "Symposium on Magnetoelectric Interaction Phenomena in Crystals," Battelle Seattle Research Center, Seattle, Washington, U.S.A., May 21–24, 1973.

magnetically ordered crystal. This discovery led them to the concept of magnetic symmetry and to the assertion that if one uses Neumann's principle (which states that the symmetry of a property tensor is that of the crystal) then the symmetry of the crystal is to be taken as that of its state, i.e., as the symmetry of the distribution of electric charge and magnetic moment. The symmetry of a magnetically ordered crystal is actually lower than that of its Hamiltonian, and this is a manifestation of the fact that such a crystal is not in a true equilibrium state even though it does appear to be in equilibrium for the purposes of most measurements. On the basis of these considerations, Landau and Lifshitz[1] predicted the possible existence of two effects which had previously been thought to be forbidden by symmetry. These new effects are piezomagnetism and ME effects, both of which can occur in magnetically ordered crystals possessing an appropriate magnetic symmetry.

The first explicit prediction of ME effects in a specific material was made by Dzyaloshinskii[2] who showed that in Cr_2O_3 these effects are allowed by the magnetic symmetry. Experimentally, an ME effect was first observed by Astrov[3] on an unoriented crystal of Cr_2O_3 and by Folen, Rado and Stalder[4] who used an oriented crystal of Cr_2O_3 ro reveal the anisotropy of the ME effect. Both of these experiments involved what is now referred to as the electrically induced ME effect [$(ME)_E$ effect]. A few months later, Rado and Folen[5] first observed the converse effect, the one which is now referred to as the magnetically induced ME effect [$(ME)_H$ effect]. From a thermodynamic standpoint, of course, it is obvious that if one of these effects exists then the other must exist also. From an experimental standpoint, however, the situation is not so simple because ME effects were found to vanish if the sample consists of equal volumes of antiferromagnetic domains which are related to each other by time reversal. The important role of antiferromagnetic domains in the observation of ME effects was, in fact, discovered in this way.[5,6] Also worth mentioning is that in the summer of 1958, when an English translation of the Landau-Lifshitz book[1] was not yet available, an unsuccessful search for non-linear as well as linear ME effects was carried out experimentally by Rado and Folen. They were motivated by their discovery[7] that the magnetocrystalline anisotropy of some ferrites is a "one-ion" type crystalline field effect and by the present writer's proposal that this effect be changed by supplementing the crystalline electric field with an applied electric field. The first atomic theory of ME effects is that proposed by Rado[8,9] and does, in fact, involve a one-ion-type mechanism.

III THERMODYNAMICS AND MAGNETIC SYMMETRY

In reviewing the thermodynamic theory of ME effects in antiferromagnets,[1,2,10] we start with a certain thermodynamic potential F which is that Legendre transform of the Helmholtz free energy in which the magnetization **M** and the electric polarization **P** are replaced as independent variables by the magnetic field **H** and the electric field **E**, respectively. Under the conditions specified below, an expansion of F to second order in the components of **E** and **H** yields

$$F = F_0 - \tfrac{1}{2}\kappa_{ij}E_iE_j - \tfrac{1}{2}\chi_{ij}H_iH_j - \alpha_{ij}E_iH_j \tag{1}$$

where F_0 is independent of the components of **E** and **H** and each of the subscripts i, j denotes any of the Cartesian axes x, y, z. Here and elsewhere in this paper, summation over repeated subscripts is understood. We assume, for simplicity, the absence of a term proportional to E_i, i.e., the absence of pyroelectricity, and we note that a term proportional to H_i is excluded because in an antiferromagnet the net magnetic moment is zero. Use of the relations

$$\mathbf{P} = -\,\partial F/\partial \mathbf{E}; \qquad \mathbf{M} = -\,\partial F/\partial \mathbf{H} \tag{2}$$

then leads to the constitutive equations

$$P_i = \kappa_{ij}E_j + \alpha_{ij}H_j \tag{3}$$

$$M_i = \chi_{ji}H_j + \alpha_{ji}E_j \tag{4}$$

and the commutativity of partial differentiation yields

$$\kappa_{ij} = \kappa_{ji}; \qquad \chi_{ij} = \chi_{ji}. \tag{5}$$

The ME susceptibility tensor $\boldsymbol{\alpha}$, which describes the linear ME effects, does not have any intrinsic symmetry. It should also be noted that if the linear ME term in Eq. (1) is written as $-\hat{\alpha}_{ij}H_iE_j$, the $\hat{\alpha}_{ij}$ equals α_{ji}. If the expansion of F is carried to third order, then one obtains terms of the form $-\frac{1}{2}\beta_{ijk}E_iH_jH_k$ and $-\frac{1}{2}\gamma_{ijk}E_iE_jH_k$ which give rise to non-linear ME effects. The property tensors κ, χ, α, β, γ must be invariant under the symmetry operations of the magnetic point group (i.e., directional symmetry) of a given crystal. If this group is known from neutron diffraction or from other experiments, then the invariance of a property tensor can be ascertained by the use of tables of the type given by Birss.[11] In his notation $\boldsymbol{\alpha}$ is an axial c-tensor of rank 2, β is a polar i-tensor of rank 3, and $\boldsymbol{\gamma}$ is an axial c-tensor of rank 3. As an example, we recall that the magnetic point group of Cr_2O_3 is $\bar{3}'m'$ (where the prime denotes the time reversal transformation) so that the matrix of $\boldsymbol{\alpha}$ has the form

$$(\alpha) = \begin{pmatrix} \alpha_\perp & 0 & 0 \\ 0 & \alpha_\perp & 0 \\ 0 & 0 & \alpha_\parallel \end{pmatrix} \tag{6}$$

provided the z-axis is chosen to be parallel to the principal axis of the crystal.

Next to be discussed is the thermodynamic theory of ME effects in ferromagnets,[12,13,11] a theory which differs significantly from that in antiferromagnets. The essential point is that the expansion embodied in Eq. (1) assumes the changes in F due to **E** and **H** to be small. In a ferromagnet, however, this assumption is not fulfilled because the changes in the magnetic structure (and hence in F) due to **H** can be quite large. The simplest situation occurs when the ferromagnet is magnetically saturated so that **M** is always parallel to **H.** In that case the magnetic symmetry depends on the orientation of **H** with respect to the crystallographic axes. No matter how **H** (and hence **M**) is rotated, however, **M** always "sees" the crystallographic rather than the magnetic symmetry.

As an example, we consider the case of $GaFeO_3$, the first ferromagnetic (actually ferrimagnetic) material found to exhibit ME effects.[12,13] We choose Cartesian axes x, y, z along the crystallographic axes a, b, c respectively, and note that b is the polar axis. The crystallographic point group is $m2m$ and the magnetic point group in the absence of applied fields is $m'2'm$. We introduce the unit vector $\mathbf{u} = \mathbf{M}/|\mathbf{M}|$ and assume that the only non-vanishing component of **E** is E_y. For any direction **u** of **M**, we expand F about that (extrapolated) spontaneous magnetization which corresponds to **u.** We keep solely the terms linear in E_y and bilinear in the components of **u** and **H.** Since this expansion is intended to be valid in saturating fields only, we put $\mathbf{H} = |H|\mathbf{u}$ and obtain

$$F = F_0 - \eta_{iiy}u_i^2 E_y - \alpha_{iiy}u_i^2|H|E_y - \beta_{iiy}u_i^2 H^2 E_y. \tag{7}$$

Next we recall the condition $u_i u_i = 1$ and introduce the abbreviations

$$\eta_0 = \eta_{zzy}; \quad \eta_1 = \eta_{xxy} - \eta_{zzy}; \quad \eta_2 = \eta_{yyy} - \eta_{zzy} \tag{8}$$

as well as analogous abbreviations for $\alpha_0, \alpha_1, \alpha_2$ and for $\beta_0, \beta_1, \beta_2$. With the use of $P_y = -\partial F/\partial E_y$, we then find that Eq. (7) yields

$$P_y = (\eta_0 + \eta_1 u_x^2 + \eta_2 u_y^2) + (\alpha_0 + \alpha_1 u_x^2 + \alpha_2 u_y^2)|H| + (\beta_0 + \beta_1 u_x^2 + \beta_2 u_y^2)H^2. \tag{9}$$

Thus P_y contains, for any orientation of **H**, a field-independent term, a term linear in $|H|$, and a term quadratic in $|H|$. The data of Figure 1 (which refer to 77°K) clearly show that if **H** is along the c (or easy) axis, then the voltage which measures P_y is proportional to $|H|$ provided $|H|$ is sufficiently large. The reason P_y is proportional to $|H|$ (rather than proportional to H, as in an antiferromagnet) is that in a ferromagnet a reversal of **H** causes a reversal of **M.** For sufficiently small values of $|H|$, however, the effects of domain structure cause the P_y *vs.* $|H|$ curve to be non-linear, as shown by the inset of Figure 1. A somewhat surprising result is that the application of **H** along the a (or intermediate) axis produces a P_y which is quadratic in $|H|$ but nevertheless quite large. The fact that this result also is a manifestation of domain structure is shown by

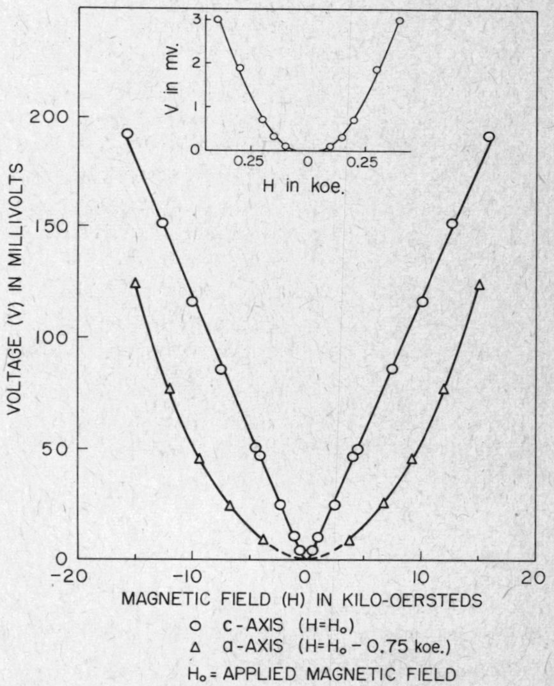

FIGURE 1 Measured dependence of the voltage V on the static magnetic field H (-15 kOe $< H <$ 15 kOe) at 77°K in $Ga_{0.92}Fe_{1.08}O_3$. Here V is induced along the b axis and H_0 is applied along the c axis (circles) or along the a axis (triangles). [After Rado, Ref. 12].

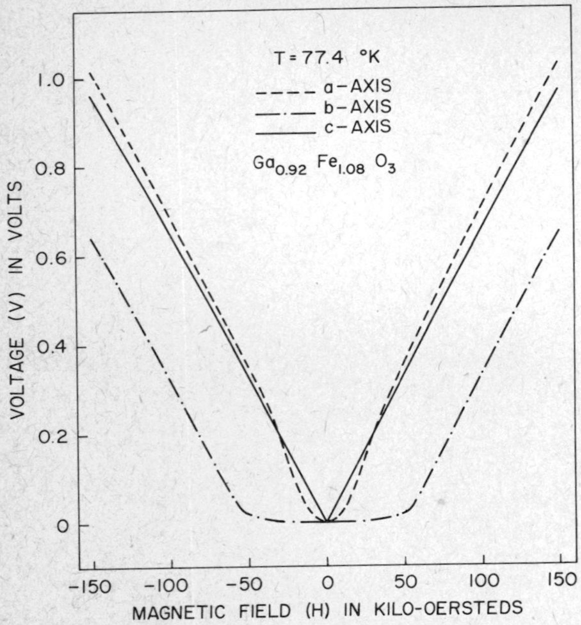

FIGURE 2 Measured dependence of the voltage V on the static magnetic field H (-150 kOe $< H <$ 150 kOe) at 77.4°K in $Ga_{0.92}Fe_{1.08}O_3$. Here V is induced along the b axis and H is applied along the a axis (dashes), along the b axis (dots and dashes), or along the c axis (full line). [After Rado, Ref. 13].

the data of Figure 2 which refer to 77°K and include $|H|$ values 10 times larger than those used in Figure 1. It is seen from Figure 2 that the linear term of Eq. (9) predominates, as expected, for all three principal crystallographic directions, and that its magnitude is essentially isotropic. Although Eq. (9) provides a good account of the ME data taken at 77°K (and also at 4.2°K), its applicability is limited by the fact that it assumes the spontaneous magnetization to be isotropic. Experiments by Schelleng and Rado[14] have shown that at higher temperatures the spontaneous magnetization of $GaFeO_3$ is highly anisotropic.

IV ATOMIC MECHANISMS OF ME EFFECTS IN TYPE 1 MATERIALS

In turning to the central topic of this paper, namely the known atomic mechanisms of ME effects, we note that all these mechanisms consist of variations of the first proposed ME mechanism[8,9] in that they involve changes due to the electric field **E** of some parameter in a spin Hamiltonian. The atomic ME mechanisms treated so far apply to two types of materials, to be referred to as type 1 and type 2, which represent limiting situations in regard to the relative importance of the crystalline electric field and the spin-orbit coupling in determining the energy levels of the magnetic ions.

The magnetic ions in type 1 materials are iron group transition metal ions whose ground state in the total crystalline electric field is an orbital singlet. It is this type of material (specifically Cr_2O_3) which is considered in the first ME theory.[8,9] An additional assumption is that the width in energy of the ground state spin multiplet is small compared to the energy separation between the orbital ground state and the first excited orbital state. The mechanism for $\boldsymbol{\alpha}$ proposed in this theory involves changes due to **E** of the quadratic axial term ("*D* term") in the spin Hamiltonian. Also introduced by this theory is the use[9] of thermodynamic perturbation theory for calculating the components of $\boldsymbol{\alpha}$ and their temperature dependences.

As an example of this theory, we briefly outline the calculation by Rado[9] of $\alpha_\perp$ for Cr_2O_3. The underlying ME mechanism[8,9] (which contributes to $\alpha_\parallel$ as well as to $\alpha_\perp$) is variously referred to as the "*D*-term," "spin-orbit," or "one-ion anisotropy" mechanism. It appears at present that in Cr_2O_3 it is this mechanism which dominates $\alpha_\perp$.

We consider a two-sublattice antiferromagnet of unit volume containing $N/2$ identical magnetic ions in its "+ sublattice" and also in its "– sublattice." Next we choose some rectangular coordinate system ξ, η, ζ, whose positive ζ direction is parallel to the magnetization of one of the sublattices. This sublattice will be referred to as the + sublattice. In addition, we find it convenient to introduce a separate rectangular coordinate system x, y, z for each of the sublattices. By definition, the positive x, y, z directions of the + sublattice coordinate system are parallel to the positive ξ, η, ζ directions, whereas the positive x, y, z directions of the – sublattice coordinate system are antiparallel to the positive ξ, η, ζ directions. We take the direction of quantization for each sublattice to be the positive z direction of its own coordinate system, and we assume that the external electric and magnetic fields, **E** and **H**, are applied parallel to the positive ξ direction.

To each magnetic ion in a sublattice (specified by the superscript + or –) we then assign the spin Hamiltonian

$$\mathcal{H}^\pm = -g\mu_B(S_xH_x^{\text{eff}\pm} + S_zH_z^{\text{eff}\pm}) \pm g\mu_B(a_\perp/2)(S_xS_z + S_zS_x)E_\xi \tag{10}$$

which contains a Zeeman term arising from the effective magnetic field $\mathbf{H}^{\text{eff}\pm}$ and a symmetrized magnetoelectric term embodying the spin–orbit mechanism. The symbols g, μ_B, and $\mathbf{S}$ denote, respectively, the spectroscopic splitting factor, the Bohr magneton, and the spin operator. On the basis of fourth-order perturbation theory, the order of magnitude of the coefficient $a_\perp$ pertaining to either sublattice is given by the typical term

$$a_\perp \approx \frac{E_z{}^c e^2 \lambda^2 \langle i|z|j\rangle\langle j|x|k\rangle\langle k|L_z|l\rangle\langle l|L_x|i\rangle}{g\mu_B \Delta_1 \Delta_2 \Delta_3} \tag{11}$$

where e, λ, and $\mathbf{L}$ denote, respectively, the electronic charge, the spin-orbit parameter, and the orbital angular momentum operator. Here $-E_z{}^c z$ is the linear portion of the crystalline electric potential, quantities of the type $\langle i|z|j\rangle$ are appropriate matrix elements, and Δ_1, Δ_2, and Δ_3 are differences between appropriate eigenvalues of the unperturbed part of $\mathcal{H}^\pm$. Since $\langle i|z|j\rangle$ vanishes except between states of different parity, one of these Δ's will usually represent an optical splitting. Although the numerical value of $|a_\perp|$ can be estimated, we prefer to treat $a_\perp$ as a parameter which is to be determined by means of suitable experiments.

The effective fields appearing in Eq. (10) will now be evaluated by means of the molecular field approximation. We define $H_x^{\text{eff}\pm}$ by

$$H_x^{\text{eff}\pm} = H_x^{\text{mol}\pm} \pm H_\xi \tag{12}$$

and use the molecular field relation

$$\mathbf{H}^{\text{mol}\pm} = -A\mathbf{M}^\mp - \Gamma\mathbf{M}^\pm \tag{13}$$

to write Eq. (12) in the form

$$H_x^{\text{eff}\pm} = AM_x{}^\mp - \Gamma M_x{}^\pm \pm H_\xi \tag{14}$$

where A and Γ are molecular field coefficients describing the inter–and intra–sublattice exchange interactions, respectively, and $\mathbf{M}^\pm$ is the sublattice magnetization. Since the applied fields are small, it is appropriate to make the usual assumption that half of the induced magnetization

$$M_\xi = M_x{}^+ - M_x{}^- \tag{15}$$

is associated with each sublattice. Thus we may put

$$M_x{}^\pm = \pm\tfrac{1}{2}M_\xi \tag{16}$$

which may be combined with Eq. (14) to give

$$H_x^{\text{eff}\pm} = \mp\tfrac{1}{2}(A+\Gamma)M_\xi \pm H_\xi. \tag{17}$$

If we now use the constitutive equation

$$M_\xi = \chi_\perp H_\xi + \alpha_\perp E_\xi \tag{18}$$

then Eq. (17) becomes

$$H_x^{\text{eff}\pm} = \pm[1 - \tfrac{1}{2}(A+\Gamma)\chi_\perp]H_\xi \mp \tfrac{1}{2}(A+\Gamma)\alpha_\perp E_\xi \tag{19}$$

which shows that ME effects cause $H_x^{\text{eff}\pm}$ to depend on E_ξ as well as on H_ξ. Defining $H_z^{\text{eff}\pm}$ by $H_z^{\text{eff}\pm} = H_z^{\text{mol}\pm}$, we then use Eq. (13) to obtain

$$H_z^{\text{eff}\pm} = (A - \Gamma)M_{0z} \tag{20}$$

where M_{0z} denotes the spontaneous magnetization of either sublattice.

Next we decompose $\mathcal{H}^\pm$ into a field-independent part $\mathcal{H}_0$ and a field-dependent part $V^\pm$ and regard $\mathcal{H}_0$ as the unperturbed Hamiltonian and $V^\pm$ as the perturbation. After substituting Eqs. (19) and (20) into Eq. (10) we have

$$\mathcal{H}^\pm = \mathcal{H}_0 + V^\pm \tag{21}$$

$$\mathcal{H}_0 = -g\mu_B(A - \Gamma)M_{0z}S_z \tag{22}$$

$$V^\pm = \pm g\mu_B\{\tfrac{1}{2}[(A+\Gamma)\alpha_\perp S_x + \alpha_\perp(S_xS_z + S_zS_x)]E_\xi - [1 - \tfrac{1}{2}(A+\Gamma)\chi_\perp]S_xH_\xi\}. \tag{23}$$

Since we work in a representation in which S_z is diagonal, the matrix elements of $V^\pm$ have the form

$$V_{nm}^\pm \equiv \langle n|V^\pm|m\rangle = \pm V_{nm}(1 - \delta_{nm}) \tag{24}$$

where V_{nm} (which we do not give explicitly) vanishes unless n is either $m+1$ or $m-1$.

In calculating the free energy F (defined in Section III), we retain only the part

$$F_2 = N\left\langle {\sum_n}' \frac{|V_{nm}|^2}{E_m^{(0)} - E_n^{(0)}} \right\rangle_{\text{av}} - \tfrac{1}{4}(A+\Gamma)M_\xi{}^2 \tag{25}$$

which is bilinear in E_ξ and H_ξ. The prime indicates $n \neq m$, the symbol $\langle\ \ \rangle_{\text{av}}$ denotes an average over the unperturbed canonical distribution, and

$$E_m^{(0)} = -g\mu_B m(A - \Gamma)M_{0z} \tag{26}$$

is the mth eigenvalue of the $\mathcal{H}_0$ given by Eq. (22). While the first term in Eq. (25) arises from thermodynamic perturbation theory, the second term ensures that none of the exchange interactions is counted twice. With the help of Eq. (18) and the equation

$$M_{0z} = \tfrac{1}{2}Ng\mu_B\langle m\rangle_{\text{av}} \tag{27}$$

we then find that Eq. (25) may be written in the form

$$F_2 = d_{HH}H_\xi{}^2 + d_{EH}E_\xi H_\xi + d_{EE}E_\xi{}^2 \tag{28}$$

where d_{HH}, d_{EH}, and d_{EE} are calculable coefficients which we do not write down in detail. The thermodynamic relation

$$\alpha_\perp = -\partial^2 F_2 / \partial E_\xi \partial H_\xi \tag{29}$$

thus leads to the result[9]

$$\alpha_\perp = -d_{EH} = -\frac{3a_\perp}{2A}\left(\frac{\langle m^2\rangle_{av} - \frac{1}{3}S(S+1)}{\langle m\rangle_{av}}\right) \tag{30}$$

where both $a_\perp$ and $\alpha_\perp$ are dimensionless in the Gaussian units used in this paper.

Evaluations of the thermal averages occurring in Eq. (30) can be carried out in at least three different ways. Perhaps the most consistent way[9] is to adhere strictly to the molecular field approximation which underlies Eqs. (19) and (20). In this way we obtain

$$\alpha_\perp = -\frac{3a_\perp SC[B_S'(u) + B_S{}^2(u) - \frac{1}{3}S(S+1)]}{2(T_N + \Theta)B_S(u)} \tag{31}$$

where

$$C = Ng^2\mu_B{}^2 S(S+1)/3k \tag{32}$$

is the Curie constant,

$$\Theta = \tfrac{1}{2}C(A + \Gamma) \tag{33}$$

is the Curie–Weiss constant,

$$T_N = \tfrac{1}{2}C(A - \Gamma) \tag{34}$$

is the Néel temperature, and $B_S'(u)$ is the derivative of the Brillouin function $B_S(u)$ with respect to its argument u. We also note the useful relations

$$M_{0z} = \tfrac{1}{2}Ng\mu_B SB_S(u) \tag{35}$$

$$T = \frac{3SC(A - \Gamma)B_S(u)}{2(S+1)u} \tag{36}$$

where T is the absolute temperature. A second way[9] to evaluate the thermal averages occurring in Eq. (30) and in the corresponding expression for $\alpha_\parallel$ is to make limited use of the experimentally determined temperature dependence of the magnetic susceptibility. But the most accurate way is probably that suggested by Hornreich and Shtrikman[15] who used the experimentally determined $\chi_\parallel$ *vs.* T relation to fully evaluate all the thermal averages appearing in the expressions for ME susceptibilities. Their method still involves the molecular field approximation but greatly decreases the "burden" placed on its validity. Accordingly, we shall base the comparison of Eq. (30) and experiment on the Hornreich and Shtrikman[15] paper rather than on Ref. 9.

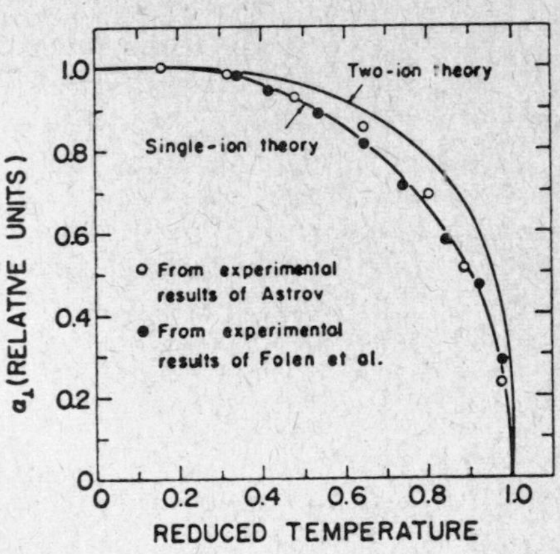

FIGURE 3 Comparison of theoretical and experimental temperature dependences of $\alpha_\perp$ in Cr_2O_3. References to the original papers are given in the text. [After Hornreich and Shtrikman, Ref. 15].

Figure 3 shows that in Cr_2O_3 the temperature dependence of $\alpha_\perp$ predicted on the basis of Eq. (30) (see the curve labeled "single-ion theory") agrees very well with that determined by two different experiments.[6,4] Furthermore, the value $|a_\perp| = 50.6 \times 10^{-3}$ required to fit[15] the predictions of the spin–orbit mechanism to the experimental values of $|\alpha_\perp|$ agrees as well as can be expected with the value $|a_\perp| = 9.7 \times 10^{-3}$ adapted[9] from the paramagnetic resonance data of Royce and Bloembergen[16] in ruby ($Al_2O_3 : Cr^{3+}$). The latter authors interpreted their results in terms of a spin Hamiltonian which is a generalized version of that proposed[8,9] for the spin–orbit mechanism of the ME effect.

The first two-ion mechanism of the ME susceptibility is due to Date, Kanamori and Tachiki[17] and is based on the change due to **E** of the isotropic exchange interaction. Symmetry considerations show that in Cr_2O_3 this mechanism involves only intra-sublattice exchange interactions and contributes to $\alpha_\parallel$ only. The temperature dependence of $\alpha_\parallel$ predicted by this exchange mechanism is proportional to $\chi_\parallel\langle S_z\rangle$ in the molecular field approximation. This is the same temperature dependence as that of a phenomenological expression[8] suggested earlier. Figure 4 shows that in Cr_2O_3 the temperature dependence of $\alpha_\parallel$ predicted on the basis of the exchange mechanism (see the curve labeled "two-ion theory") agrees well with that determined experimentally[6] provided the theoretical $\alpha_\parallel$ curve is displaced along the ordinates by a temperature-independent amount which is given by the g-factor mechanism[16,18] mentioned below.

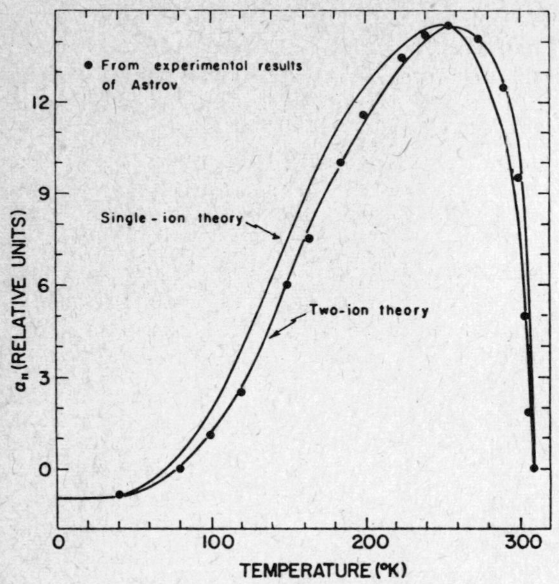

FIGURE 4 Comparison of theoretical and experimental temperature dependences of $\alpha_{\parallel}$ in Cr_2O_3. References to the original papers are given in the text. [After Hornreich and Shtrikman, Ref. 15].

Later extensions of the atomic theory of $\boldsymbol{\alpha}$ in type 1 materials consist in incorporating into the original formalism[9] the changes due to **E** of the g-factor,[16,18] antisymmetric exchange,[15] and biquadratic exchange.[19] In fact, the curve labeled "two-ion theory" in Figure 3 is based on the antisymmetric exchange mechanism and the curve labeled "single-ion theory" in Figure 4 is based on the spin–orbit and g-factor mechanisms. Hornreich and Shtrikman[15] have concluded from their detailed comparisons (Figures 3 and 4) between the extended theory of $\boldsymbol{\alpha}$ and experimental results that in Cr_2O_3 the mechanism which best describes the temperature dependence of $\alpha_{\perp}$ is of the "single-ion" (namely spin–orbit) type and that the mechanism which best describes the temperature dependence of $\alpha_{\parallel}$ is of the "two-ion" (namely isotropic exchange) type at high temperatures and of the "single-ion" (namely g-factor) type at low temperatures. Similar conclusions were reached by Mercier *et al.*[19] with respect to the temperature dependence of $\alpha_{\perp}$ and $\alpha_{\parallel}$ in $LiMnPO_4$.

We wish to point out that it is difficult to decide conclusively which mechanisms contribute appreciably to a specified component of $\boldsymbol{\alpha}$ in a given material. One reason for this difficulty is the unknown magnitude of the error in the calculation of the thermal averages occurring in the theory. The introduction of Green's function methods[20,21] is, in principle, an improvement over the originally used molecular field methods. In practice, however, the results obtained by means of Green's function methods are not yet sufficiently reliable. Figure 4 of Ref. 21 clearly shows that the differences between the $\alpha_{\perp}$ values predicted on the basis of two versions of the Green's function theory (namely the "random phase approximation" and the "Callen decoupling") are quite large. These differences are, in fact, comparable to the differences existing between the predictions of two ME mechanisms which are evaluated in the molecular field approximation or in some other specified approximation. Another reason for the above-mentioned difficulty of deciding between competing ME mechanisms is that various parameters (such as $a_{\perp}$) contained in the theory represent complicated sums of perturbation terms which cannot be calculated in practice and thus must be estimated on the basis of appropriate spectroscopic or paramagnetic resonance data. In several cases, moreover, the relevant atomic parameter (such as that describing the antisymmetric exchange contribution to $\alpha_{\perp}$ in Cr_2O_3) has not even been estimated.

V ATOMIC MECHANISMS OF ME EFFECTS IN TYPE 2 MATERIALS

The magnetic ions in type 2 materials are rare-earth ions, and it is the changes due to **E** of a quantity similar to the g-factor of rare-earth ions which are involved in the first mechanism proposed for $\boldsymbol{\alpha}$ in such materials. This mechanism, due to Rado,[22] is probably the most effective one suggested so far because it is based on the lowest possible order of perturbation theory. We now describe this mechanism and its use for calculating $\boldsymbol{\alpha}$ in the Ising-like antiferromagnet $DyPO_4$.

Magnetic symmetry considerations show that the magnetic point group $4'/m'mm'$ of antiferromagnetic $DyPO_4$ allows linear ME effects and that

$$F_{\mathrm{ME}} = -\alpha(E_x H_x - E_y H_y) \tag{37}$$

is the ME contribution to the F defined in Section III. The Cartesian axes x, y, z are parallel, respectively, to the tetragonal axes a, a, c of the crystal, and

$$(\alpha) = \begin{pmatrix} \alpha & 0 & 0 \\ 0 & -\alpha & 0 \\ 0 & 0 & 0 \end{pmatrix} \tag{38}$$

is the matrix of $\boldsymbol{\alpha}$ in the x, y, z coordinate system. In constructing an atomic mechanism of α, however, we find it convenient to use a Cartesian coordinate system ξ, η, ζ which is related to the x, y, z system by a $\frac{1}{4}\pi$ rotation about the $\zeta \equiv z$ axis. Thus we replace Eq. (37) by

$$F_{\mathrm{ME}} = -\alpha(E_\xi H_\eta + E_\eta H_\xi) \tag{39}$$

and the matrix (38) by

$$(a) = \begin{pmatrix} 0 & \alpha & 0 \\ \alpha & 0 & 0 \\ 0 & 0 & 0 \end{pmatrix}. \tag{40}$$

Since the ξ, η, ζ axes pass through the Dy^{3+} sites and are, in fact, the symmetry axes associated with these sites, the use of the ξ, η, ζ coordinates causes the crystalline potential at the Dy^{3+} sites to be somewhat simplified.

We assume that the unperturbed Hamiltonian $\mathcal{H}_0$ contains contributions describing the free Dy^{3+} ion (including spin–orbit coupling), the even part of the crystalline potential energy, and the (axial) exchange and dipolar interactions. Because of the three latter contributions, the $J = \frac{15}{2}$ manifold of the ${}^6\mathrm{H}_{15/2}$ ground state of the free Dy^{3+} ion is decomposed into eight exchange- and dipole-split Kramers doublets which are *not* pure m_J states. The ground doublet, for example, is expected to contain not only $m_J = \pm\frac{15}{2}$ but also admixtures of $m_J = \pm\frac{7}{2}, \mp\frac{1}{2}$, and $\mp\frac{9}{2}$ due to the tetragonal potential as well as admixtures due to that part of the exchange interaction which is non-diagonal in the low-lying crystal field states. Now we calculate the value of F_{ME} at 0°K by considering the combined action of three perturbations, namely the odd part $eV^{(u)}$ of the crystalline potential energy, the interaction $-e\mathbf{E}\cdot\mathbf{r}$ with the applied electric field, and the Zeeman energy $\mu_B(\mathbf{L} + 2\mathbf{S})\cdot\mathbf{H}$ due to the applied magnetic field. Comparison of the resulting expression with Eq. (39) yields

$$\alpha = \frac{4e^2\mu_B}{v} \times \sum_{k,l} \frac{\langle 0|V^{(u)}|l\rangle\langle l|\xi|k\rangle\langle k|L_\eta + 2S_\eta|0\rangle}{(W_k - W_0)(W_l - W_0)}, \tag{41}$$

where the W's are appropriate eigenvalues of $\mathcal{H}_0$ and the factor 4 denotes the number of Dy^{3+} ions in the unit cell of volume $v = a^2c$. It should be noted that the two sublattices are related to each other by the product of space inversion and time reversal and thus contribute equally to Eqs. (39) and (41), and that Eq. (41) does satisfy the requirement of invariance under the symmetry operations of the magnetic point group of the material. [The combined Eqs. (30) and (11) also satisfy this requirement.] Another point worth mentioning is that the ME mechanism embodied in Eq. (41) involves essentially the change due to $\mathbf{E}$ of a quantity similar to the spectroscopic splitting factor in a magnetically ordered material containing rare-earth ions. This change is given by a fourth-order process[18] in the case of Cr_2O_3 but by a third-order process [Eqs. (39) and (41)] in the case of $DyPO_4$.

Next we estimate the order of magnitude of $|\alpha|$. We use a point charge model and the site symmetry $\bar{4}2m$ to express $V^{(u)}$ by the third-order term

$$V^{(u)} = 15\xi\eta\zeta \sum_n (Q_n u_n v_n w_n / R_n{}^7), \tag{42}$$

because in $DyPO_4$ (unlike in Cr_2O_3) the first-order terms vanish by symmetry. Here $\mathbf{r} = \mathbf{i}\xi + \mathbf{j}\eta + \mathbf{k}\zeta$ and $\mathbf{R}_n = \mathbf{i}u_n + \mathbf{j}v_n + \mathbf{k}w_n$ denote, respectively, the position of a $4f$ electron and of a neighboring ion (having charge Q_n) with respect to the nucleus of a central Dy^{3+} ion. The unit vectors $\mathbf{i}, \mathbf{j}, \mathbf{k}$ are parallel to the coordinate axes ξ, η, ζ, respectively. We approximate the $(PO_4)^{3-}$ complexes by charges $-3|e|$ located at the P^{5+} nuclei and neglect the (unknown) shielding of the $4f$ electrons of Dy^{3+}. Using $\xi \approx \eta \approx \zeta \approx 1$ Å and the lattice parameters $a = 6.917$ Å and $c = 6.053$ Å, we thus obtain $|eV^{(u)}| \approx 3.5 \times 10^4$ cm^{-1} after summing over four neighboring Dy^{3+} ions and four neighboring $(PO_4)^{3-}$ complexes. For $W_l - W_0$, which is the (unknown) separation between the ground level W_0 and a level W_l of opposite parity, we assume the value 10^5 cm^{-1}, while for $W_k - W_0$, which may be taken as the separation between the lowest-lying Kramers doublet and the nearest excited level, we use the measured[23] value of ≈ 70 cm^{-1}. According to the Wigner–Eckart theorem,[24] the matrix element $\langle k|L_\eta + 2S_\eta|0\rangle$ equals $g_J\langle k|J_\eta|0\rangle$, where g_J is the Landé g factor. Using $g_J = \frac{4}{3}$ and assuming, for simplicity, that the states $|0\rangle$ and $|k\rangle$ contain only $m_J = \frac{15}{2}$ and $m_J = \frac{13}{2}$, respectively, we obtain $|\langle k|L_\eta + 2S_\eta|0\rangle| \approx 2.6$. The final result of our rough estimate is the value $|\alpha| \approx 4 \times 10^{-2}$, which is probably too high. This value may be compared with the value[25] $|\alpha| \approx 1.2 \times 10^{-3}$ measured on a crystal which had been "ME-annealed" (see Section VII and hence probably consisted of a single antiferromagnetic domain.

The present numerical estimate of $|\alpha|$ differs from the estimate[22] obtained at a time when the above-mentioned energy separation of 70 cm^{-1} had not yet been measured. We wish to point out, furthermore, that terms in which $W_k - W_0$ represent the splitting of the lowest-lying Kramers doublet make a relatively small contribution to the α of Eq. (41).

As shown in the Appendix, this limitation follows from certain properties of the time reversal operator θ and depends ultimately on the fact that for a Kramers ion θ^2 is -1. For a non-Kramers ion, θ^2 is $+1$ so that the limitation is inapplicable. Under appropriate circumstances, therefore, a crystal containing non-Kramers ions may exhibit a smaller "usable" value of $W_k - W_0$, and hence a larger actual value of $|\alpha|$, than a crystal having the same structure but containing Kramers ions. These considerations may provide a partial explanation of the experimental result[26] that the non-Kramers ion Tb^{3+} leads to a considerably larger $|\alpha|$ in $TbPO_4$ than the Kramers ion Dy^{3+} in $DyPO_4$. In fact, the absolute value of a measured component of $\boldsymbol{\alpha}$ in $TbPO_4$ is found[26] to be larger by at least an order of magnitude than any component of $\boldsymbol{\alpha}$ previously reported for any material.

Next we discuss the temperature dependence of α in $DyPO_4$ and demonstrate that ME effects provide a new and convenient method for measuring the critical behavior of the sublattice magnetization M_T. We begin by showing theoretically that in the case of $DyPO_4$ (and also in analogous cases) the quantities M_T and α have the same temperature dependence. The theory which follows is due to Rado[25] and is rigorous within the assumption that the energy levels are very nearly of the "one-ion" type. This is partly due to the result[22] that the physical origin of α in $DyPO_4$ is predominantly a "one-ion" mechanism, and partly due to the experimental fact that the lowest-lying energy levels of $DyPO_4$ constitute an exchange–and dipole–split Kramers doublet whose separation from the nearest excited level is rather large[23] ($\approx 70\ \text{cm}^{-1}$), being about ten times the doublet splitting. Even at temperatures T as high as the Néel temperature T_N ($=3.39°$K), therefore, $DyPO_4$ may be described accurately by a fictitious spin S$'$ of $\frac{1}{2}$. Focusing our attention on one of the two sublattices, we denote the effective magnetic quantum number by m and use W_m to label the corresponding energy per Dy^{3+} ion in the absence of applied magnetic or electric fields. This "unperturbed" energy is taken to be $W_{1/2}$ for the ground level and $W_{-1/2} = -W_{1/2}$ for the other level of the essentially isolated doublet under consideration. We let W_m^{ME} be the magnetoelectric perturbation energy per Dy^{3+} ion of the chosen sublattice and recall that both sublattices contribute equally to α_0, the value of α at $T = 0°$K. Next we note that, in the x, y, z coordinate system, $W_{1/2}^{\text{ME}}$ is given by $-\alpha_0 E_x H_x/N$ for the case $E_y = H_y = 0$. By using Eq. (41) and the change of matrix elements upon time reversal[27] of the states, we obtain $W_{-1/2}^{\text{ME}} = -W_{1/2}^{\text{ME}}$ provided we assume temporarily that all the Kramers doublets which are connected by the various perturbation paths have the same unperturbed splitting.

Use of the above considerations in conjunction with the unperturbed canonical distribution

$$\rho_m = \exp(-W_m/kT)/\sum_m \exp(-W_m/kT) \tag{43}$$

easily leads to the equations

$$M_T = \tfrac{1}{2}Ns_\parallel\mu_B \sum_m m\rho_m = -\tfrac{1}{4}Ns_\parallel\mu_B \tanh(W_{1/2}/kT), \tag{44}$$

$$\tfrac{1}{2}\alpha = -\tfrac{1}{2}N\partial^2 \sum_m W_m^{\text{ME}} \rho_m/\partial E_x \partial H_x = -\tfrac{1}{2}\alpha_0 \tanh(W_{1/2}/kT), \tag{45}$$

where $s_\parallel$ is the effective parallel spectroscopic splitting factor for $S' = \frac{1}{2}$. Equations (44) and (45) yield the crucial result

$$\alpha/\alpha_0 = M_T/M_0, \tag{46}$$

where $M_0 = \frac{1}{4}Ns_\parallel\mu_B$ is the value of M_T at $T = 0°$K. If we now drop the temporary assumption mentioned above, then we find that in the molecular field approximation $W_{-1/2}^{\text{ME}}$ equals $-W_{1/2}^{\text{ME}}(1 + 2\Delta_0 M_T/M_0)$, where $2\Delta_0$ is essentially the ratio of the unperturbed splitting of the ground Kramers doublet at $T = 0°$K to the separation between this doublet and the first excited doublet. A straight-forward calculation now gives

$$\alpha/\alpha_0 = (M_T/M_0)[1 - \Delta_0(1 - M_T/M_0)], \tag{47}$$

which shows that the largest fractional effect due to Δ_0 occurs at $T = T_N$. Since in $DyPO_4$ we have $2|\Delta_0| < 0.1$, we can interpret the measurements with sufficient accuracy by using Eq. (46) rather than Eq. (47). We wish to emphasize, furthermore, that Eq. (46) applies to both sublattices and that its derivation does not involve any assumption about the nature of the temperature-dependent (and intrinsically negative) quantity $W_{1/2}$. In particular, $W_{1/2}$ need not contain a molecular field even though the very concept of $W_{1/2}$ is based on the assumption that the energy levels are very nearly of the "one-ion" type. It is seen, therefore, that Eq. (46) [in contrast to Eq. (47)] does not make use of the molecular field approximation. By actually measuring α/α_0 and using Eq. (46), we are thus able to obtain what may be regarded as a measured M_T/M_0. We then compare this latter M_T/M_0 with a calculated M_T/M_0 obtained on the basis of some suitable statistical theory.

In comparing the experimental α/α_0 with a theoretical M_T/M_0, we make use of the series expansion[28] for the reduced spontaneous magnetization of the Ising ferromagnet of spin $\frac{1}{2}$ on the diamond lattice. This expansion converges at all temperatures up to T_N, remains valid for the antiferromagnetic[29] M_T/M_0, and may be expected to apply closely to the lattice of $DyPO_4$. The critical behavior[28,30] of M_T/M_0 has the form

$$M_T/M_0 = B(1 - T/T_N)^{\beta}, \tag{48}$$

where B and β are constants whose most recently calculated values[30] are $B = 1.661 \pm 0.001$ and $0.307 \leqslant \beta \leqslant 0.314$. Adopting the values $B = 1.661$ and $\beta = 0.314$, which are within the uncertainty of the theory, we combine Eqs. (46) and (48) to obtain

$$(\alpha/1.661\alpha_0)^{(1/0.314)} = 1 - T/T_N. \tag{49}$$

Use of the measured values of α/α_0 in the left-hand side of Eq. (49) then yields the data plotted in Figure 5 as a function of temperature. Also shown in Figure 5 is a straight line which fulfills the

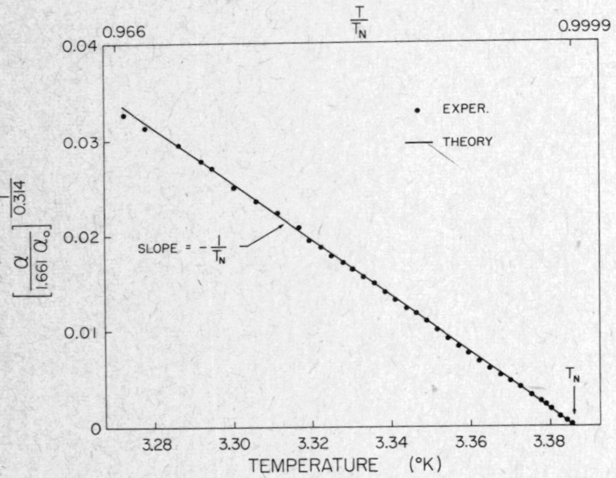

FIGURE 5 Critical behavior of the experimental α/α_0 in $DyPO_4$. The theoretical line is explained in the text. [After Rado, Ref. 25].

theoretical requirement [see Eq. (49)] of passing through the experimental points and having a slope that equals the negative reciprocal of the intercept on the T axis. The agreement between theory and experiment is seen to be excellent and to extend over the range $0.966 \lesssim T/T_N \lesssim 0.9999$. The value $T_N = 3.39°K$ deduced from Figure 5 agrees with the values obtained from specific heat[31] and magnetic susceptibility[31] measurements which also showed excellent agreement with the Ising model. We note that the use of a linear graph (such as Figure 5) permits a more stringent test of the theory than the use of a log–log graph.

Figure 6 shows data on the measured α/α_0 at low as well as high values of T/T_N. It also shows that the predictions of the molecular field theory (with spin $\frac{1}{2}$) for the quantity M_T/M_0 strongly disagree with experiment. This disagreement is not limited to the critical region but extends throughout the entire temperature range. The predictions of the Ising

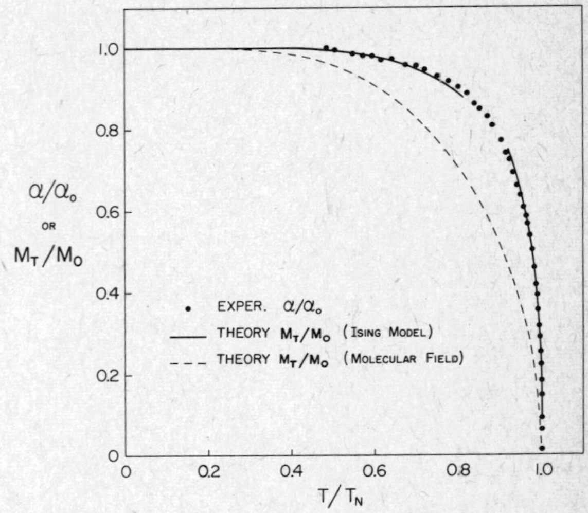

FIGURE 6 Comparison of the temperature dependence of the experimental α/α_0 in $DyPO_4$ with the temperature dependences of the theoretical M_T/M_0 according to the molecular field and Ising models. [After Rado, Ref. 25].

model theory, on the other hand, are seen to be in excellent agreement with experiment. This agreement obtains not only in the critical region, as already demonstrated in Figure 5, but also at lower temperatures, where we used the series expansion[28] of M_T/M_0. In a small range of intermediate temperatures, of course, M_T/M_0 cannot yet be predicted because the number of presently known coefficients of this expansion is inadequate. Thus we conclude that the excellent agreement between theory and experiment, which involves no adjustable parameters, extends essentially over the whole temperature range of magnetic ordering. This conclusion is confirmed by the experiments of Holmes *et al*[32] which show that the ME properties of $DyAlO_3$ are analogous to those of $DyPO_4$.

VI ME EFFECTS ARISING FROM THE INTERPLAY OF PIEZOELECTRICITY AND PIEZOMAGNETISM

A possible mechanism of ME effects in a ferromagnet is the combined action of piezoelectricity and piezomagnetism.[12,13] This mechanism will now be demonstrated by means of a thermodynamic calculation.[13] We extend, for this purpose, our thermodynamic treatment of ME effects in ferromagnets (see Section III) by including the stress tensor $\boldsymbol{\sigma}$ as an additional independent variable and letting the direction of **E** be arbitrary. Of particular interest is the contribution to F given by

$$\Delta F = -\alpha_{ijk}u_iH_jE_k - d_{ijk}E_i\sigma_{jk} - p_{ijkm}u_iH_j\sigma_{km} - \tfrac{1}{2}s_{ijkm}\sigma_{ij}\sigma_{km} \tag{50}$$

which contains a linear ME term, a piezoelectric term, a piezomagnetic term, and an elastic term. Here d_{ijk}, p_{ijkm}, and s_{ijkm} denote, respectively, the piezoelectric, piezomagnetic and elastic compliance coefficients. The relations $P_i = -\partial F/\partial E_i$ and $e_{mi} = -\partial F/\partial \sigma_{mi}$ then lead to

$$P_i = \alpha_{jki}u_jH_k + d_{ijk}\sigma_{jk} \tag{51}$$

$$e_{mi} = d_{kmi}E_k + p_{jkmi}u_jH_k + s_{jkmi}\sigma_{jk} \tag{52}$$

where the e_{mi} are components of the strain tensor **e**. Since we are again concerned with a saturated ferromagnet, we simplify Eqs. (51) and (52) with the use of $\mathbf{u} = \mathbf{M}/M = \mathbf{H}/|H|$. Next we assume $\mathbf{e} = 0$ and solve the 6 equations (52) for the 6 stress components σ_{jk}. Substitution of the σ_{jk} into Eq. (51) then yields

$$P_y{}^H = (\alpha_{kky} + \Delta\alpha_{kky})u_k{}^2|H| = \alpha^*_{kky}u_k{}^2|H| \tag{53}$$

for that part of P_y which is proportional to $|H|$. Here α^*_{kky} denotes the ME susceptibility at zero strain, i.e., the ME susceptibility which would result from a quantum mechanical calculation based on a rigid lattice. The quantity α_{kky}, on the other hand, is the ME susceptibility at zero stress, i.e., the ME susceptibility measured experimentally. Eq. (53) shows that α_{kky} is given by

$$\alpha_{kky} = \alpha^*_{kky} - \Delta\alpha_{kky} \tag{54}$$

where $\Delta\alpha_{kky}$ is the contribution due to the combined action of piezoelectricity and piezomagnetism. The quantity $\Delta\alpha_{kky}$ is a complicated function of the coefficients d_{ijk}, p_{ijkm} and s_{ijkm} and will not be presented explicitly. We note, however, that omission of all subscripts leads to the order-of-magnitude estimate

$$\Delta\alpha \sim pd/s. \tag{55}$$

In the case of $GaFeO_3$, we use the measured value[33] $d \sim 10^{-7}$ and assume the typical value $s \sim 10^{-12}$. Since p has not yet been measured in $GaFeO_3$, we can only state that if p had the (not unreasonable) value 10^{-9}, then $\Delta\alpha$ would be 10^{-4}, which is comparable to an experimentally measured[12] value of α. It appears, in fact, that the ME mechanism embodied in $\Delta\alpha$ is probably significant in $GaFeO_3$ near the Curie point. More generally, we conclude that this mechanism must exist in all unclamped, piezoelectric ferromagnets which are magnetically saturated.

VII OTHER TOPICS

a) A theoretical upper bound for the magnitudes of the components of $\boldsymbol{\alpha}$ in a given material was proposed by Brown, Hornreich and Shtrikman[34] on the basis of thermodynamic perturbation theory. They assumed that **E** has only the component E_j ($j = 1, 2, 3$) and **H** only the component H_k ($k = 1, 2, 3$) along the Cartesian axes 1, 2, 3 fixed with respect to any particular crystal, and they considered only those situations which involve either $j = k$ ($\mathbf{E} \parallel \mathbf{H}$) or $j \neq k$ ($\mathbf{E} \perp \mathbf{H}$). Their result may be written in the form

$$\alpha_{kj} < (\chi^p_{kk}\kappa_{jj})^{1/2} \tag{56}$$

where χ^p_{kk} is the paramagnetic contribution to the magnetic susceptibility χ_{kk}. Since, for all practical purposes, χ^p_{kk} equals χ_{kk}, the inequality (56) essentially states that each element of $\boldsymbol{\alpha}$ must be smaller than the geometric mean of appropriate elements of $\boldsymbol{\chi}$ and $\boldsymbol{\kappa}$. More recently, the inequality (56) was derived by Yatom[35] on the basis of a purely thermodynamic method.

b) In any given sample of material, the magnetic domain structure strongly influences the measured values of the ME susceptibilities. As noted in Section III, the domain structure of a ferromagnetic sample can be eliminated by simply biasing the sample with a sufficiently strong magnetic field. To eliminate the domain structure of an antiferromagnetic sample, on the other hand, it is usually necessary to use the method of ME annealing.[36] This method was originally proposed in connection with polycrystalline antiferromagnets but it is applicable to monocrystalline antiferromagnets as well. In fact, a monocrystalline antiferromagnet must be ME-annealed if the magnitudes of its experimental and theoretical ME susceptibilities are to be compared in a meaningful way. A polycrystalline antiferromagnet, of course, may not exhibit any ME effect at all unless it has been ME-annealed. Existing formulas[36,37] relating the $\boldsymbol{\alpha}$ of a polycrystal to that of the corresponding monocrystal are based on averaging procedures which implicitly assume that each crystallite contains only one antiferromagnetic domain and that possible interactions between domains are negligible.

c) Nonlinear ME effects arising from a term of the form $-\frac{1}{2}\gamma_{ijk}E_iE_jH_k$ in the thermodynamic potential F (defined in Section III) were calculated by Cardwell[38] for ferrimagnetic yttrium iron garnet. He used a formalism similar to that described in Section IV but omitted a term [corresponding to the second term in Eq. (25)] which would have prevented the "double counting" of the exchange interactions. A careful theoretical treatment of certain non-linear ME effects arising from a term of the form $-\frac{1}{2}\beta_{ijk}E_iH_jH_k$ in F was given by Hou and Bloembergen[39] for $NiSO_4 \cdot 6H_2O$. This material is a piezoelectric paramagnet and the effects treated are known as "paramagnetoelectric effects." Their physical origin is analogous to that of $\alpha_\perp$ in Cr_2O_3 in that it is dominated by changes due to $\mathbf{E}$ of the D term in the spin Hamiltonian.

d) ME effects arising solely from the relative motion of a medium with respect to an observer can occur in any medium (even in a nonmagnetic one) and are not treated in the present paper. A discussion of such relativistic effects is given by O'Dell[40] along with a review of various electrodynamic properties of ME media. The recently proposed reciprocity relations[41] for susceptibilities and fields in ME antiferromagnets are additional examples of such properties.

Appendix

It remains to justify our assertion (see Section V) that terms in which $W_k - W_0$ represent the splitting of the lowest-lying Kramers doublet make a relatively small contribution to the α of Eq. (41). This particular contribution will be denoted by $\delta\alpha$.

We let θ be the time reversal operator so that $|\theta 0\rangle$ signifies the Kramers conjugate of the ground state $|0\rangle$. Thus the states $|k\rangle$ of Eq. (41) now represent the singlet state $|\theta 0\rangle$ and the energies W_k represent the single energy $W_{\theta 0}$. Each pair of states $|l\rangle$, $|l+1\rangle$ represents some Kramers doublet (of parity opposite to that of the ground doublet $|0\rangle$, $|\theta 0\rangle$) and will be labeled $|\lambda\rangle$, $|\theta\lambda\rangle$. Hence Eq. (41) leads to

$$\delta\alpha = \frac{4e^2\mu_B \langle\theta 0|L_\eta + 2S_\eta|0\rangle (S_{v\xi} + S_{\xi v} + S_{v\xi}^* + S_{\xi v}^*)}{v(W_{\theta 0} - W_0)} \tag{A1}$$

where $S_{v\xi}$ is given by

$$S_{v\xi} = \sum_\lambda \frac{\langle 0|V^{(u)}|\lambda\rangle\langle\lambda|\xi|\theta 0\rangle}{W_\lambda - W_0} + \sum_\lambda \frac{\langle 0|\xi|\theta\lambda\rangle\langle\theta\lambda|V^{(u)}|\theta 0\rangle}{W_{\theta\lambda} - W_0} \tag{A2}$$

and $S_{\xi v}$ is obtained from $S_{v\xi}$ by interchanging $V^{(u)}$ and ξ. In Eq. (A1) and in what follows, the asterisk denotes the complex conjugate.

Before proceeding, we require the change of matrix elements upon time reversal of the states. As is easily shown,[27] Eqs. (26.11), (26.12) and (26.8) of Wigner[42] yield

$$\langle\theta\psi_a|Q|\theta\psi_b\rangle = \pm\langle\theta\psi_a|\theta|Q\psi_b\rangle = \pm\langle\psi_a|Q|\psi_b\rangle^* \tag{A3}$$

where ψ_a and ψ_b are arbitrary states and where the + or − sign is to be used depending on whether the operator Q commutes or anticommutes with θ.

Next we transform the numerator of the second term of Eq. (A2). Noting that the operators $V^{(u)}$ and ξ are Hermitian and that they commute with θ, we use Eq. (A3) to obtain

$$\langle\theta\lambda|V^{(u)}|\theta 0\rangle = \langle\theta 0|V^{(u)}|\theta\lambda\rangle^* = \langle 0|V^{(u)}|\lambda\rangle \tag{A4}$$

$$\langle 0|\xi|\theta\lambda\rangle = \langle\theta\lambda|\xi|0\rangle^* = \langle\theta^2\lambda|\xi|\theta 0\rangle. \tag{A5}$$

The crucial point is that for a Kramers ion the eigenvalue of θ^2 is −1 so that substitution of Eqs. (A4) and (A5) into Eq. (A2) gives

$$S_{v\xi} = \sum_\lambda \left[\langle 0|V^{(u)}|\lambda\rangle\langle\lambda|\xi|\theta 0\rangle \left(\frac{1}{W_\lambda - W_0} - \frac{1}{W_{\theta\lambda} - W_0}\right)\right] \tag{A6}$$

Since the factor

$$f = (W_{\theta\lambda} - W_\lambda)/(W_\lambda - W_0) \tag{A7}$$

is usually small compared to unity, Eq. (A6) becomes

$$S_{v\xi} = \sum_{\lambda} \frac{\langle 0|V^{(u)}|\lambda\rangle\langle\lambda|\xi|\theta 0\rangle f}{W_\lambda - W_0}. \tag{A8}$$

By transforming $S_{\xi v}$, $S^*_{v\xi}$, and $S^*_{\xi v}$ in a manner analogous to that used for $S_{v\xi}$, we see from Eq. (A1) that the order of magnitude of $\delta\alpha$ is smaller by a factor f in the case of a Kramers ion than in the case of a non-Kramers ion possessing a similar energy level structure. For $DyPO_4$, we estimate f to be smaller than $7\ cm^{-1}/10^5\ cm^{-1} \approx 10^{-4}$, so that the assertion made in Section V and at the beginning of this Appendix is indeed justified.

REFERENCES

1. L. D. Landau and E. M. Lifshitz, *Electrodynamics of Continuous Media* (Addison-Wesley Publishing Company, Inc., Reading, Massachusetts, 1960).
2. I. E. Dzyaloshinskii, *J. Exptl. Theoret. Phys.* (*U.S.S.R.*) **37**, 881 (1959) [Translation: *Soviet Phys.–JETP* **10**, 628 (1960)].
3. D. N. Astrov, *J. Exptl. Theoret. Phys.* (*U.S.S.R.*) **38**, 984 (1960) [Translation: *Soviet Phys.–JETP* **11**, 708 (1960)].
4. V. J. Folen, G. T. Rado and E. W. Stalder, *Phys. Rev. Letters* **6**, 607 (1961).
5. G. T. Rado and V. J. Folen, *Phys. Rev. Letters* **7**, 310 (1961).
6. D. N. Astrov, *J. Exptl. Theoret. Phys.* (*U.S.S.R.*) **40**, 1035 (1961) [Translation: *Soviet Phys.–JETP* **13**, 729 (1961)].
7. G. T. Rado and V. J. Folen, *Bull. Am. Phys. Soc.* **1**, 132 (1956); V. J. Folen and G. T. Rado, *J. Appl. Phys.* **29**, 438 (1958).
8. G. T. Rado, *Phys. Rev. Letters* **6**, 609 (1961).
9. G. T. Rado, *Phys. Rev.* **128**, 2546 (1962).
10. Reference 9, Appendices A and B.
11. R. R. Birss, *Symmetry and Magnetism* (North Holland Publishing Company, Amsterdam, 1964).
12. G. T. Rado, *Proceedings of the International Conference on Magnetism, Nottingham*, 1964 (The Institute of Physics and The Physical Society, London, 1965), p. 361. Part of this work appeared also in *Phys. Rev. Letters*, **13**, 335 (1964).
13. G. T. Rado, *J. Appl. Phys.* **37**, 1403 (1966), and paper in preparation.
14. J. H. Schelleng and G. T. Rado, *Phys. Rev.* **179**, 541 (1969); *Phys. Letters* **A28**, 318 (1968). See also G. T. Rado, *Phys. Rev.* **176**, 644 (1968); **184**, 606 (E) (1969).
15. R. Hornreich and S. Shtrikman, *Phys. Rev.* **161**, 506 (1967). See also R. Hornreich and S. Shtrikman, *Phys. Rev.* **159**, 408 (1967): *Phys. Rev.* **166**, 598 (E) (1968); H. Shaked and S. Shtrikman, *Solid State Commun.* **6**, 425 (1968).
16. E. B. Royce and N. Bloembergen, *Phys. Rev.* **131**, 1912 (1963).
17. M. Date, J. Kanamori and M. Tachiki, *J. Phys. Soc. Japan* **16**, 2589 (1961).
18. S. Alexander and S. Shtrikman, *Solid State Commun.* **4**, 115 (1966).
19. M. Mercier, E. F. Bertaut, G. Quezel and P. Bauer, *Solid State Commun.* 7, 149 (1969).
20. H. Yatom and R. Englman, *Phys. Rev.* **188**, 793 (1969).
21. R. Englman and H. Yatom, *Phys. Rev.* **188**, 803 (1969).
22. G. T. Rado, *Phys. Rev. Letters* **23**, 644 (1969); **23**, 946 (E) (1969).
23. G. A. Prinz, J. L. Lewis, and R. J. Wagner, in *Magnetism and Magnetic Materials–1972*, AIP Conference Proceedings No. 10, edited by C. D. Graham, Jr. and J. J. Rhyne (American Institute of Physics, New York, 1973), p. 1299.
24. M. Tinkham, *Group Theory and Quantum Mechanics* (McGraw-Hill, New York, 1964), p. 132.
25. G. T. Rado, *Solid State Commun.* **8**, 1349 (1970); Erratum: **9**, No. 2, vii (1971).
26. G. T. Rado and J. M. Ferrari, in *Magnetism and Magnetic Materials–1972*, AIP Conference Proceedings No. 10, edited by C. D. Graham, Jr. and J. J. Rhyne (American Institute of Physics, New York, 1973), p. 1417, and paper in preparation.
27. V. J. Folen, R. L. White, and G. T. Rado (unpublished work, 1969). Their result is contained in Eq. (A3) of the present paper.
28. J. W. Essam and M. F. Sykes, *Physica* **29**, 378 (1963).
29. L. P. Kadanoff, W. Götze, D. Hamblen, R. Hecht, E. A. S. Lewis, V. V. Palciauskas, M. Rayl, and J. Swift, *Rev. Mod. Phys.* **39**, 395 (1967).
30. G. A. Baker and D. S. Daunt, *Phys. Rev.* **155**, 545 (1967).
31. J. H. Colwell, B. Mangum, D. D. Thornton, J. C. Wright, and H. W. Moos, *Phys. Rev. Letters* **23**, 1245 (1969).
32. L. M. Holmes, L. G. Van Uitert and G. W. Hull, *Solid State Commun.* **9**, 1373 (1971).
33. D. L. White, *Bull. Am. Phys. Soc.* **5**, 189 (1960). See also the citation in J. P. Remeika, *J. Appl. Phys.* **31**, 263 S (1960).
34. W. F. Brown, R. M. Hornreich and S. Shtrikman, *Phys. Rev.* **168**, 574 (1968).
35. H. Yatom, *Solid State Commun.* **12**, 1117 (1973).
36. S. Shtrikman and D. Treves, *Phys. Rev.* **130**, 986 (1963).
37. R. M. Hornreich, *J. Appl. Phys.* **41**, 950 (1970) and paper to be published.
38. M. J. Cardwell, *Phys. Stat. Sol.* (*b*) **45**, *597 (1971).*
39. S. L. Hou and N. Bloembergen, *Phys. Rev.* **138**, A1218 (1965).
40. T. H. O'Dell, *The Electrodynamics of Magneto-electric Media* (North-Holland, Amsterdam, 1970).
41. G. T. Rado, *Phys. Rev.* B 8, 5239 (1973).
42. E. P. Wigner, *Group Theory and its Application to the Quantum Mechanics of Atomic Spectra* (Academic Press, New York, 1959).

Discussion

T. P. SRINIVASAN Do the ME crystals having non-Kramers ions show larger value of ME susceptibility than those having Kramers ions?

G. T. RADO As mentioned in my present paper, it is an experimental fact that the absolute value of the largest ME susceptibility in RPO_4 is larger by an order of magnitude if R represents the non-Kramers ion Tb^{3+} than if R represents the Kramers ion Dy^{3+}. Although I did interpret this fact theoretically, I wish to emphasize that my arguments were based not only on time reversal considerations but also on the arrangement of the energy levels of the magnetic ions in $TbPO_4$ and $DyPO_4$. In general, because of the role of the arrangement of energy levels, I do not expect a crystal containing non-Kramers ions to be necessarily more strongly magnetoelectric than a crystal having the same structure but containing Kramers ions. This expectation agrees with the experimental results [summarized in Table I of R. M. Hornreich, *IEEE Transactions on Magnetics*, **MAG-8**, 584 (1972)] which show that $LiCoPO_4$, which contains the Kramers ion Co^{2+}, has a considerably larger ME susceptibility than $LiNiPO_4$ which contains the non-Kramers ion Ni^{2+}.

T. P. SRINIVASAN Could you comment as to why the fourth-order process of single ion mechanism and the third order process of two ion mechanism lead to the same temperature dependence of ME susceptibility?

G. T. RADO If the calculation of the thermal averages occurring in the theory is based partly or entirely on the molecular field approximation, then the two processes you mentioned lead to similar but nevertheless non-identical temperature dependences of the ME susceptibility. This result is discussed in my present paper in connection with Cr_2O_3 and illustrated by the work of Hornreich and Shtrikman (see my reference 15) shown in my Figures 3 and 4. Moreover, the results of Englman and Yatom (see my Ref. 21) indicate that the two processes you mentioned lead to similar but non-identical temperature dependences even if the molecular field approximation is replaced by the same Green's function scheme for both processes.

W. OPECHOWSKI Concerning the proportionality between α and M, do I understand you correctly that this proportionality is linked with the existence of Kramers doublets, and therefore would not be expected in the case of non-Kramers crystals?

G. T. RADO What I have shown is that to a good approximation the temperature dependence of the magnetoelectric susceptibility α is given by that of the sublattice magnetization M provided (1) the ground state is an "essentially isolated" Kramers doublet and (2) the physical origin of α is described by a certain third order mechanism. Thus the theory applies to $DyPO_4$ and apparently also to $DyAlO_3$. Because of the non-existence of a comparably accurate theory which applies to other situations, I do not know whether the proportionality between α and M can or cannot occur in other crystals containing Kramers ions or in crystals containing non-Kramers ions.

LOW TEMPERATURE THEORIES OF MAGNETO-ELECTRIC EFFECTS†

R. ENGLMAN and H. YATOM
Soreq Nuclear Research Centre, Yavne, Israel

After describing a number of intuitive approaches to the theory of magneto-electric effects, we present and relate to experiments (in Cr_2O_3) the results of Green function theories for two-sublattice antiferromagnets. These theories have strict validity in the temperature region much below the Néel temperature; however, by consideration of the ratio (magneto-electric susceptibility/sublattice magnetization) the reliability and the significance of the theories are extended, even up to the critical region.

INTRODUCTION

The interpretation of magneto-electric phenomena rests on two pillars. The first of these is the mathematical theory of interacting spin systems in the presence of magnetic and electric fields. Here the main line of development follows the advances in the Green's function theories (GFT)[1] as developed by Callen[2] and others.[3–6]

The second mainstay involved the search for and evaluation of physical effects which could be held responsible for magneto-electric phenomena. A cumulative histroical development has here taken place focusing attention, in turn, on one-ion[7–8] and two-ion exchange[9] effects, on the *g*-factor term[10] and on the Dzyaloshinskii interaction term (arising from the anisotropic interaction between ions).[11–12] A considerable portion of these developments was tied to the magneto-electric susceptibility in anti-ferromagnetic Cr_2O_3 (Néel temperature $T_N \simeq 307^\circ K$) which are due to the theory of Dzyaloshinskii[13] and to the experiments of Astrov.[14] Subsequently, magneto-electric effects were observed also in several other compounds, to which references are made in other of the papers presented at the Symposium on Magnetoelectric Interaction Phenomena (published in the *International Journal of Magnetism*). In addition, three very useful books on the subject have seen light,[15–17] which delve deeply in the symmetry-aspect of the theory.

The present review leans strongly on the formulae and results of the papers of the authors,[18–19] but includes also several newer developments. It dispenses with much of the mathematics (formulae, proofs, etc.) and emphasizes the connection between theory and experiment in a (hopefully) pragmatic fashion. It is mainly aimed at two-sublattice antiferromagnets.

EMPIRICAL APPROACHES

In comparing the experimental and theoretical standing of magneto-electric effects it seems a fair point to make that whereas experimentally a new set of data is obtained by the combination of electrical and magnetic measurements, elementary explanation can be found for these data which are no more complicated than, and indeed follow from, the theory of the magnetic properties. It is worth our while to emphasize the successes of these elementary theories for two reasons. First, since in the majority of this review we shall be concerned with (essentially) the refinement to the elementary theories. Second, the good semi-quantitative agreement of these theories with the data provides us with the assurance that the fundamental physical concepts in the phenomena have been properly recognized. For experimentalists such assurance is invaluable.

---

† Presented at the Symposium on Magnetoelectric Interaction Phenomena in Crystals, Seattle, Washington, U.S.A., May 21–24, 1973.

Rado

The first of these "mini-theories" is Rado's phenomenological result[7] for Cr_2O_3

$$\chi^{ME}_{\|,\perp} \propto \chi^{M}_{\|,\perp}\, \bar{S} \tag{1}$$

or, in words, that the parallel and perpendicular magneto-electric susceptibilities are proportional tc the respective magnetic susceptibility and to the sublattice magnetization $\bar{S}$. Since $\bar{S}$ as a function of temperature shows below the transition point the behaviour characteristic of order parameters, whereas $\chi^M_\|$ and $\chi^M_\perp$ are monotonically increasing and approximately constant, respectively, we see that the product in Eq. (1), reproduces qualitatively the actually observed behaviour of the magneto-electric susceptibilities in Cr_2O_3. ($\chi^{ME}_\|$ will be shown later in Figure 3). The low temperature behaviour of $\chi^{ME}_\|$ is incorrectly predicted, as regards the sign from Eq. (1), but numerically this is a relatively small error and could safely wait another five years.[10]

The constant of proportionality in Eq. (1), is in Gaussian units about 0.5 for the parallel and 0.05 for the perpendicular susceptibilities.[8,12] The difference reflects the relative strengths of the mechanisms operative in the magneto-electric effects, the magnetic exchange (for the parallel case) and the spin-orbit coupling (for the perpendicular effect). The physical meaning of the constant 0.05 is that (if $\bar{S}$ is taken as unity) an electric field intensity of 20 gaussian units (or 0.7×10^5 V/m) is equivalent in its effect on the free energy to a magnetic field intensity of one oersted. We thus see that the magneto-electric effect is a weak phenomenon.

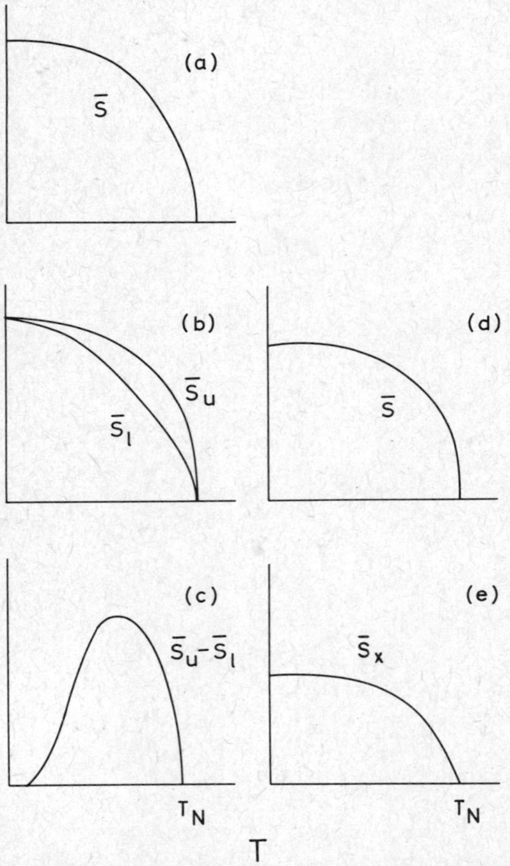

FIGURE 1 Elementary visual derivation of the magneto-electric susceptibilities. (All vertical units arbitrary.) (a) Sublattice magnetization versus temperature. (b) When an electric field E_z is applied the upper end lower sublattice magnetizations differ. (c) The difference yields the parallel magneto-electric susceptibility $\chi^{ME}_\|$. (d) Applying a transverse field H_x cants the magnetizations to the x-direction. (e) The magnetizations depend also on the electric field E_x, yielding $\chi^{ME}_\perp$.

O'Dell

Another explanation at an elementary level was given by O'Dell in his book.[17] He considered the behaviour (as function of the temperature) of the sublattice magnetizations in any two-sublattice antiferromagnet as an electric field is applied externally. Depending on whether the magnetic *and* electric fields are along or transverse to the magnetization axis one gets the changes in the sublattice magnetizations and the magneto-electric susceptibilities as in Figures 1(b), 1(c) and in Figure 1(e). It is clear from the figure that in the parallel effect the two sublattices become inequivalent, while they retain their equivalence (i.e., acquire canting angles which are equal and opposite) in the perpendicular geometry.

Hornreich and Shtrikman

There is also another procedure which cuts across the theoretical intricacies and refinements, and gets extremely good agreement for the magneto-electric susceptibilities. We mean the use of phenomenological susceptibility-derived magnetization in the expressions for the magneto-electric susceptibilities, due to Hornreich and Shtrikman.[12] This procedure is based on the result that in the molecular field approximation the thermal averages $\langle m^r \rangle$ of the moments of the local spin S^z are functionally interrelated through the Brillouin function and its derivatives. Deriving, e.g. $\langle m^2 \rangle - \langle m \rangle^2$, from the experimental magnetic susceptibilities, any given function of the moments $\langle m^r \rangle$ is numerically directly given. Such experimental computations have been very successful for the magnetizations in a number of antiferromagnets (MnF_2, $CuCl_2 . 2H_2O$, FeF_2, Cr_2O_3)[20] as well as for the magneto-electric susceptibilities. It has already been pointed out[18] that even if the moments were uniquely available, semi-empirically, one would still have to make a choice between the various theoretical approximations which predict different functional dependence (of, e.g., the susceptibilities) on the moments. It is nevertheless true that the simple artifact of Hornreich and Shtrikman have reproduced, in a broad spectrum of cases, the observed results better than the more sophisticated theories. In the sequel we turn to these.

GREEN FUNCTION THEORIES

The Hamiltonian

For the sake of consistency we write the Hamiltonian in the same notation as in our paper.[18] Heuristic derivations for the Hamiltonian have been given before.[11–12] We refer to a two-sublattice antiferromagnet which may be regarded as an idealization of Cr_2O_3 and write the Hamiltonian with the magnetic and electric fields in the x–z plane:

$$H = \underbrace{-\mu \sum_j \mathbf{H} \cdot \mathbf{S}_j}_{\text{(i)}} + \underbrace{\mu^2/J_d(a_{\|}{}^g H_z E_z + a_{\perp}{}^g H_x E_x) \sum_j \epsilon_{(j)} S_j{}^z}_{\text{(ii)}}$$

$$+ \underbrace{\mu\, a_{\|}^{LS} E_z \sum_j \epsilon_{(j)} (S_j{}^z)^2 + \mu\, a_{\perp}^{LS} E_x \sum_j \epsilon_{(j)} \tfrac{1}{2}(S_j{}^x S_j{}^z + S_j{}^z S_j{}^x)}_{\text{(iii)}}$$

$$+ \underbrace{\tfrac{1}{2}\mu E_z \sum_{jl} \tfrac{1}{2}(\epsilon_{(j)} + \epsilon_{(l)}) a_{\|}{}^J (j,l)\, \mathbf{S}_j \cdot \mathbf{S}_l + \tfrac{1}{2}\mu E_x \sum_{jl} \tfrac{1}{2}(\epsilon_{(j)} + \epsilon_{(l)}) a_{\perp}{}^J (j,l) \mathbf{S}_j \cdot \mathbf{S}_l}_{\text{(iv)}}$$

$$+ \underbrace{\mu E_x \sum_{jl} \tfrac{1}{2}(\epsilon_{(j)} - \epsilon_{(l)})\, \mathbf{a}_{jl}^D\, (\mathbf{S}_j \times \mathbf{S}_l)}_{\text{(v)}} + \underbrace{\tfrac{1}{2} \sum_{jl} J_{jl}\, \mathbf{S}_j \cdot \mathbf{S}_l}_{\text{(vi)}} - \underbrace{\tfrac{1}{2} I_a \sum_j (S_j{}^z)^2}_{\text{(vii)}}$$

$$- \underbrace{\tfrac{1}{2} \chi_{VV} H^2}_{\text{(viii)}}$$

All coupling coefficients (a^g, a^{LS}, a^D) are dimensionless. The terms have the following meaning:

i) Zeeman term.

ii) g-factor term. $a_{\parallel}{}^g, a_{\perp}{}^g$ represent the change in the g-factor due to the electric field E parallel and perpendicular to the magnetic axis (z) of the anti-ferromagnet.

iii) Single ion term, arising from spin-orbit coupling.

iv) Two ion term due to the exchange integral J_{lj}

$a^J(j,l) = a^J(l,J) = a^J(|j-l|)$.

v) Dzyaloshinskii term, due to vectorial interaction between angular momenta on different ions. This term appears only in the perpendicular case $\mathbf{a}_{jl}^D = \mathbf{a}_{lj}^D = \mathbf{a}^D(|j-l|)$.

vi) Isotropic exchange interaction.

vii) Anisotropic term, $I_a > 0$.

viii) Constant Van Vleck term.[21]

Further,

$\mu = \mu_B g$, where μ_B is the Bohr magneton

$\epsilon_{(j)} = \pm 1$ for the upper and lower sublattices, respectively

j = index of atom

(j) = index of the sublattice of the atom j

$\mathbf{S}_j$ = spin operator on atom j

$J_d = \sum_j J_{jl}$, where the l summation is over cations in the nearest co-ordination polyhedron, all having different spins from j.

Thermodynamics

It seems appropriate to give a brief reconsideration of the thermodynamic theory,[18] if only to specify at which point the Green Function theories enter. We would not attempt to elaborate on these here; they are treated at length in our paper.[18]

Starting with the fundamental definitions

$$\chi_{\parallel}^{ME} = -\frac{\partial^2 F}{\partial H_z \partial E_z}$$

$$\chi_{\perp}^{ME} = -\frac{\partial^2 F}{\partial H_x \partial E_x}$$

expressing the free energy F in terms of the Hamiltonian introduced earlier and working in the weak field ($\mathbf{H} \to 0, \mathbf{E} \to 0$) limit we obtain the following expressions for χ^{ME}

$$\chi_{\parallel}^{ME} = -\frac{\mu^2 a_{\parallel}{}^g}{J_d} \sum_j \epsilon_{(j)} \langle S_j{}^z \rangle_0 - \beta\mu^2 a_{\parallel}^{LS} \sum_{jk} \epsilon_{(j)} \langle S_k{}^z (S_j{}^z)^2 \rangle_0$$

$$- \tfrac{1}{2}\beta\mu^2 \sum_k \sum_{jl:s} \epsilon_{(j)} a_{\parallel}{}^J(jl) \langle S_k{}^x (\mathbf{S}_j \cdot \mathbf{S}_l) \rangle_0$$

$$\chi_{\perp}^{ME} = -\frac{\mu^2 a_{\perp}{}^g}{J_d} \sum_j \epsilon_{(j)} \langle S_j{}^z \rangle_0 - \tfrac{1}{2}\beta\mu^2 a_{\perp}^{LS} \sum_{jk} \epsilon_{(j)} \langle S_k{}^x (S_j{}^x S_j{}^z + S_j{}^z S_j{}^x) \rangle_0$$

$$- \beta\mu^2 \sum_{jkl} \tfrac{1}{2}(\epsilon_{(j)} - \epsilon_{(l)})\, a_{jl}^{D(y)} \langle S_k{}^x (S_j{}^z S_l{}^x - S_j{}^x S_l{}^z) \rangle_0$$

$(\beta = 1/kT)$

$\chi_{\perp}^{ME}$

The perpendicular susceptibilities are derived relatively simply, using the algebra of spin operators and by applying an approximating linearization procedure of the GFT. The result is

$$\chi_{\perp}^{ME} = -N\mu^2 J_d^{-1} a_{\perp}{}^g \bar{S} - \chi_{\perp}{}^M \bar{S}\{a_{\perp}^{LS} B_0 + a^D B_d\}$$

where N is the number of magnetic ions per sublattice. Note that the second term is the product of the magnetic susceptibility and of $\bar{S}$ (just as in Rado's phenomenological result) as well as of the coefficients **a** and of the averages B which represent the correlations. These are defined as

$$B = 1 - \alpha\langle S_l^- S_j^+\rangle_0$$

where $\langle\rangle_0$ denotes the thermal average at zero magnetic and electric fields. $l = j$ for B_0, l is on the neighbouring site to j for B_d and could be a farther neighbours in terms corresponding to interaction between neighbours. It is clear that B measures the correlation. The coefficient α is characteristic of the decoupling procedure used:

$\alpha = 0$ in RPA (random phase approximation)

$= (2S^2)^{-1}$ in CD (Callen decoupling approximation)[2–3]

$= \dfrac{S-1}{2S^2(S+1)} + \dfrac{(\bar{S})^2}{S^4(S+1)}$ MCS (modified Callen decoupling approximation)[6]

$\bar{S}$

The correlation (and the decoupling) enters in an important way through yet another quantity S, the (positive) sublattice magnetization. It is essential to follow through its dependence on the spin wave energies $w_{\mathbf{k}}$ (**k** is the wave-vector). These are given by the GFT self-consistently as

$$w_{\mathbf{k}} = \bar{S}[\lambda_{\mathbf{k}}{}^2 - (J_d B_d \gamma_{d\mathbf{k}})^2]^{\frac{1}{2}}$$

$$\lambda_{\mathbf{k}} = J_d B_d - J_s B_s(1 - \gamma_{s\mathbf{k}}) + I_a B_0$$

wherein appear the sums over nearest and next-nearest neighbours, respectively:

$$\gamma_{d\mathbf{k}} = \frac{1}{Z_1}\sum_j e^{i\mathbf{k}(\mathbf{i}-\mathbf{j})a_1}$$

$$\gamma_{s\mathbf{k}} = \frac{1}{Z_2}\sum_j e^{i\mathbf{k}(\mathbf{i}-\mathbf{j})a_2}$$

$Z_1 Z_2$ being the number of cations on the first and second coordination polyhedron round the magnetic ion and a_1 and a_2 are the corresponding distances.

The spectrum of energies is seen in Figure 2. The single energy characterizing the molecular field approximation (MFA) is also shown.

We now show the string of formulae defining $\bar{S}$. These involve the **k**-averages of the form

$$\langle R_{\mathbf{k}}\rangle_{\mathbf{k}} = 2/N \sum_k R_{\mathbf{k}} \coth \tfrac{1}{2}\beta w_{\mathbf{k}}$$

$$\bar{S} = \frac{1}{2}\left[(2S+1)\frac{(X_0+1)^{2S+1} + (X_0-1)^{2S+1}}{(X_0+1)^{2S+1} - (X_0-1)^{2S+1}} - X_0\right]$$

$$X_0 = \bar{S}\{(J_d B_d + I_a)[\langle 1\rangle_{\mathbf{k}} + \alpha\bar{S}^2 J_s(\langle\gamma_{s\mathbf{k}}\rangle^2 - \langle 1\rangle_{\mathbf{k}}\langle\gamma_{s\mathbf{k}}^2\rangle_{\mathbf{k}})] - J_s\langle 1 - \gamma_{s\mathbf{k}}\rangle_{\mathbf{k}}\}/D$$

$$D = 1 + \alpha\bar{S}^2[I_a\langle 1\rangle_{\mathbf{k}} + J_s\langle\gamma_{s\mathbf{k}}(1-\gamma_{s\mathbf{k}})\rangle_{\mathbf{k}}] + \alpha^2\bar{S}^4 J_s I_a[\langle\gamma_{s\mathbf{k}}\rangle_{\mathbf{k}}{}^2 - \langle 1\rangle_{\mathbf{k}}\langle\gamma_{s\mathbf{k}}^2\rangle_{\mathbf{k}}]$$

In spite of the elaborate expressions and the rather involved simultaneous algebraic equations (arising from the GFT) one should not leave out sight that the sublattice magnetization $\bar{S}$ is the key quantity for the perpendicular

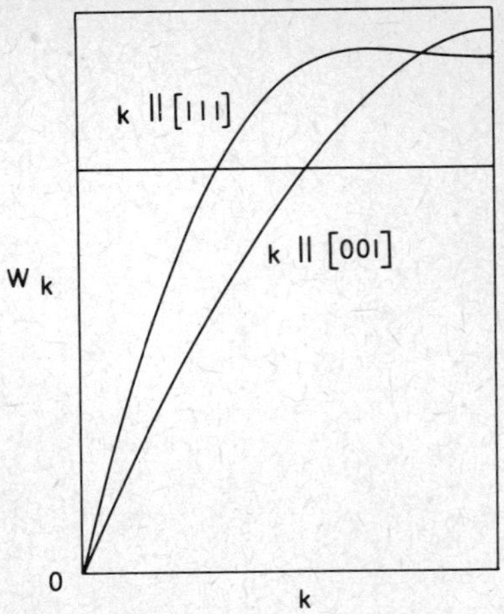

FIGURE 2 Spin-wave (or magnon) energies along $k \parallel [111]$ and $k \parallel [001]$ in a cubic approximation to Cr_2O_3. The constant line is the MFA result. Horizontal scale: k-vector up to the Brillouin-zone edge. Vertical scale $w_k/\bar{S}$.

susceptibilities. In a sense this was implicit in Hornreich and Shtrikman's empirical method. In the MFA (or the Weiss theory) one has simply

$$X_0 = \coth J_0 \bar{S}/2kT$$

so that the string of simultaneous equations reduces to a single self-consistent equation for $\bar{S}$ (J_0 is the usual coupling coefficient or exchange integral).

$\chi_{\parallel}^{ME}$

The parallel susceptibilities are not so simple at all. They involve the boson distribution function $N(w_{\mathbf{k}}) = (e^{\beta w_{\mathbf{k}}} - 1)^{-1}$ and the quantities $A(\bar{S}, \beta)$ and C which are shown in turn.

$$\chi_{\parallel}{}^{M} = \chi_{VV} + N\mu^2 A(\bar{S},\beta)\left\{\left\langle\frac{dN}{dw_{\mathbf{k}}}\right\rangle_{\mathbf{k}} + 2\alpha\bar{S}^2 J_s\left(\left\langle\gamma_{s\mathbf{k}}\frac{dN}{dw_{\mathbf{k}}}\right\rangle_{\mathbf{k}}^2 - \left\langle\frac{dN}{dw_{\mathbf{k}}}\right\rangle_{\mathbf{k}}\left\langle\gamma_{s\mathbf{k}}^2\frac{dN}{dw_{\mathbf{k}}}\right\rangle_{\mathbf{k}}\right)\right\}\Big/ C$$

$$\chi_{\parallel}^{ME} = -N^2 J_d^{-1} a_{\parallel}{}^{g}\bar{S}$$

$$-(\chi_{\parallel}{}^{M} - \chi_{VV})\bar{S}\left\{2a_{\parallel}^{LS}B_0 + a_{\parallel}{}^{J}B_s \frac{\left\langle(1-\gamma_{s\mathbf{k}})\frac{dN}{dw_{\mathbf{k}}}\right\rangle_{\mathbf{k}}}{\left\langle\frac{dN}{dw_{\mathbf{k}}}\right\rangle_{\mathbf{k}} + 2\alpha\bar{S}^2 J_s\left(\left\langle\gamma_{s\mathbf{k}}\frac{dN}{dw_{\mathbf{k}}}\right\rangle_{\mathbf{k}}^2 - \left\langle\frac{dN}{dw_{\mathbf{k}}}\right\rangle_{\mathbf{k}}\left\langle\gamma_{s\mathbf{k}}^2\frac{dN}{dw_{\mathbf{k}}}\right\rangle_{\mathbf{k}}\right)}\right\}$$

$$C = 1 + 2\alpha\bar{S}^2\left[I_a\left\langle\frac{dN}{dw_{\mathbf{k}}}\right\rangle_{\mathbf{k}} + J_s\left\langle\gamma_{s\mathbf{k}}(1-\gamma_{s\mathbf{k}})\frac{dN}{dw_{\mathbf{k}}}\right\rangle_{\mathbf{k}}\right]$$

$$+ 4\alpha^2\bar{S}^4 I_a J_s\left[\left(\gamma_{s\mathbf{k}}\frac{dN}{dw_{\mathbf{k}}}\right)_{\mathbf{k}} - \left\langle\frac{dN}{dw_{\mathbf{k}}}\right\rangle_{\mathbf{k}}\left\langle\gamma_{s\mathbf{k}}^2\frac{dN}{dw_{\mathbf{k}}}\right\rangle_{\mathbf{k}}\right] + \left[J_d B_d - I_a(2B_0 - 1)\right]$$

$$\times\left[\left\langle\frac{dN}{dw_{\mathbf{k}}}\right\rangle_{\mathbf{k}} + 2\alpha\bar{S}^2 J_s\left(\left\langle\gamma_{s\mathbf{k}}\frac{dN}{dw_{\mathbf{k}}}\right\rangle_{\mathbf{k}}^2 - \left\langle\frac{dN}{dw_{\mathbf{k}}}\right\rangle_{\mathbf{k}}\left\langle\gamma_{s\mathbf{k}}^2\frac{dN}{dw_{\mathbf{k}}}\right\rangle_{\mathbf{k}}\right)\right] A(\bar{S},\beta) + J_s(2B_s - 1)\left\langle(1-\gamma_{s\mathbf{k}})\frac{dN}{dw_{\mathbf{k}}}\right\rangle_{\mathbf{k}} A(\bar{S},\beta)$$

$$A(\bar{S},\beta) = 4(2S+1)^2\frac{(X_0{}^2 - 1)^{2S}}{[(X_0+1)^{2S+1} - (X_0-1)^{2S+1}]^2} - 1$$

As has been pointed out[4,18] the reason for the contrast with the perpendicular case is that there we had to calculate averages of S^x, $(S^x)^2$ etc. which are small quantities, whereas now we are concerned with deviations from the finite quantities S^z, $(S^z)^2$, etc.

Lastly we list the magneto-electric susceptibilities in the various approximations as obtained by us. Some of these have already been given before[18] but in our paper they appear with errors which we have now corrected.

$$\chi_{\|}^{ME} = -\frac{N\mu^2}{J_d} a_{\|}{}^{g} \langle m \rangle - (\chi_{\|}{}^{M} - \chi_{VV}) \left[a_{\|}^{LS} \frac{\langle m^3 \rangle - \langle m \rangle \langle m^2 \rangle}{\langle m^2 \rangle - \langle m \rangle^2} + a_{\|}{}^{J} \langle m \rangle \right] \qquad \text{(MFA)}$$

$$= -\frac{N\mu^2}{J_d} a_{\|}{}^{g} \langle m \rangle - (\chi_{\|}{}^{M} - \chi_{VV}) \langle m \rangle \left\{ 2a_{\|}^{LS} + a_{\|}{}^{J} \frac{\left\langle (1-\gamma_{s\mathbf{k}}) \dfrac{dN_{\mathbf{k}}}{dw_{\mathbf{k}}} \right\rangle_{\mathbf{k}}}{\left\langle \dfrac{dN_{\mathbf{k}}}{dw_{\mathbf{k}}} \right\rangle_{\mathbf{k}}} \right\} \qquad \text{(RPA)}$$

$$= -\frac{N\mu^2}{J_d} a_{\|}{}^{g} \langle m \rangle - (\chi_{\|}{}^{M} - \chi_{VV}) \langle m \rangle \left\{ a_{\|}^{LS} B_0 \right.$$

$$\left. + a_{\|}{}^{J} B_s \frac{\left\langle (1-\gamma_{s\mathbf{k}}) \dfrac{dN_{\mathbf{k}}}{dw_{\mathbf{k}}} \right\rangle_{\mathbf{k}}}{\left\langle \dfrac{dN_{\mathbf{k}}}{dw_{\mathbf{k}}} \right\rangle_{\mathbf{k}} + J_s \dfrac{\langle m \rangle^2}{S^2} \left[\left\langle \gamma_{s\mathbf{k}} \dfrac{dN_{\mathbf{k}}}{dw_{\mathbf{k}}} \right\rangle_{\mathbf{k}}^2 - \left\langle \dfrac{dN_{\mathbf{k}}}{dw_{\mathbf{k}}} \right\rangle_{\mathbf{k}} \left\langle \gamma_{s\mathbf{k}}^2 \dfrac{dN_{\mathbf{k}}}{dw_{\mathbf{k}}} \right\rangle_{\mathbf{k}} \right]} \right\}$$

(Callen Decoupling Approximations)

B_0 as below for CD and MCD.

$$\chi_{\perp}^{ME} = -\frac{N\mu^2}{J_d} a_{\perp}{}^{g} \langle m \rangle - \chi_{\perp}{}^{M} \langle m \rangle \left\{ \tfrac{3}{2} a_{\perp}^{LS} \left[\langle m^2 \rangle - \tfrac{1}{3} S(S+1) \right] + \right.$$

$$\left. + \frac{1}{2} \frac{N\mu}{J_d} (\chi_{\perp}^{M-1} - \Gamma) a^D \right\} \qquad \text{(MFA)}$$

$$= -\frac{N\mu^2}{J_d} a_{\perp}{}^{g} \langle m \rangle - \chi_{\perp}{}^{M} \langle m \rangle \left\{ a_{\perp}^{LS} + \frac{1}{2} \frac{N\mu}{J_d} (\chi_{\perp}^{M-1} - \Gamma) a^D \right\} \qquad \text{(RPA)}$$

$$= -\frac{N\mu^2}{J_d} a_{\perp}{}^{g} \langle m \rangle - \chi_{\perp}{}^{M} \langle m \rangle \left\{ \frac{a_{\perp}^{LS}}{B_d} B_0 + \frac{1}{2} \frac{N\mu}{J_d} (\chi_{\perp}^{M-1} - \Gamma) a^D \right\}$$

(Callen Decoupling Approximations)

where

$$B_0 \equiv 1 - \alpha \bar{S} X_0$$

$$= \frac{S-1}{2S} + \frac{\langle m^2 \rangle}{2S^2} \qquad \text{(CD)}$$

$$= \frac{S+1}{2S} + \frac{(S-1)\langle m^2 \rangle}{2S^2(S+1)} - \frac{\langle m \rangle^2}{S^3} \left(1 - \frac{\langle m^2 \rangle}{S(S+1)} \right) \qquad \text{(MCD)[6]}$$

Comparison with Experiments

These formulae formed the basis for our computations which resulted in a set of diagrams testing the relevancy of the theoretical effort to the experimental data. The question that we have asked ourselves is "Do different theoretical methods differ sufficiently from each other (and from experiment) to justify the effort that goes into

TABLE I

Susceptibilities	Term	Sensitivity to approximation
$\chi_{\parallel}^{M}$		Yes
$\chi_{\perp}^{M}$		Yes
$\chi_{\parallel}^{ME}$	Single ion	Yes
	Double ion	No
$\chi_{\perp}^{ME}$	Single ion	Yes
	Double ion	No
	Dzyaloshinskii	No

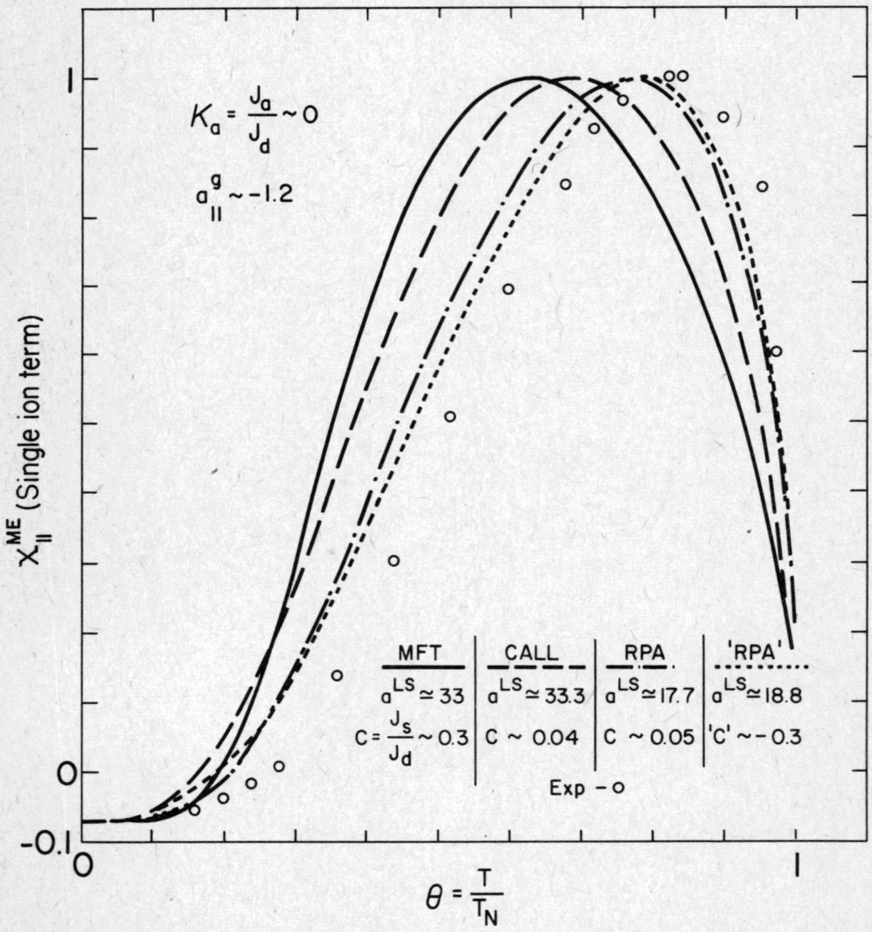

FIGURE 3 Single ion term (and a small *g*-factor term) contribution to $\chi_{\parallel}^{ME}$ versus *T*. Arbitrary vertical units.

them?" Our conclusions are shown in Table I and a case where a sensitivity is indicated with respect to the theoretical schema is shown in Figure 3.

Unfortunately, the experimental data points do not uniquely discriminate between theories, as long as the coupling coefficients (the *a*'s) are treated as phenomenological parameters. Nor can it be said that the complicated GFT are necessarily better than some of the simple-minded theories noted earlier. On the contrary, the Hornreich and Shtrikman approach conforms to the data probably better than, but certainly as well as, the GFT.

The Ratio $\chi^{ME}/\bar{S}$

It seems that a different presentation of both the theoretical and experimental results is called for if we are to sharpen the discriminatory capacity of our data. In Figures 4, 5 and 6 we have plotted the ratio of $\chi^{ME}/\bar{S}$ as function of temperature. We shall presently see the theoretical advantages involved in this approach, but for the time we note that the different theories significantly diverge over the temperature range $0 < T < T_N$ and afford useful comparison with the observed data with respect to both the microscopic mechanism and the decoupling approximation adopted. As regards the experimental values, the drawn curves (marked e) are not very reliable near the transition temperature, since they were derived by interpolation between experimental points which are subject to large errors. It appears that in the future some effort ought to be made to produce reliable experimental values for the ratio $\chi^{ME}/\bar{S}$. This is, of course, not an easy assignment since arduous experimental procedures are required. In cases where these are unattainable, there appears to be a way which involves the critical exponent ϵ of the magneto-electric susceptibilities. It is suggested that this be derived by experiment, as e.g. in Refs. 26–28, and then a plot of

$$(T_N - T)^{\epsilon}\chi^{ME}$$

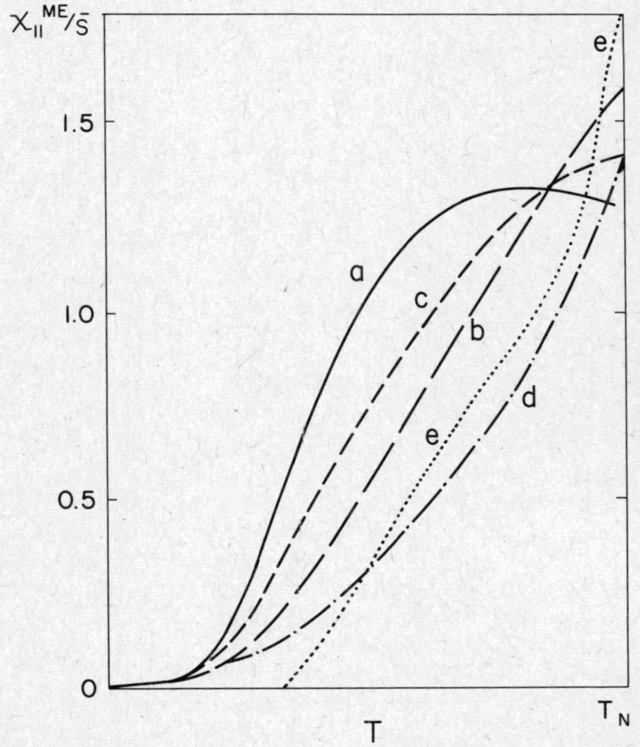

FIGURE 4 $\chi_{\parallel}^{ME}/\bar{S}$ due to the single ion term versus temperature in different approximations. (a) Molecular field approximation (MFA). (b) Random phase approximation (RPA). (c) Callen decoupling (CD). (d) Modified Callen decoupling (MCD).[6] (e) Experiment. Arbitrary vertical units.

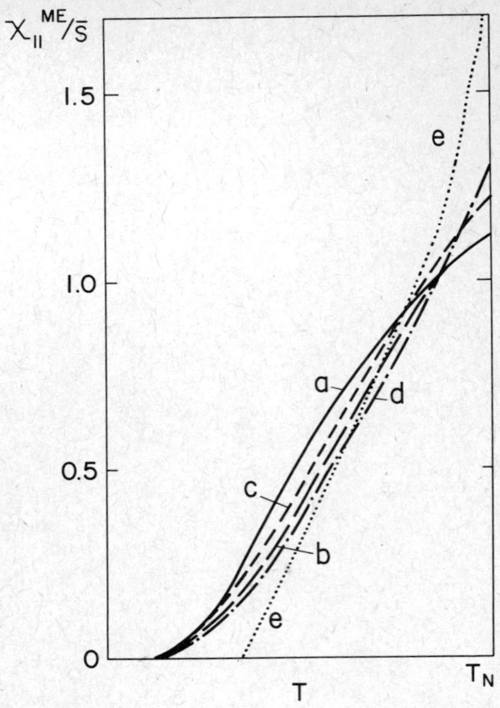

FIGURE 5 $\chi_{\parallel}^{ME}/\bar{S}$ due to the double ion term versus temperature in different approximations. (a)–(e) as in Figure 4.

is made. The behaviour of this should resemble that of the ratio $\chi^{ME}/\bar{S}$, as we shall go on to argue in the next section, and is expected to be a more useful quantity to compare with theory than χ^{ME}.

GFT in the Critical Region

It is generally known that the linearizing (decoupling) approximation is unsatisfactory in the critical region, i.e. near the transition temperature T_N. It is for this reason that we have called the GFT "Low temperature theories". The insufficiency of the theories is shown up by the incorrect transition temperatures that they predict,[6] compared to the Padé approximant results.[22–23] Another deficiency is found in the critical exponent, which is e.g. 1/2 for the order parameter according to MFA, whereas values lower than this were generally found for the magnetization.[24–25] For the magneto-electric susceptibilities the value of $\epsilon = 0.35$ has been measured in Cr_2O_3[26] and the surprising value of $\epsilon \approx 0.5$ for $GdVO_4$.[27] The reason for this failure of the GFT is of course that the linearization does not allow for the correlation to be properly taken into account, although the various decoupling procedures[1–6] represent different attempts to do this. Obviously the GFT are not single particle theories and they do embrace collective interactions, however they fail for critical phenomena. [Having made the verbal distinction between collective effects and critical effects we may perhaps elaborate on this by noting that if we had a finite but large crystal then the effects (e.g., the behaviour of the specific heat near T_N) which varied markedly with the size of the crystal could be called "critical", whereas effects which did not vary significantly with size, but depended on the collective nature of the forces could be termed "collective". The low temperature expansion of the spin wave energies includes "collective" effects.]

The GFT having been declared suspect in the critical region, we would do well to employ some means to remedy or to eliminate this deficiency. It appears that if we considered a quantity which, unlike χ^{ME}, possessed an innocuous behaviour near T_N we would rid ourselves of the difficulties inherent in the GFT. It is for this reason that we have chosen to consider in detail the ratio $\chi^{ME}/\bar{S}$, the implication being that the denominator and the numerator have the same critical exponents.

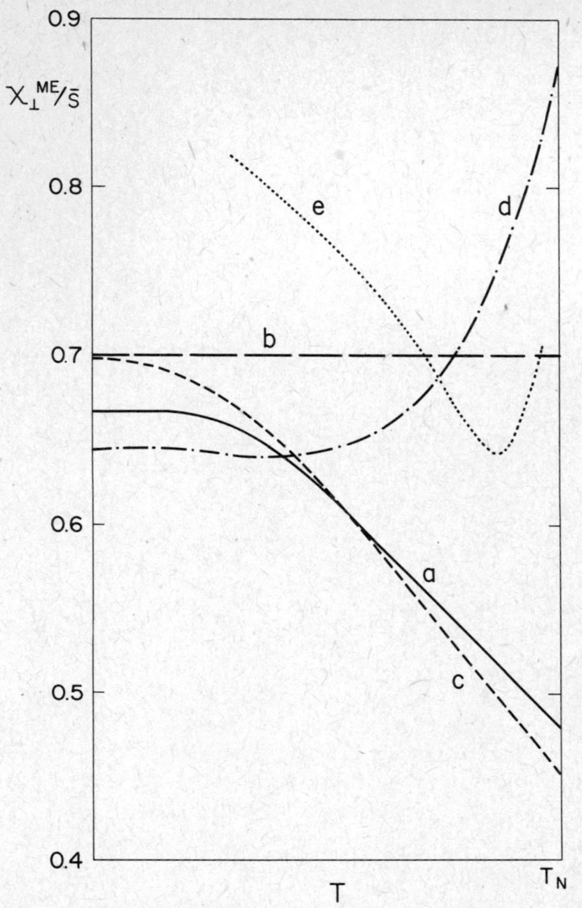

FIGURE 6 $\chi_\perp^{ME}/\bar{S}$ due to the single ion term versus temperature in different approximations. (a)–(e) as in Figure 4. The Dzyaloshinskii term adds a constant to $\chi_\perp^{ME}/\bar{S}$ in (a)–(d).

Is that really so? Experimentally, in Cr_2O_3, this was found to be the case.[20,26] In $DyPO_4$ experimentally, and theoretically for particular forms of the magneto-electric coupling this was shown to hold.[28] In the general case one can only quote the hypothesis by M.E. Fisher[29] that "the nature of the [magnetic] transition and the critical point remains ideal if observed at fixed [electric] force". On the basis of this, we assume that the ratio $(\partial^2 F/\partial H \partial E)/(\partial F/\partial H)$ will be regular (in the mathematical sense) near T_N.

THE COUPLING CONSTANTS

The fitting of theoretical curves to the experimental susceptibilities affords the determination of the coupling constants (the a's) in the Hamiltonian. This has been done in Refs. 8 and 12 and more recently by us[19] in our notation $a_\parallel{}^J \gg -a_\parallel^{LS} \gg -a_\parallel{}^g$.

Naturally one would like to obtain confirmation of these values from other physical phenomena. In the past comparisons have been sought with the values obtained from the Stark effect on ESR-lines[30] and, as regards the single ion coupling parameter a^{LS}, with the zero-field anisotropy splitting parameter in the spin-Hamiltonian (usually denoted by D).[10] For the latter the value derived from χ^{ME} are smaller by 1–2 orders of magnitude than the corresponding value obtained from ESR. However, it appears that the way the electric field affects D is capable of accounting for the discrepancy.

Magnon–Phonon Interaction

Another physical phenomenon of considerable interest, both in its own right and by the way it sheds light on magneto-electric coupling, is phonon–magnon coupling, particularly its effect on the spectra of phonons and magnons. One is concerned with the terms in the Hamiltonian having the form

$$b_k^- \sum_{jl} C_{j-l}(\mathbf{k})\,(S_j^z S_l^+ + S_j^+ S_l^z)e^{-i\mathbf{k}\cdot l} + C\cdot C$$

where b_k^- is the destruction operator for a phonon of wave-vector **k** and the sum goes over the same ($j = l$) or neighbouring ($j \neq l$) ions.

It appears form symmetry considerations[16] that, e.g. for Cr_2O_3, a long wave-length ($k \to 0$) transverse acoustic phonon can indeed interact with magnon. Effects of this type are called magneto-elastic effects. The coefficients $C(\mathbf{k})$ are not immediately related to the a's of the magneto-electric coupling. They would be, if a long wavelength optical phonon was involved (as is the case in $FeCl_2 \cdot 2H_2O$[32]). On the other hand the optical phonons in Cr_2O_3 are about twice as energetic[33] as the top magnons (Figure 2). Little information is available for short wave-length acoustic phonons in Cr_2O_3, such that e.g. two sublattice vibrate against each other. If the frequencies of such vibrations would come near to the corresponding magnon frequencies (a question whose resolution seems at present to belong to the experimental domain), then important insight into magneto-electric coupling may be gained.

FUTURE DIRECTIONS

Theory

This review has emphasized the deficiency of the currently used theories (GFT) in the critical region and the close connection between the theories of the magneto-electric effect and those of (anti-ferro) magnetism. The great and steady advances of the latter theory in the critical region, in particular those related to critical exponents, will no doubt leave their impact on the former theories.

Experiments

As regards the (all-important) experimental stimulus to the theory, one expects that newer (yet newer) magneto-electric materials will be discovered and more m.e. data will there be to interpret. For some of these (e.g., the class of two sublattice antiferromagnets) the present theoretical apparatus may be sufficient when extended to farther neighbour interactions. In those fascinating cases (e.g. some boracites) where the magnetic and electric polarizations occur simultaneously, novel thinking and development may be required.

The spin-wave spectra and their coupling with phonons have already been alluded to. Experimental work is awaited here.

Applications

With magneto-electric materials receiving practical application new theoretical challenges will no doubt arise. One expects that the problems will, at least partly, parallel those associated with magnetic devices, e.g. stability, hardness, switching-time, etc. A good deal will have to be done.

ACKNOWLEDGEMENTS

Our thanks are due to Professor Hornreich for prepublication copies of his works and to Dr. J. Imry for discussions.

REFERENCES

1. S. V. Tyablikov, *Ukr. Math. Zh.* **11**, 287 (1959).
2. H. B. Callen, *Phys. Rev.* **130**, 890 (1963).
3. F. B. Anderson and H. B. Callen, *Phys. Rev.* **136**, A1068 (1964).
4. K. H. Lee and S. H. Liu, *Phys. Rev.* **159**, 390 (1967).
5. M. E. Lines, *Phys. Rev.* **B3**, 1749 (1971).
6. R. H. Swendsen, *Phys. Rev.* **B5**, 116 (1972).
7. G. T. Rado, *Phys. Rev. Letters,* **6**, 609 (1961).

8. G. T. Rado, *Phys. Rev.* **128**, 2546 (1962).
9. M. Date, J. Kanamori and M. Tachiki, *J. Phys. Soc. Japan* **16**, 2589 (1961).
10. S. Alexander and S. Shtrikman, *Solid State Commun.* **4**, 115 (1966).
11. E. F. Bertaut, *J. Phys. Radium,* **23**, 460 (1962).
12. R. Hornreich and S. Shtrikman, *Phys. Rev.* **161**, 506 (1967).
13. I. E. Dzyalohinskii, *Soviet Phys. – JETP* **10**, 628 (1960).
14. D. N. Astrov, *Soviet Phys. – JETP,* **11**, 708 (1960).
15. R. R. Birss, *Symmetry and Magnetism* (North Holland, Amsterdam, 1964).
16. S. Bhagavantam, *Crystal Symmetry and Physical Properties* (Academic, London, 1966).
17. T. H. O'Dell, *The Electrodynamics of Magneto-Electric Media* (North Holland, Amsterdam, 1970).
18. H. Yatom and R. Englman, *Phys. Rev.* **188**, 793 (1969).
19. R. Englman and H. Yatom, *Phys. Rev.* **188**, 803 (1969).
20. H. Shaked and S. Shtrikman, *Solid State Commun.* **6**, 425 (1968).
21. S. D. Silberstein and I. S. Jacobs, *Phys. Rev. Letters* **12**, 670 (1964).
22. R. Tahir-Kheli, *Phys. Rev.* **132**, 689 (1963).
23. G. A. Baker, Jr., H. E. Gilbert, J. Eve and G. S. Rushbrooke, *Phys. Rev.* **164**, 800 (1967).
24. M. E. Fisher, *Repts. Progr. Phys.* **30**, 615 (1967).
25. P. Heller, *Repts. Progr. Phys.* **30**, 731 (1967).
26. E. Fischer, G. Gorodetsky and S. Shtrikman, *J. de Phys.* **32**, C1, 479 (1971).
27. G. Gorodetsky, R. M. Hornreich and B. M. Wanklyn (1972, private communication).
28. G. T. Rado, *Solid State Commun.* **8**, 1349 (1970).
29. M. E. Fisher, *Phys. Rev.* **176**, 257 (1968).
30. E. B. Royce and N. Bloembergen, *Phys. Rev.* **131**, 1912 (1963).
31. B. A. Auld, *Proc. I.E.E.* **53**, 1517 (1965).
32. J. B. Torrance and J. C. Slonczewski, *Phys. Rev.* **B5**, 4648 (1972).
33. R. Marshall, S. S. Mitra, P. J. Gielisse, J. N. Plendl and L. S. Mansur, *J. Chem. Phys.* **43**, 2893 (1965).

PARAMAGNETOELECTRIC EFFECT IN CRYSTALS†‡

S. L. HOU

Addressograph Multigraph Corp., Graphics R&D Center, Warrensville Heights, Ohio 44128

and

N. BLOEMBERGEN

Harvard University, Cambridge, Massachusetts 02138

Magnetoelectric effect exists in piezoelectric paramagnetic crystals. This effect is called the paramagnetoelectric effect which has a term in thermodynamic free energy $F_{\text{PME}} = -\xi_{ijk}\varepsilon_i H_j H_k$. We have observed the PME effect in $NiSO_4.6H_2O$ which belongs to point group symmetry D_4. $F_{\text{PME}} = -\xi(\varepsilon_a H_b - \varepsilon_b H_a) H_c$. The temperature dependence of the PME effect shows a peak at 3.0°K, changes sign at −1.38°K, and follows approximately T^{-2} for $5°K < T < 77°K$. The microscopic theory may be explained by linear variation in an applied electric field of all parameters in the spin Hamiltonian of the four Ni^{++} ions in the unit cell. The dominant contribution comes from the electric field change in the crystal field splitting D. Effect due to exchange interaction is also discussed. Theory may be applicable to most piezoelectric paramagnetic crystals.

I INTRODUCTION

Ever since Astrov[1] and Rado[2] experimentally verified the existence of a term in free energy proportional to EH in antiferromagnetic Cr_2O_3 predicted from thermodynamics by Landau and Lifshitz[3] and Dzyaloshinski,[4] many physicists have studied the origin of magnetoelectric effect (ME) and have explored the possibilities for the existence of the ME effect in various magnetic systems. Hou and Bloembergen[5] first observed and explained the ME effect in the paramagnetic system (or the paramagnetoelectric effect (PME)), and Rado[6] observed the ME effect in the ferromagnetic (or ferrimagnetic) system. The induced ME effect was observed by O'Dell,[7] and the higher order ME effects based on the point group symmetry were derived later by Ascher[8] and Schmid.[9]

The paramagnetoelectric (PME) effect exists in any piezoelectric paramagnetic crystal. It has a term in free energy proportional to $\varepsilon_i H_j H_k$ as pointed out by Bloembergen.[10] He arrived at this conclusion as a result of his work on the linear electric field effect.[11] The shift (or splitting) in magnetic resonance spectra is linear with respect to the applied electric field, whenever the magnetic ion or nuclear spin is located at a site without inversion symmetry. In this case all parameters in the spin Hamiltonian become linear functions of the applied electric field. Because the magnetic susceptibility is a function of these parameters in the spin Hamiltonian, it in turn becomes a linear function of electric field.

The electrostatic perturbation Hamiltonian $-e\varepsilon.\mathbf{r}$ can have a non-vanishing matrix element only if the wave functions of the ground state or the excited state have mixed parity. The relative shift due to electric field can be estimated as $e\varepsilon a/\Delta W = 1:10^8 \sim 1:10^9$ per volt/cm where a is Bohr radius, and $\Delta W = 2 \sim 10$ eV the energy necessary to excite a valence electron. We expect the PME effect to be of the same order of magnitude at low temperatures. This paper reviews the PME effect, explains the microscopic mechanism, and discusses the effect due to exchange interactions.

† This research, which was carried out at Harvard University, was supported jointly by the U.S. Office of Naval Research, the Signal Corps of the U.S. Army and the U.S. Air Force. A preliminary work was presented in *Bull. Am. Phys. Soc.* **9**, 13 (1964). A paper was published in *Phys. Rev.* **138**, A1218 (1965). Some detailed information may be found in the Ph.D. thesis of S. L. Hou, Harvard University, 1964 (unpublished).

‡ Part of this paper was presented at the Symposium on Magneto-Electric Interaction Phenomena in Crystals, May 21, 1973, at Battelle Seattle Research Center, Seattle, Washington.

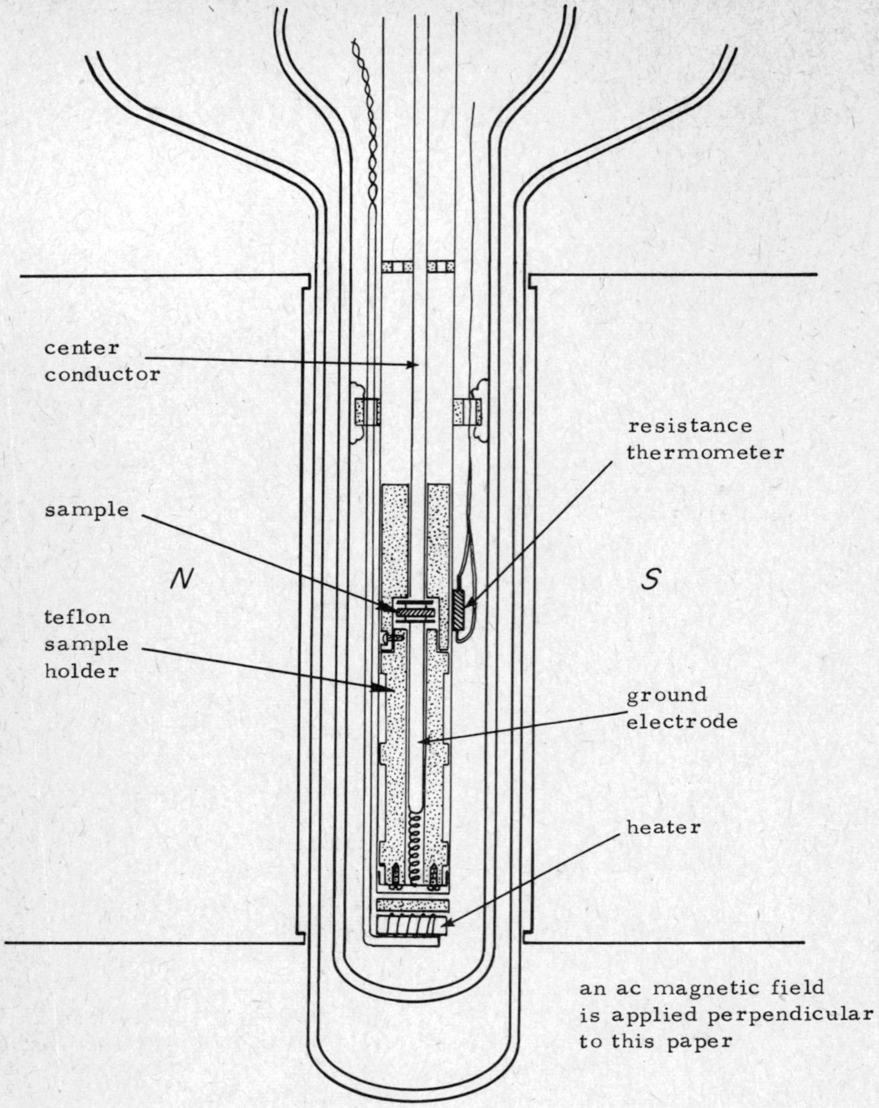

FIGURE 1 Apparatus for measuring $(\mathrm{PME})_P$ effect.

II EXPERIMENTAL AND RESULTS

The PME effect can generally be derived from a thermodynamic potential

$$F_{\mathrm{PME}} = -\xi_{ijk}\,\varepsilon_i\,H_j\,H_k \tag{1}$$

where ξ_{ijk} is a third-rank tensor which has the same symmetry properties as the third rank piezoelectric tensor. $NiSO_4 . 6H_2O$ belongs to the crystal class with point group symmetry D_4 or 422.[12] The only non-vanishing elements in the third-rank tensor are $-\xi_{14} = \xi_{25} = \xi$. The PME thermodynamic potential takes the form

$$F_{\mathrm{PME}} = -\xi(\varepsilon_a H_b - \varepsilon_b H_a)\,H_c = -\xi\varepsilon h H_c \tag{2a}$$

where c is the crystal tetragonal axis, $\boldsymbol{\varepsilon}$ and $\mathbf{h}$ are in the *ab*-plane and $\boldsymbol{\varepsilon} \perp \mathbf{h}$. The electric polarization induced by magnetic fields is called the $(\mathrm{PME})_P$ effect.

$$P = -\frac{\partial}{\partial\varepsilon}(F_{\mathrm{PME}}) = \xi h H_c \tag{2b}$$

The magnetic moment induced by applying a magnetic field and an electric field is called the $(PME)_M$ effect.

$$M = -\frac{\partial}{\partial h}(F_{\mathrm{PME}}) = \xi \varepsilon H_c \qquad (2c)$$

Crystals for the $(PME)_P$ effect were cut to the size of 0.077 in. × 0.352 in. × 0.350 in., where the last dimension is along the c-axis. Two broad surfaces of the crystal were silver painted. This crystal was placed in a low-capacitance coaxial tube with a dc magnetic field and an ac magnetic field applied at 90° to each other on the plane of the broad surfaces. The apparatus was shown in Figure 1. The electrical signal due to electric polarization between the painted surfaces was detected by a Keithley #603 electrometer amplifier, measured by a lock-in amplifier, and was recorded on a recorder. If the sample is rotated in a place where $H_{dc} \perp h\cos\omega_M t$, a component of the $(PME)_P$ effect, which has the

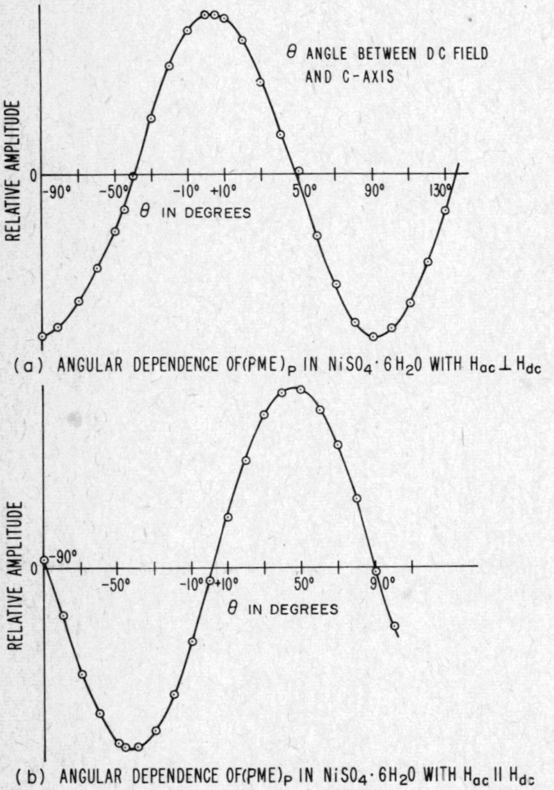

FIGURE 2 Angular dependence of $(PME)_p$ effect in $NiSO_4 . 6H_2O$.

same modulation frequency, has an angular dependence of $\xi h H_{dc} \cos\omega_m t \cos 2\theta$ where θ is the angle between H_{dc} and the c-axis of the sample. If the sample is rotated in a place where $H_{dc} \| h\cos\omega_m t$, the component of the $(PME)_P$ effect, which has the same

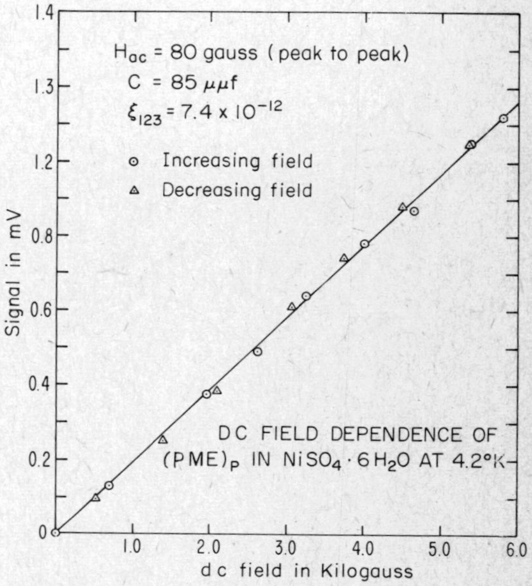

FIGURE 3 Determination of the coefficient of (PME) effect.

modulation frequency f_m, has an angular dependence of $\xi h H \cos\omega_m t \sin 2\theta$. The measured angular dependences for both cases are shown in Figure 2. Furthermore, no signal was observed in an experiment with a fused quartz of the same dimensions to replace the sample. The vanishing signal and the angular dependences as that shown in Figure 2 confirm that the detected signal was indeed a $(PME)_p$ effect.

Following the Gauss theorem, an induced ac polarization **P** may be considered as a constant current source. If the modulation frequency f_m is high enough so that $\omega_m(C_s + C_d) \gg (R_s^{-1} + R_d^{-1})$, the detected signal should be independent of frequency, as verified experimentally,

$$V_s = \frac{i\omega AP}{i\omega(C_s + C_d)} = \frac{\xi h H_c A}{C_s + C_d}\cos\omega_m t,$$

where A is the sample area, C_s and C_d are capacitance of sample and the input capacitance of the detection system, respectively. By measuring the slope of the $(PME)_P$ effect versus dc magnetic field as shown in Figure 3, the coefficient of the $(PME)_P$ effect at 4.2°K was determined as

$\xi(4.2°K) = 7.4 \times 10^{-12}$ ergs $(G)^{-2}(V/cm)^{-1}$ or

$\xi(T = 4.2°K) = (2.2 \pm 0.2) \times 10^{-9}$ e.s.u.

Measurements for temperatures between 1.28°K to 77°K revealed that the $(PME)_P$ effect has a peak at 3.0°K, changes sign at 1.39°K, and has a temperature dependence approximately proportional to T^{-n} for $77°K > T > 5°K$, where $n = 1.8 \pm 0.1$.

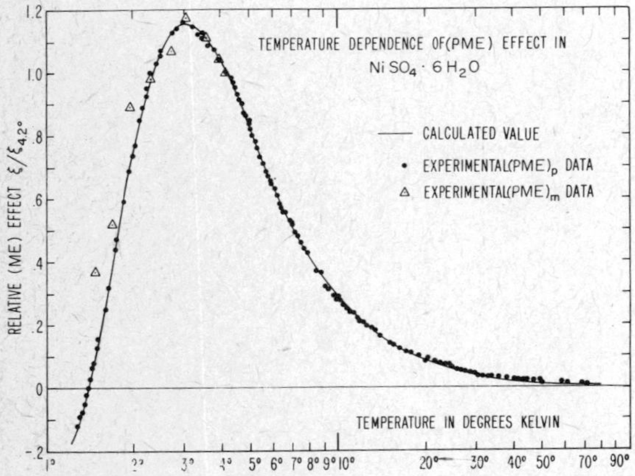

FIGURE 4 Experimental results of the temperature dependence of (PME) effect in $NiSO_4 . 6H_2O$.

The $(PME)_M$ effect was measured by applying a dc magnetic field along the *c*-axis and an ac electric field on the *ab*-plane. The induced magnetic moment $M = \xi h H$ was detected by a pair of identical 8000-turn pick-up coils. Similar result as the $(PME)_P$ effect was measured. Both the $(PME)_P$ and the $(PME)_M$ effects versus temperature were shown in Figure 4.

III MAGNETIC PROPERTIES OF $NiSO_4 . 6H_2O$

Understanding of magnetic properties is essential for studies on the (PME)-effect. A detailed review of magnetic properties of $NiSO_4 . 6H_2O$ has been given by Stout and Hadley.[13] The crystal structure as a whole has a point group symmetry of D_4 or 422, as examined by Beevers and Lipson.[12] Each Ni^{++} site has a C_2 symmetry. The degeneracy of spin triplet is totally lifted. The two-fold axis for sites I and III is along the (1,1,0) direction, and for sites II and IV along $(1,\bar{1},0)$. According to Pryce,[14] the spin Hamiltonian for $S = 1$ may be described as follows, if the principal axes are chosen to diagonalize the Λ tensor.

$$\mathscr{H}_S = -\lambda^2 \sum_\alpha \Lambda_\alpha S_\alpha^2 + \beta \sum_\alpha g_\alpha H_\alpha S_\alpha - \beta^2 \sum_\alpha \Lambda_\alpha H_\alpha^2 \tag{3}$$

where

$$\Lambda_\alpha = \sum_i \frac{\langle 0|L_\alpha|i\rangle\langle i|L_\alpha|0\rangle}{W_i - W_0} \tag{4}$$

and

$$g_\alpha = 2(1 - \lambda\Lambda_\alpha) \tag{5}$$

$|i\rangle$ and W_i are states and energy levels of the orbital triplet 3_{T_2}. The conventional spin Hamiltonian for $S = 1$ may be written as

$$\mathscr{H}_S = D[S_z^2 - \tfrac{1}{3}S(S+1)] + E(S_x^2 - S_y^2) + \mathbf{S}.\mathbf{g}.\mathbf{H} + C \tag{6}$$

Comparing Eq. (3) with Eq. (6), we have the following identifications.

$$\mathrm{D} = \tfrac{1}{2}\lambda^2(\Lambda_x + \Lambda_y - 2\Lambda_z) \tag{7}$$

$$\mathrm{E} = -\tfrac{1}{2}\lambda^2(\Lambda_x - \Lambda_y) \tag{8}$$

$$\mathrm{C} = -\tfrac{2}{3}\lambda^2(\Lambda_x + \Lambda_y + \Lambda_z) = -\frac{8\lambda^2}{10\mathrm{Dq}} \tag{9}$$

Hence

$$\left.\begin{aligned} g_x &= 2\left[1 - \frac{4\lambda}{10Dq} - \frac{1}{3\lambda}(D - 3E)\right] \\ g_y &= 2\left[1 - \frac{4\lambda}{10Dq} - \frac{1}{3\lambda}(D + 3E)\right] \\ g_z &= 2\left[1 - \frac{4\lambda}{10Dq} + \frac{2D}{3\lambda}\right] \end{aligned}\right\} \tag{10}$$

The parameters of spin Hamiltonians have been determined by a number of experiments. The low-temperature specific heat measured by Hadley and Stout[13] determines accurately $D = 4.77$ cm^{-1} and $E = 0.27$ cm^{-1}. The isotropic part of the g tensor is determined from the powder susceptibility at high temperatures, $\langle g \rangle_{av} = \frac{1}{3}(g_x + g_y + g_z) = 2.203 \pm 0.004$. Since the cubic splitting between T_2 and A_2 in the ground multiplet ^3F of a Ni^{++} ion is $10Dq = 8600$ cm^{-1}, from Eq. (5), $g_x = 2.216$, $g_y = 2.212$, and $g_z = 2.181$, if $\lambda = -285$ cm^{-1} is used. The magnetic susceptibility along each principal axis under the weak magnetic field approximation may be given as

$$\chi_x = \frac{2Ng_x^2\beta^2}{D-E}\frac{1-e^{-(D-E)/KT}}{1+e^{-(D-E)/KT}+e^{-(D+E)/KT}} + \frac{2N\beta^2}{\lambda}(1-g_x/2)$$

$$\chi_y = \frac{2Ng_y^2\beta^2}{D+E}\frac{1-e^{-(D+E)/KT}}{1+e^{-(D-E)/KT}+e^{-(D+E)/KT}} + \frac{2N\beta^2}{\lambda}(1-g_y/2)$$

$$\chi_z = \frac{2Ng_z^2\beta^2}{2E}\frac{e^{-(D-E)/KT}-e^{-(D+E)/KT}}{1+e^{-(D-E)/KT}+e^{-(D+E)/KT}} + \frac{2N\beta^2}{\lambda}(1-g_z/2) \quad (11)$$

Stout and Hadley[13] have derived from the observed magnetic anisotropy that the two-fold axis at the Ni^{++} site cannot be the z-axis. If this axis is chosen as the x-axis, the z-axis of the Λ tensor makes an angle $\phi = 40.2°$ with the tetragonal c-axis. Summation over the four different sites gives the macroscopic susceptibilities

$$\chi_c = 4(\chi_y \sin^2\phi + \chi_z \cos^2\phi)$$
$$\chi_{ab} = 2\chi_x + 2\chi_y \cos^2\phi + 2\chi_z \sin^2\phi \quad (12)$$

At extremely low temperatures, the exchange interaction $-2Jz\mathbf{S}.\langle\mathbf{S}\rangle$ should be included in Eq. (6). The magnetic susceptibility is modified into

$$\chi_{ex} = \frac{\chi}{1-2Jz/Ng^2\beta^2} \quad (13)$$

The best fit to low-temperature magnetic susceptibility measurements by Watanabe[15] and Hou[15] indicated that $2Jz = 0.42\ \text{cm}^{-1}$ (ferromagnetic). The results are shown in Figure 5. Comparison of Figures 4 and 5 shows the pronounced difference in temperature dependence of χ and ξ. In the next three sections, we shall study how each term in the spin Hamiltonian varies as a linear function of electric field ε, and in turn the derivation of the coefficient of the PME effect.

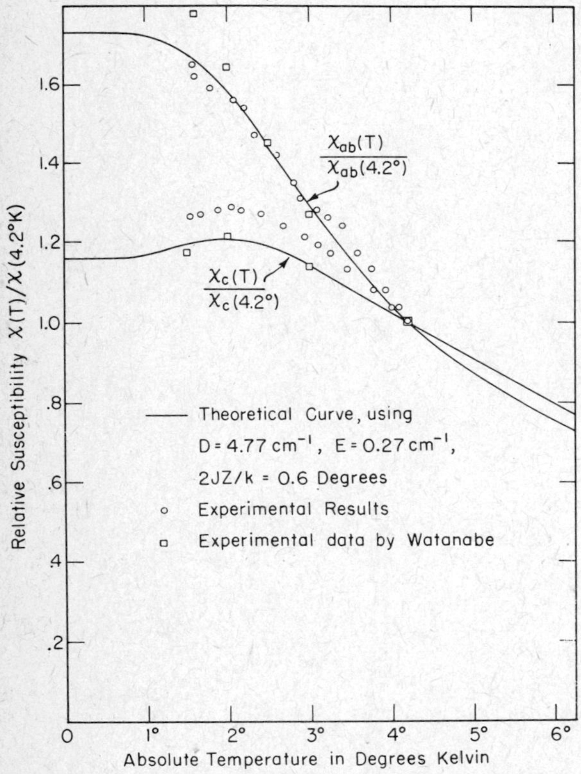

FIGURE 5 Low-temperature magnetic susceptibilities of $NiSO_4.6H_2O$.

IV GENERAL THEORY OF THE (PME) EFFECT

Since the Ni^{++} sites lack inversion symmetry, there will be components in the crystal-field potential with odd parity. This odd part of the potential V^u_{cr} will admix some odd parity configurations, such as $3d^7\,4p$, to $3d^8$ configuration. The admixture will be small because the $4p$ configuration is an estimated 60,000 cm^{-1} above the ^3Fg ground multiplet. Let the perturbing Hamiltonian be $\mathscr{H}' = \mathscr{H}_1 + \mathscr{H}_2$, where $\mathscr{H}_1 = \lambda\mathbf{L}.\mathbf{S} + \beta\mathbf{L}.\mathbf{H}$ and $\mathscr{H}_2 = V^u_{cr} - e\varepsilon.\mathbf{r}$. The higher-order corrections to the orbital ground singlet $^3A_{2g}$ may be written as

$$(\Delta\mathscr{H})_{\text{orbital}} = -\sum_m \frac{\langle ^3A_{2g}|\mathscr{H}_1|m\rangle\langle m|\mathscr{H}_1|^3A_{2g}\rangle}{W_m - W_0} - \sum_u \frac{\langle ^3A_{2g}|\mathscr{H}_2|u\rangle\langle u|\mathscr{H}_2|^3A_{2g}\rangle}{W_u - W_0} + \sum_{m,u} \frac{\langle ^3A_{2g}|\mathscr{H}_1|m\rangle\langle m|\mathscr{H}_2|u\rangle\langle u|\mathscr{H}_2|^3A_{2g}\rangle}{(W_m-W_0)(W_u-W_0)} - \sum_{m,n,u} \frac{\langle ^3A_{2g}|\mathscr{H}_1|m\rangle\langle m|\mathscr{H}_2|u\rangle\langle u|\mathscr{H}_2|n\rangle\langle n|\mathscr{H}_1|^3A_{2g}\rangle}{(W_m-W_0)(W_u-W_0)(W_n-W_0)}$$
$$+ \text{higher order perturbations} \quad (14)$$

Here W_0 is the energy of the orbital ground state $^3A_{2g}$, m and n are states of the $^3T_{2g}$ orbital triplet of the $3d^8 3F$ configuration, and u is a state with opposite parity configuration $3d^7 4p$. The first term is identical to Eq. (3) which is the spin Hamiltonian derived by Pryce.[14] The second term produces a shift to $^3A_{2g}$ linearly proportional to the applied electric field. Similar shift may occur to the excited states. The difference of shifts between the excited state and the ground state constitutes the shift of optical spectrum linear with respect to electric field. For example, splitting on R line of Cr^{3+} in ruby linear with respect to the electric field is a result of the shift of the ground state $^4A_{2g}$ and that of the excited state ^2E_g of two inequivalent Cr^{3+} sites.[16] However, since all spin multiplicities of the orbital ground state shift equally, it contributes nothing to the ME effect. The third term connects the external applied field with one spin operator or a magnetic field with an electric field. This term could exist in ferromagnetic or antiferromagnetic bodies, as it changes sign on time reversal. It also contributes to the paramagnetic system at extremely low temperatures where the exchange interaction can no longer be ignored. This term will be discussed further in section VI. The fourth term is the term which shifts Λ-tensor elements linearly with respect to the electric field.

If we temporarily ignore the linear electric field effect to the exchange interaction in a piezoelectric paramagnetic system, the modified spin Hamiltonian may be written as

$$\mathscr{H}_{s,\varepsilon} = \mathscr{H}_s - \lambda^2 \sum R_{Jkl}\,\varepsilon_J S_k S_l - \lambda\beta \sum R_{Jkl}\,\varepsilon_J S_k H_l - \beta^2 \sum R_{Jkl}\,\varepsilon_J H_k H_l \tag{15}$$

where κ_s is given in Eq. (3), and

$$\Delta_\varepsilon \Lambda_{kl} = \sum_J R_{Jkl}\,\varepsilon_J \tag{16}$$

The third rank R tensor has components given by

$$R_{Jkl}\,\varepsilon_J = 2 \sum_{m,n,u} \frac{\langle {}^3A_{2g}|L_k|m\rangle\langle m|V_{cr}^u|u\rangle\langle u|-e\boldsymbol{\varepsilon}.\mathbf{r}|n\rangle\langle n|L_l|{}^3A_{2g}\rangle}{(W_m - W_0)(W_u - W_0)(W_n - W_0)} \tag{17}$$

The factor 2 is added because the real perturbations V_{cr}^u and $-e\varepsilon.\mathbf{r}$ can be taken in different order. In general, the first nine R-toner elements change the lengths of the principal axes of the Λ tensor as given in Eq. (16). Following the relationships given in Eqs. (7), (8), and (9), the shifts in $10Dq$, and in the parameters of spin Hamiltonian D and E are therefore obtained. The second nine R-tensor elements produce a rotation of the Λ tensor, if the original Λ tensor is diagonalized. For example, the elements R_{14}, R_{24}, and R_{34} describe a rotation in the yz-plane around the x-axis produced by ε_x, ε_y, and ε_z. From Eq. (16), we have $\Delta\Lambda_{yz} = R_{14}\varepsilon_x + R_{24}\varepsilon_y + R_{34}\varepsilon_z$. The magnitude of the rotation may be obtained from a rediagonalization of the Λ tensor

$$\begin{Vmatrix} \Lambda_x & 0 & 0 \\ 0 & \Lambda_y & \Delta\Lambda_{yz} \\ 0 & \Delta\Lambda_{yz} & \Lambda_z \end{Vmatrix}$$

The tensor is rotated through an angle θ_x given by

$$\tan 2\theta_x \approx 2\theta_x = \frac{2(R_{14}\,\varepsilon_x + R_{24}\,\varepsilon_y + R_{34}\,\varepsilon_z)}{\Lambda_z - \Lambda_y} \tag{18}$$

The change in the magnetic free energy caused by this rotation is

$$-\tfrac{1}{2}(\chi_2 - \chi_y)\sin 2\theta_x H_y H_z \approx -(\chi_z - \chi_y)\frac{R_{14}\,\varepsilon_x + R_{24}\,\varepsilon_y + R_{34}\,\varepsilon_z}{\Lambda_z - \Lambda_y} H_y H_z \tag{19}$$

Hence the eighteen R tensor elements are reduced into six irreducible parameters, namely, the electrical shifts of D, E, and $10Dq$ (or the isotropic part of g tensor), and three magnetic anisotropies caused by rotations around principal axes of Λ tensor. The PME free energy of the crystal as a whole is the summation of the PME free energies in each site created by the six irreducible parameters.

V THE PME-EFFECT IN $NiSO_4.6H_2O$

Since Ni^{++} is in a site of C_2 symmetry with z-axis as twofold axis, 18 R-tensor elements are reduced into 8 non-vanishing elements.

$$\begin{Vmatrix} R_{11} & R_{12} & R_{13} & R_{14} & 0 & 0 \\ 0 & 0 & 0 & 0 & R_{25} & R_{26} \\ 0 & 0 & 0 & 0 & R_{35} & R_{36} \end{Vmatrix}$$

According to Eqs. (5), (7), (8), and (9), we have he following identifications.

$$\left.\begin{aligned} \Delta D &= \tfrac{1}{2}\lambda^2(R_{11}+R_{12}-2R_{13})\,\varepsilon_x \\ \Delta E &= -\tfrac{1}{2}\lambda^2(R_{11}-R_{12})\,\varepsilon_x \\ \frac{8\lambda}{(10Dq)^2}\Delta(10Dq) &= -\tfrac{2}{3}\lambda(R_{11}+R_{12}+R_{13})\,\varepsilon_x \end{aligned}\right\} \quad (20)$$

$$\left.\begin{aligned} \Delta g_x &= \frac{8\lambda}{(10Dq)^2}\Delta(10Dq) - \tfrac{2}{3}\lambda^{-1}(\Delta D - 3\Delta E) \\ \Delta g_y &= \frac{8\lambda}{(10Dq)^2}\Delta(10Dq) - \tfrac{2}{3}\lambda^{-1}(\Delta D + 3\Delta E) \\ \Delta g_z &= \frac{8\lambda}{(10Dq)^2}\Delta(10Dq) + \frac{4}{3\lambda}\Delta D \end{aligned}\right\} \quad (21)$$

The electrical shift in $10Dq$ is the electrical shift in the isotropic part of the g tensor. The shifts in the anisotropic part of g-tensor may be combined with that of D and E. The paramagnetoelectric free energy for the Ni^{++} ions in site I is found to be

$$F^{\mathrm{I}}_{\mathrm{PME}} = \frac{1}{2}\sum_{\alpha=x,y,z}\left\{\frac{\partial\chi_\alpha}{\partial g_\alpha}\frac{\partial g_\alpha}{\partial\varepsilon_x} + \frac{\partial\chi_\alpha}{\partial D}\frac{\partial D}{\partial\varepsilon_x} + \frac{\partial\chi_\alpha}{\partial E}\frac{\partial E}{\partial\varepsilon_x}\right\} H_\alpha^{\,2}\,\varepsilon_x$$

$$- \frac{\chi_z-\chi_y}{\Lambda_z-\Lambda_y} R_{14}\,\varepsilon_x H_y H_z$$

$$- \frac{\chi_y-\chi_x}{\Lambda_y-\Lambda_x}(R_{26}\,\varepsilon_y + R_{36}\,\varepsilon_z) H_x H_y$$

$$- \frac{\chi_x-\chi_z}{\Lambda_x-\Lambda_z}(R_{25}\,\varepsilon_y + R_{35}\,\varepsilon_z) H_z H_x \quad (22)$$

The paramagnetoelectric free energy of the entire crystal as a whole is the summation of the PME free energies of four sites. The orthorhombic two-fold axis of these sites are related to one another by rotation of $\pi/2$ around the c-axis. The tetragonal axis is at right angle to the x-axis and makes an angle $\phi = 40.2°$ with the z-axis at each site. On summation various terms cancel and the only remaining terms in the PME free energy of the crystal are those occurring in Eq. (2a). The temperature dependence of the PME effect is controlled by six functions in front of six irreducible parameters. The coefficient of the PME effect of entire crystal ξ is given as follows:

$$\begin{aligned} \xi = \sin 2\phi\{&[f_D(T)+f_{gD}(T)](R_{11}+R_{12}-2R_{13})\,\lambda^2/2 \\ &+ [f_E(T)+f_{gE}(T)](R_{11}-R_{12})\,\lambda^2/2 \\ &+ f_{10Dg}(T)\,2\lambda^2(R_{11}+R_{12}+R_{13})\} \\ &+ 2\lambda^2\cos 2\phi f_A(T)(R_{14}) \\ &+ 2\lambda^2\cos\phi f_B(T)(R_{25}\cos\phi - R_{35}\sin\phi) \\ &+ 2\lambda^2\sin\phi f_C(T)(R_{26}\cos\phi - R_{36}\sin\phi) \end{aligned} \quad (23)$$

where

$$f_D(T) = \frac{\partial\chi_y}{\partial D} - \frac{\partial\chi_z}{\partial D}$$

$$f_E(T) = -\frac{\partial\chi_y}{\partial E} + \frac{\partial\chi_z}{\partial E}$$

$$f_{gD}(T) = -\frac{2}{3\lambda}\left(\frac{\partial\chi_y}{\partial g_y} + 2\frac{\partial\chi_z}{\partial g_z}\right)$$

$$f_{gE}(T) = -\frac{2}{3\lambda}\frac{\partial\chi_y}{\partial g_y}$$

$$f_A(T) = (\chi_z-\chi_y)/(D-E)$$

$$f_B(T) = -(\chi_x-\chi_z)/(D+E)$$

$$f_C(T) = (\chi_x-\chi_y)/2E$$

and

$$f_{10Dq}(T) = -\frac{1}{3\lambda}\left(\frac{\partial\chi_y}{\partial g_y} - \frac{\partial\chi_z}{\partial g_z}\right)$$

$$\approx -\frac{1}{3\lambda}\frac{2}{g}(\chi_y-\chi_z) = \frac{2}{g}\frac{D-E}{3\lambda}f_A(T) \quad (24)$$

The temperature dependence due to electric shift of the isotropic part of g-tensor (or the electric shift of $10Dq$) $f_{10D}(T)$ is accidentally identical to that due to a rotation of Λ-tensor around the x-axis $f_A(T)$. The magnitude is, however, only $(D-E/3\lambda) \approx 10^{-2}$ that of $f_A(T)$. The temperature dependence of all functions given in Eq. (24) are listed in Table I.

In the high temperatures, all functions follow a T^{-2} dependence except those due to electric shift in

TABLE I
Temperature dependence of five irreducible parameters

	$F_D(T)$		$F_E(T)$		$F_R(T)$
	$f_D(T)$	$f_{gD}(T)$	$f_E(T)$	$f_{gE}(T)$	$f_A(T), f_B(T), f_C(T)$
T = 0°K	$-\dfrac{2Ng^2\beta^2}{(D+E)^2}$	$\dfrac{8Ng\beta^2}{3\lambda(D+E)}$	$\dfrac{2Ng^2\beta^2}{(D+E)^2}$	$\dfrac{-8Ng\beta^2}{\lambda(D+E)}$	$\dfrac{2Ng^2\beta^2}{D^2-E^2}$
At high temp.	$\dfrac{2Ng^2\beta^2}{6k^2T^2}$	$\dfrac{8Ng\beta^2}{2\lambda kT}$	$\dfrac{2Ng^2\beta^2}{6k^2T^2}$	$\dfrac{-8Ng\beta^2}{3\lambda kT}$	$\dfrac{2Ng^2\beta^2}{6k^2T^2}$

the anisotropic part of g-tensor $f_{gD}(T)$ and $f_{gE}(T)$, which follow a T^{-1} dependence. The relative contributions of $f_{gD}(T)$ and $f_{gE}(T)$ are $(D+E)/\lambda \cong 10^{-2}$ times smaller than those of $f_D(T)$ and $f_E(T)$ at low temperature. It is the small contribution due to electric shift of anisotropic part of g-tensor which makes the temperature dependence of PME effect slightly deviated from T^{-2} for $5°\text{K} < T < 70°\text{K}$.

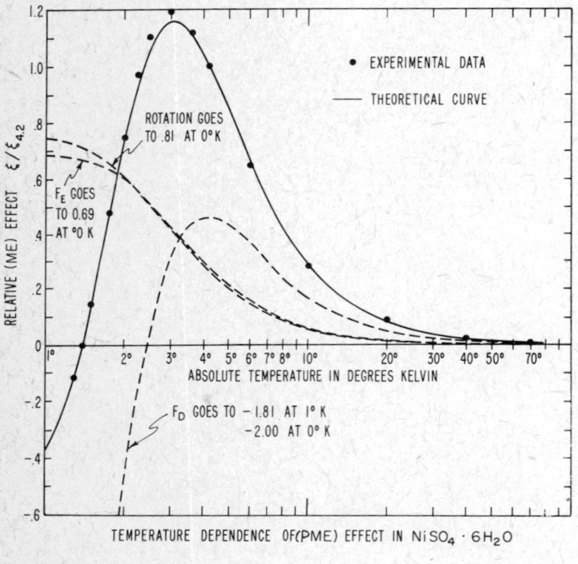

FIGURE 6 Temperature dependence of (PME) effect in $NiSO_4 . 6H_2O$.

At $T = 0°\text{K}$, all functions approach their asymptotic values of the same sign, except that caused by the linear electric shift of $D, f_D(T)$, which approaches its asymptotic value with an opposite sign. This is the only mechanism which changes the sign of PME effect at low temperatures.

Let $F_D(T) = f_D(T) + f_{gD}(T)$ be the overall effect due to electric shift of D, $F_E(T) = f_E(T) + f_{gE}(T)$ be the overall effect due to electric shift of E, and $F_R(T) = \frac{1}{3}[f_{10Dq}(T) + f_A(T) + f_B(T) + f_C(T)]$ be the overall effect due to rotations around principal axes of Λ-tensor, the theoretical temperature dependences of $F_D(T)$, $F_E(T)$ and $F_R(T)$ are plotted in Figure 6.

After multiplying by appropriate weight factors to $F_D(T)$, $F_E(T)$, and $f_R(T)$ as shown in Figure 5, the sum represented by the full drawn curve approximates closely the experimentally observed temperature dependence of ξ. The sign reversal at low temperature is due to variation of D with electric field. The best fit with the experimental data is obtained for the following values of coefficients of Eq. 20,

$$\frac{\lambda^2}{2}(R_{11} + R_{12} - 2R_{13}) = 7.86 \times 10^{-8}\,\text{cm}^{-1}\,[V/\text{cm}]^{-1} \tag{27}$$

$$\frac{\lambda^2}{2}(R_{11} - R_{12}) = 2.60 \times 10^{-8}\,\text{cm}^{-1}\,[V/\text{cm}]^{-1} \tag{28}$$

Overall contribution due to rotations

$$= 2.5 \times 10^{-8}\,\text{cm}^{-1}\,(V/\text{cm})^{-1} \tag{29}$$

and

$$\left|\frac{1}{D}\frac{\partial D}{\partial \varepsilon_x}\right| = (1.65 \pm 0.15) \times 10^{-8}\,(V/\text{cm})^{-1} \tag{30}$$

VI CONTRIBUTIONS DUE TO EXCHANGE INTERACTION

At extremely low temperatures, the exchange interaction can no longer be ignored from the spin Hamiltonian given in Eq. (6). The modified magnetic susceptibility may be obtained in Eq. (13) and is shown in Figure 5. The existence of the electric shift in exchange interaction has been discussed in the third term of Eq. (14). This term is a third order perturbation which consists of one orbital angular momentum operator, one odd parity crystalline field and a linear electric field operator. Hence, the magnetoelectric effect in magnetically order system will be larger in crystals with $3d$ transition metal ions than those with $4f$ rare earth ions. The correction to the PME effect due to shift of exchange interaction may be described by

$$\frac{\partial \chi_{ex}}{\partial \varepsilon} = \frac{\dfrac{\partial \chi}{\partial \varepsilon} + \chi^2[2zN^{-1}g^{-2}\beta^{-2}]\dfrac{\partial J}{\partial \varepsilon}}{[1-(2Jz/Ng^2\beta^2)\chi]^2} \tag{31}$$

The contribution to the PME effect due to $\partial J/\partial \varepsilon$ has a temperature dependence of T^{-2} at high temperature and does not change sign at absolute zero. Such a temperature dependence is quite similar to those of $f_E(T)$ and $f_R(T)$. Although J is small in most of the paramagnetic systems and the effect of $\partial J/\partial \varepsilon$ does not affect the contribution of $F_D(T)$, the relative contributions of the terms $F_E(T)$ and $f_R(T)$ in Figure 6 and Eqs. (25) and (26) may require a considerable correction. The various contributions to the PME effect cannot be unravelled from the measurement of $\xi(T)$ along. Perhaps the best method to reveal the importance of $\partial J/\partial \varepsilon$ is studying the electric shift of magnetic resonance spectra of ion pairs. The experiment is feasible in ruby, but is not informative in $NiSO_4.6H_2O$ because of the excessive line width in the EPR spectrum.

The contribution of $\partial J/\partial \varepsilon$ to the magnetoelectric effect may be significant in the magnetically ordered system. The possible contribution from the electric shift in exchange interaction between two Cr^{3+} ion pairs to the magnetoelectric effect in antiferromagnetic Cr_2O_3 has been discussed in detail by Pratt.[17] The electric field variations of the exchange interaction between $3d$-transition metal ions may be important for the magneto-electric effect of the weakly ferromagnetic piezoelectric crystal of gallium ferrite,[6] and the ferroelectric and weakly ferromagnetic boracites.[18]

VII DISCUSSION AND CONCLUSION

The microscopic mechanism of the PME effect has been discussed in detail in this paper. An excellent agreement has been obtained between the theoretically derived temperature dependence and the experimentally measured results. The sign reversal of the PME effect in $NiSO_4.6H_2O$ at low temperature is dominated by the variation of D with electric field.

In the extremely high magnetic field approximation where $g\beta H \gg D$, the discussions in the previous sections no longer holds. The spin Hamiltonian given in Eqs. (3) and (15) is dominated by Zeeman splitting. One must diagonalize the spin Hamiltonian with respect to $g\beta \mathbf{S}.\mathbf{H}_{dc}$. Magnetic moments should be calculated instead of magnetic susceptibilities. The paramagnetoelectric effect becomes quite small. One should observe the PME effect with a T^{-1} dependence and without a change in signs at low temperatures.

Throughout this paper, we have ignored the strain effect of the piezoelectric crystal. However, one should not misinterpret the observed paramagnetoelectric effect as the piezoelectric strain-induced magnetoelectric effect, because the typical effect due to piezoelectric strain is only a few percent of that due to the influence of the electric field on the spin resonance spectra as measured in a number of dilute paramagnetic crystals.[22] The overall temperature dependence of the observed effect is dominated by the variation of D with electric field. The best method to separate the variation of spin Hamiltonian parameters due to electric field may be obtained from the linear electric field shift on the spin resonance spectra to be discussed later by Dr. Kiel.[19]

Although $NiSO_4.6H_2O$ is the material discussed in detail in this paper, the microscopic mechanism and physics are general enough for most of $3d$ transition metal ions. The microscopic basis of the PME effect is essentially the same as for the magnetoelectric effect in antiferromagnetic Cr_2O_3 as proposed by Rado.[20] Since the R tensor defined in Eqs. (14) and (17) equally weighs the spin-orbit interaction and the odd crystalline field, piezoelectric crystals consisting of $3d$ transition metal ions and those consisting of $4f$ rare earth ions are equally favorable for the (PME) effect. Similar paramagnetoelectric effect may be observed in piezoelectric crystals $NiSeO_4.6H_2O$, $Nd(BrO_3)_3.9H_2O$ and the rare earth ethylsulfates.[21]

REFERENCES

1. D. N. Astrov, *JETP*, **11**, 708 (1960).
2. V. J. Folen, G. T. Rado and E. W. Stalder, *Phys. Rev. Letters* **6**, 607 (1961); G. T. Rado and V. J. Folen, *Phys. Rev. Letters* **7**, 310 (1961).
3. L. D. Landau and E. M. Lifshitz, *Electrodynamics of Continuous Media* (Addison-Wesley Publishing Co., Inc., Reading, Mass., 1930), pp. 116–119.
4. I. F. Dzyaloshinskii, *JETP*, **10**, 628 (1960).

5. N. Bloembergen and S. L. Hou, *Bull. Amer. Phys. Soc.* **9**, 13 (1964); S. L. Hou and N. Bloembergen, *Phys. Rev.* **138**, A1218 (1965).
6. G. T. Rado, *Phys. Rev. Letters* **13**, 335 (1964).
7. T. H. O'Dell, *Philosophical Magazine* **16**, 487 (1967).
8. E. Ascher, *Philosophical Magazine* **17**, 149 (1968).
9. H. Schmid, Meeting on Ferroelectricity, Orleans, France, September, 1972.
10. N. Bloembergen, *Proceedings of the International Conference on High Magnetic Fields*, ed. B. Lax (John Wiley & Sons, Inc., 1962), p. 454.
11. N. Bloembergen, *Science* **133**, 1363 (1961); N. Bloembergen, *Proceedings of the XI Colloque Ampere*, ed. J. Smidt (North-Holland Publishing Co., Amsterdam 1963), p. 39.
12. C. A. Beevers and H. Lipson, *Z. Krist.* **83**, 123 (1932).
13. J. W. Stout and W. B. Hadley, *J. Chem. Phys.* **40**, 55 (1964).
14. M. H. L. Pryce, *Proc. Roy. Soc.* (London) **A214**, 237 (1952).
15. T. Watanabe, *J. Phys. Soc. Japan* **17**, 1356 (1962).
16. W. Kaiser, S. Sugano and D. L. Wood, *Phys. Rev. Letters* **6**, 605 (1961); M. B. Cohen and N. Bloembergen, *Phys. Rev.* **135**, A950 (1964).
17. T. Isuyama and G. W. Pratt, *J. Appl. Phys.* **34**, 1226 (1963).
18. H. Schmid, *Phys. Stat. Sol.* **37**, 209 (1970); Ascher, H. Rieder, H. Schmid and H. Stossel, *J. Appl. Physics* **37**, 1404 (1966).
19. A. Kiel, This Conference, Seattle (1973).
20. G. T. Rado, *Phys. Rev. Letters* **6**, 609 (1961). G. T. Rado, *Phys. Rev.* **128**, 2546 (1964).
21. Tsuyoshi Murao, *Prog. Theor. Phys.* **37**, 1038 (1967).
22. F. A. Collins and N. Bloembergen, *J. Chem. Phys.* **40**, 3479 (1964).

MICROSCOPIC ORIGINS OF PIEZOMAGNETISM AND MAGNETOELECTRICITY†

ROBERT L. WHITE

Stanford Electronic Laboratories, Stanford University, Stanford, California 94305

The sensitivity to microscopic strains of various magnetic parameters, particularly g factors, have been determined in investigations whose primary objective was the understanding of piezomagnetic constants. These same microscopic strain coefficients may be used, in conjunction with estimates of local strains produced by an applied electric field, to estimate magnetoelectric coefficients assuming that microscopic displacements play an important role in magnetoelectricity. Making such a model, we obtain magnetoelectric coefficients of the same order of magnitude as those actually observed. We therefore conclude that atomic displacements are important in the magnetoelectric effect, and that a "frozen lattice" calculation of the atomic origins of magnetoelectricity might well be inadequate.

INTRODUCTION

In a piezomagnetic material, an applied stress produces a magnetic moment because, at a microscopic level, magnetic parameters such as the g values or exchange interactions are altered by the local atomic displacements. In a magnetoelectric material, an applied electric field produces a magnetic moment because, at a microscopic level, magnetic parameters such as the g values or exchange interactions are altered.[1] It is not known with any degree of certainty whether these changes in magnetic parameters in the magnetoelectric case are caused by direct action of the electric field upon the charge distributions of the atoms (the overall lattice remaining rigid), or whether the electric field induces atomic displacements (conceptually similar to the displacements under stress) and the atomic displacements in turn produce the alteration of magnetic parameters. That atomic displacements might be implicated in the magnetoelectric effect was suggested as long ago as 1964 by Rado,[2] but little quantitative data has been brought to bear on the subject in the intervening years. The purpose of this article is to point out that data taken during investigations of the microscopic organs of piezomagnetism supply some of the missing quantitative links. We produce here this data and argue that the piezomagnetic data, combined with known or estimated ionic displacements in magnetoelectric crystals, lead to predicted values of α, the magnetoelectric coefficient, consistent with observed values.

† This paper has been presented in the Symposium on Magnetoelectric Interaction Phenomena in Crystals on May 21, 1973, at Battelle Seattle Research Center, Seattle, Washington.

THE PIEZOMAGNETIC DATA

The relevant microscopic data on piezomagnetism and magnetostriction were generated primarily by Phillips, Townsend, and White in a series of studies[3] relating the strain dependence of atomic parameters, especially $F_{ijkl} = \partial g_{ij}/\partial \epsilon_{kl}$, obtained from EPR and optical studies under uniaxial stress, with the macroscopic piezomagnetic and magnetostriction constants of isomorphous magnetic compounds of the same structure. Typical data for the dominant components of the F tensor for various magnetic ions, and estimates of the associated piezomagnetic constants of crystals composed of such materials, $P = N\beta SF$ (where N is the number of ions per cc, β is the Bohr magneton and S is the ionic spin) are listed in Table I.

TABLE I

Ion	Mn^{2+}, Fe^{3+}	Co^{2+}	Fe^{2+}	Yb^{3+}	Er^{3+}
$F = \dfrac{\partial g}{\partial \epsilon}$ (dimensionless)	1	60	150	150	300
$P_\epsilon = N\beta SF$ (Gaussian units)	2×10^2	3×10^3	1.5×10^4	6×10^3	1×10^4

We note that the S state ions Fe^{3+} and Mn^{2+} are one to two orders of magnitude less effective in magnetoelastic phenomena than the other magnetic ions. It is also worthy of note that for all piezomagnetic and magnetostrictive materials studied the two-ion mechanisms dominated the effect and single-ion effects were inadequate to explain the data. (In the earliest studies the variation of g with strain for S state ions was assumed negligible, and the magnetoelastic effects attributed to single-ion effects; in all cases this assumption was later shown to be wrong and two-ion effects dominant.) The moral for magnetoelectricity is obvious; two-ion effects are probably dominant.

APPLICATION OF THE PIEZOMAGNETIC DATA TO MAGNETOELECTRICITY

We now wish to show that ionic displacements occurring in magnetoelectric materials due to the applied electric field are of sufficient magnitude to produce, through essentially piezomagnetic mechanisms, the observed magnetic moments. This argument proceeds especially easily in materials which are piezoelectric as well as magnetoelectric.

Of the 58 magnetic crystal classes which are magnetoelectric, 38 are also piezoelectric and piezomagnetic. However, for only one such crystal, $Ga_{2-x}Fe_xO_3$, do we have enough data to test the hypothesis that the magnetoelectric effect is mediated by strains. The piezoelectric constants of $Ga_{2-x}Fe_xO_3$ have been measured[4] to be $d_{12} = d_{23} = 1.4 \times 10^{-7}$ esu, so the strain ϵ is given by $\epsilon = 1.4 \times 10^{-7}E$. The piezomagnetic contants of $Ga_{2-x}Fe_xO_3$ have not been measured, but we may take as a lower bound the piezomagnetic constants of αFe_2O_3,[5] which contains the S state Fe^{3+} ion in a relatively undistorted octahedral site. For αFe_2O_3, $P_{14} = 3 \times 10^{-10}$ Gauss/(dyne/cm^2). Taking a representative value of the elastic stiffness constants of such an oxide $c = 10^{12}$, we have

$$H = PcdE = 3 \times 10^{10} \times 10^{12} \times 1.4 \times 10^{-7}E$$
$$= 5 \times 10^{-5}E$$

or $\alpha = 5 \times 10^{-5}$. This is one order of magnitude smaller than the magnetoelectric constant observed[6] for $Ga_{2-x}Fe_xO_3$. On the other hand we have made a very conservative estimate of the piezomagnetic constant. The small value is posited on the iron in $Ga_{2-x}Fe_xO_3$ being Fe^{3+} in an S state. There is reason to doubt this assumption. The susceptibility of $Ga_{2-x}Fe_xO_3$ above its Curie temperature cannot be fit by the usual Fe^{3+} S-state moment,[7] and the anisotropic spontaneous magnetization at 0°K requires an anisotropic g-tensor not characteristic of the S-state ion.[8] If the S-state configuration of the Fe^{3+} ion has been compromised, the piezomagnetic constant of $Ga_{2-x}Fe_xO_3$ may easily be an order of magnitude larger than that of αFe_2O_3, as can be seen from comparing any other ion in Table I with Fe^{3+}. We conclude that for $Ga_{2-x}Fe_xO_3$, the magnetoelectric effect may well be explained by a piezomagnetic microscopic phenomenon.

We now must face the problem of magnetoelectric crystals which are not piezoelectric. Do we expect that local ionic displacements by an electric field will be less in a material that is not piezoelectric than they typically are for piezoelectric materials? Not at all. The principal difference in the two cases is that for piezoelectric crystals the microscopic strains are cumulative giving rise to a detectable deformation, while in non-piezoelectric crystals the microscopic strains are not cumulative, and do not produce observable macroscopic changes. The studies by Kiel and Mims on the changes in EPR spectra of ions subject to electric fields[9] clearly reveal that ionic displacements are significant. Given that the microscopic strains in all materials are similar to those observed in piezoelectric crystals, the microscopic piezomagnetic constants of Table I lead to α of the observed magnitude for all crystals.

CONCLUSION

We conclude that microscopic strains play an important role in the magnetoelectric effect, and must be included in any quantitative calculation of the microscopic origins of the magnetoelectric effect.

REFERENCES

1. G. T. Rado, *Phys. Rev. Letters* **6**, 609 (1961); G. T. Rado, *Phys. Rev.* **128**, 2546 (1962); M. Date, J. Kanamori and M. Tachiki, *J. Phys. Soc. Japan* **16** 2589 (1961); E. B. Royce and N. Bloembergen, *Phys. Rev.* **131**, 1912 (1963); S. Alexander and S. Shtrikman, *Solid State Commun.* **4**, 115 (1966).
2. G. T. Rado, *Proceedings of the International Conference on Magnetism, Nottingham, 1964* (The Institute of Physics and the Physical Society, London, 1965), p. 361.
3. T. G. Phillips and R. L. White, *Phys. Rev.* **153**, 616 (1967); *J. Appl. Phys.* **38**, 1222 (1967); *Phys. Rev.* **160**, 316 (Fig. 7); T. G. Phillips, R. L. Townsend, Jr. and R. L. White, *Phys. Rev. Lett.* **18**, 646 (1947); *Phys. Rev.* **160**, 316 (1967); R. L. Townsend, Jr., T. G. Phillips and R. L. White, *Solid State Comm.* **6**, 603 (1968).
4. D. L. White, *Bull. Am. Phys. Soc.* **5**, 189 (1960) or cited in J. P. Remeika, *J. Appl. Phys.* **31**, 263 (1960).
5. V. P. Andratskii and A. S. Borovik-Romanov, *JETP* **51**, 1030 (1966) [*Soviet Physics JETP* **24**, 687 (1967)].
6. G. T. Rado, *Phys. Rev. Lett.* **13**, 335 (1964). Proc. Internatl. Conf. on Magnetism, Nottingham 1964, p. 361 (London, Inst. of Physics, 1965).
7. C. H. Nowlin and R. V. Jones, *J. Appl. Phys.* **34**, 1262 (1963).
8. G. T. Rado, *Phys. Rev.* **176**, 644 (1968).
9. W. B. Mims, *Phys. Rev.* **140**, A531 (1965); A. Kiel, *Phys. Rev.* **148**, 247 (1966); A. Kiel and W. B. Mims, *Phys. Rev.* **B1**, 2935 (1970).

Discussion

S. L. HOU Although strain induced magneto-electric effect may contribute to the paramagneto-electric (PME) effect, it is unlikely that PME effect is 100% due to the piezoelectrically induced magneto-electric effect. What is your opinion?

R. L. WHITE Since the data does not exist which would allow us to separate rigid lattice effects from ion motion effects, I can only offer a statement of prejudice here. For what it is worth, my opinion is that microscopic strain will be the dominant mechanism for ions like Fe^{2+}, Yb^{3+} and Er^{3+}, which are very strain sensitive, but that for S state ions like Fe^{3+}, Mn^{2+} and Gd^{3+}, no sensible guess can be made and several effects might be of equal importance.

J. M. TROOSTER What do you mean by your statement that Fe in $GaFeO_3$ is not Fe^{3+}? Mössbauer data give no indication that the valence state in this case is different from that of Fe in, for instance, Fe_2O_3.

R. L. WHITE The evidence suggesting that Fe in $GaFeO_3$ is not Fe^{3+} is (a) Nowlin and Jones found, fitting the paramagnetic susceptibility of $GaFeO_3$ above its Curie temperature a moment per ion where upper bound was $g^2S(S+1) = 7$ rather than the value of 35 one would expect from Fe^{3+} (b) the anisotropy of the magnetization of $GaFeO_3$ strongly suggests an anisotropy of g which is much greater than typical of Fe^{3+}. However, if Mössbauer data show that the iron is trivalent, I would consider that a much more binding and specific determination than either of the above evidence. The piezomagnetism of $GaFeO_3$ might nonetheless still be much larger than that of α-Fe_2O_3 because of the large site distortion (evidenced in the Mössbauer data as a quadrupole splitting) in $GaFeO_3$ as compared to α-Fe_2O_3.

S. SHTRIKMAN How do your results compare with the measurements of Borovik-Romanov?

R. L. WHITE For CoF_2, Borovik-Romanov found $P_{14} = 2.1 \times 10^{-3}$ and $P_{36} = 0.8 \times 10^{-3}$ both in $G/(\mathrm{kg/cm^2})$. We calculate, using our EPR data on the strain-dependence of g, the values $P_{14} = 3.2 \times 10^{-3}$ and $P_{36} = 2.6 \times 10^{-3}$. Considering our simplistic approach, we consider this to be good agreement. See: T. G. Phillips, R. L. Townsend and R. L. White, *Phys. Rev. Lett.* **18**, 646 (1967).

S. IIDA As you have mentioned, Tb^{3+} is a very anomalous ion. In my measurements on $Tb_3Fe_5O_{12}$. I found that λ_{111} is very large whereas λ_{100} is very small, so that ΔK, expected from these constants at 78 K is larger than ΔK, observed. Would you please present you knowledge concerning this point in relation to the magneto-electric effects.

R. L. WHITE Terbium is an even-electron non Kramers ion, so one has to be careful about characterizing it with g factor or q tensor. If Tb^{3+} in the garnets has a pair of nearly degenerate low lying states, one can still use an effective Hamiltonian approach. In this effective Hamiltonian, the tensor will be axial, that is, it has only one non-zero principal axis entry when diagonalized. The ion is magnetically uni-axial, so all its magnetic properties are very anisotropic. I cannot make a detailed connection between this observation and the observed large λ_{111} small λ_{100} except that I am not surprised to find some very preferred directions in $Tb_3Fe_5O_{12}$.

Part II

Symmetry and Phenomenological Theory

MAGNETOELECTRIC SYMMETRY†

W. OPECHOWSKI
Department of Physics, University of British Columbia, Vancouver, Canada

A review is given of the fundamental definitions and postulates which are necessary for a consistent discussion of the symmetries characterizing the electromagnetic properties of crystalline media. Only very elementary group theory is used. Some historical remarks are included.

1 INTRODUCTION

In this paper I present a review of the fundamental mathematical definitions and physical postulates which are necessary for a meaningful discussion of the symmetries characterizing the electromagnetic properties of crystalline media, with, in particular, magnetoelectric effects in mind. There had been considerable confusion concerning these definitions and postulates until only several years ago. Even now some hesitation remains (see in particular Section 8, below); and the terminology is still far from being uniform. That is why the historical remarks included in this paper may have more than just an historical interest.

The appropriate language for describing symmetries is the group-theory language. That language will be used from the outset, but only in its most elementary form.

2 MAXWELL'S EQUATIONS AND CONSTITUTIVE RELATIONS

For a macroscopic medium the fundamental equations that the electromagnetic fields (em-fields) **E**, **B** and **D**, **H** satisfy consist of two sets of equations: the Maxwell equations,

$$\operatorname{curl} \mathbf{E} + \frac{1}{c}\frac{\partial \mathbf{B}}{\partial t} = 0, \qquad \operatorname{div} \mathbf{B} = 0$$

$$\operatorname{curl} \mathbf{H} - \frac{1}{c}\frac{\partial \mathbf{D}}{\partial t} = \frac{4\pi}{c}\mathbf{J}, \qquad \operatorname{div} \mathbf{D} = 4\pi\rho \qquad \text{(M.Eqs)}$$

and the Constitutive Relations

$$\mathbf{E} = f(\mathbf{H}, \mathbf{D}), \quad B = g(\mathbf{H}, \mathbf{D}) \qquad \text{(CRs)}$$

which characterize the electromagnetic properties of the medium. (I do not include relations involving the current density **J**, because they will not be used in what follows.) Functions f and g depend in general on temperature, stresses, and velocity of the medium (and even possibly on time explicitly). For example, the CRs are often, in sufficient approximation, of the form:

$$E_j = E_j^0 + \lambda_{jk} H_k + \eta_{jk} D_k \qquad (1)$$

$$B_j = B_j^0 + \mu_{jk} H_k + \kappa_{jk} D_k; \qquad (2)$$

here the usual summation convention has been adopted. This form describes, in particular, the linear magnetoelectric effect if κ_{jk} and λ_{jk} are not all zero, and $B_j^0 = E_j^0 = 0$. It also describes the case of vacuum, that is, the case of **E** = **D**, **B** = **H**.

3 ELECTROMAGNETIC SYMMETRY GROUP (EMSG) OF A MEDIUM

Quite generally the *electromagnetic symmetry group* (*EMSG*) of a medium is a subgroup of the symmetry group of the Maxwell equations, which is the *Poincaré group* $\mathscr{P}$ (also called the *full inhomogeneous Lorentz group*). In what follows only the case of a medium that is at rest relative to the observer will be considered. The EMSG is then a subgroup of a group $\mathscr{E}$ which is itself a subgroup of $\mathscr{P}$. Some elements of $\mathscr{P}$ depend on the velocity of the medium; by omitting them one obtains the group $\mathscr{E}$ which will explicitly be defined in Section 4, and be called the *Newton group*. However, the logical steps leading to the definition of the phrase "symmetry group of the

† This is a somewhat expanded text of a review paper given at the "Symposium on Magnetoelectric Interaction Phenomena in Crystals", Battelle Seattle Research Center, May 21–24, 1973. In particular, Section 8 has been rewritten and expanded as a result of some useful questions and remarks of Dr. G. R. Rado at the Symposium.

Maxwell equations" are the same in the case of $\mathscr{P}$ and that of $\mathscr{E}$. Therefore I will briefly indicate these steps for the more general case.

The Poincaré group $\mathscr{P}$ is defined as the group of all those linear transformations P of the Minkowski space-time which leave the square of the "distance"

$$c^2(t_1 - t_2)^2 - (x_1 - x_2)^2 - (y_1 - y_2)^2 - (z_1 - z_2)^2$$

between two space-time points, (x_1, y_1, z_1, t_1) and (x_2, y_2, z_2, t_2), unchanged.

Having defined the Poincaré group, one first of all introduces a postulate concerning the *action* of the Poincaré group on an arbitrary em-field and charge and current density field. A transformation P of the Minkowski space-time which belongs to the Poincaré group $\mathscr{P}$, acting on an em-field

$$\mathbf{E}(x, y, z, t), \quad \mathbf{B}(x, y, z, t), \quad \mathbf{D}(x, y, z, t), \quad \mathbf{H}(x, y, z, t) \tag{3}$$

gives then rise to a new em-field

$$(P\mathbf{E})(x, y, z, t), \quad (P\mathbf{B})(x, y, z, t), \quad (P\mathbf{D})(x, y, z, t),$$

$$(P\mathbf{H})(x, y, z, t) \tag{4}$$

uniquely defined for each P. This postulate is usually stated by saying that **E** and **B** on the one hand, and **D** and **H** on the other hand, are four-tensor fields in the Minkowski space-time; and that ρ and **J** is a four-vector field; and by explicitly specifying the action of the discrete space-time subgroup $\mathscr{U}$ of $\mathscr{P}$ (see Section 4). If the fields **E** and P**E** are identical, one says that the field **E** is *invariant* under the Poincaré transformation P, and the same definition is adopted for other fields.

Next one shows that if the field (3) is a solution of M.Eqs with functions $\rho(x, y, z, t)$ and $\mathbf{J}(x, y, z, t)$, then the field (4) is, for each P, a solution of M.Eqs with the functions $(P\rho)(x, y, z, t)$ and $(P\mathbf{J})(x, y, z, t)$ instead of the functions $\rho(x, y, z, t)$ and $\mathbf{J}(x, y, z, t)$. The fact that this is so for any choice of the four-vector field $\rho, \mathbf{J}$ is referred to as the *covariance* or *invariance of M.Eqs under the Poincaré group.* Both terms "covariance" and "invariance" are used in this connection in the literature. If we adopt the term "covariance" rather than the term "invariance," then we may reserve the use of the latter term for the case where the field $\rho, \mathbf{J}$ appearing in M.Eqs is invariant, in the sense defined in the preceding paragraph, under some elements of the Poincaré group. In this terminology, on the one hand M.Eqs are covariant under the Poincaré group $\mathscr{P}$ for any choice of the field $\rho, \mathbf{J}$; on the other hand, for each specific choice of the field $\rho, \mathbf{J}$, M.Eqs are invariant under a subgroup of $\mathscr{K}$ of $\mathscr{P}$ consisting of all those Poincaré transformations K for which $K\rho = \rho$ and $KJ = \mathbf{J}$. In other words, invariance of M.Eqs under $\mathscr{K}$ characterizes a specific physical situation to the extent the latter manifests itself in the symmetry properties of the field $\rho, \mathbf{J}$, while covariance characterizes the symmetry properties of M.Eqs themselves whatever physical situation they may describe.

Finally, one shows that there are no other transformations of the Minkowski space-time under which M.Eqs are covariant in an analogous sense, and that is the meaning of the statement that the Poincaré group is the *symmetry group of the M.Eqs* (if $\mathbf{J} = \rho = 0$, and $\mathbf{E} = \mathbf{D}$, $\mathbf{B} = \mathbf{H}$, then a larger group, the *conformal group*, is the symmetry group of the M.Eqs).

The EMSG of a medium is then defined as consisting of all those Poincaré transformations P for which the em-field (4) satisfies the CRs if the em-field (3) does, the functions f and g remaining unchanged. In other words, the EMSG of a medium consists of all those P's which leave the CRs unchanged. Any subgroup of the EMSG of a medium is an *electromagnetic invariance group* of the medium.

This reduction of symmetry when vacuum is replaced by a medium was stressed long ago, in 1894, by Pierre Curie.[1] It is the absence of certain elements of the symmetry group of M.Eqs in vacuum which makes it possible for certain non-trivial forms of CRs (that is, of functions f and g) to be invariant. More generally, as Curie concisely expressed it in an often quoted sentence: "C'est la dissymétrie qui crée le phénomène." For example, the absence of time inversion and space inversion are necessary for a linear magnetoelectric effect to exist.

The case of a medium in uniform motion is of some interest in connection with the question, under what condition can the EMSG of a medium at rest be equivalent to the EMSG of another medium in the state of uniform motion; see O'Dell.[2] I may mention that "relativistic symmetries" of a medium, in a quite different sense, have recently been considered by Ascher,[3] where also references will be found to earlier work by Ascher and Janner on the relativistic symmetry groups of certain solutions of M.Eqs in vacuum, like plane waves or homogeneous electric and magnetic fields. Such topics in which the Poincaré group $\mathscr{P}$, and not only its subgroup $\mathscr{E}$, plays an essential role will not further be discussed in this paper.

4 THE NEWTON GROUP, AND THE EMSG OF A MEDIUM REDEFINED

Returning now to the case of a medium at rest, all that has been said about the steps leading to the definition of the EMSG of a medium remains true "mutatis mutandis" if one considers instead of the Poincaré group $\mathscr{P}$ its subgroup $\mathscr{E}$. The group $\mathscr{E}$ consists of:

i) All proper rotations about a point; they form a subgroup $\mathscr{R}_+$ of $\mathscr{E}$; the action of a proper rotation is defined by the assumption that **E**, **B** and **D**, **H** and **J** are "vector" fields; and ρ is a "scalar" field.

ii) All space translations (subgroup $\mathscr{T}_s$) and all time translations (subgroup $\mathscr{T}_t$); the two kinds of translations commute, and all pairs consisting of one space translation and one time translation constitute the direct product group

$$\mathscr{T} = \mathscr{T}_t \times \mathscr{T}_s \tag{5}$$

of all space and time translations.† The action of a space or time translation on an em-field is again implied by the assumed vector character of the latter.

iii) The four transformations of the Minkowski space which constitute the *discrete space-time group* $\mathscr{U}$:

identity E: $(x, y, z, t \rightarrow x, y, z, t)$,
time inversion E': $(x, y, z, t \rightarrow x, y, z, -t)$,
space inversion I: $(x, y, z, t \rightarrow -x, -y, -z, t)$,
and total inversion I': $(x, y, z, t \rightarrow -x, -y, -z, -t)$.

The group $\mathscr{U}$ has three non-trivial subgroups:

$$\mathscr{A} = \{E, E'\}, \quad \mathscr{I} = \{E, I\}, \quad \mathscr{J} = \{E, I'\} \tag{6}$$

† Here are a few standard definitions used in this paper.
Suppose that $\mathscr{K}$ and $\mathscr{L}$ are two subgroups of a group $\mathscr{G}$ that have no element in common except the identity element, and furthermore suppose that, for each element G of $\mathscr{G}$,

$$G = KL$$

where K belongs to $\mathscr{K}$ and L to $\mathscr{L}$. Then, if $\mathscr{K}$ and $\mathscr{L}$ are normal in $\mathscr{G}$, the group $\mathscr{G}$ is called the *direct product of its subgroups* $\mathscr{K}$ *and* $\mathscr{L}$:

$$\mathscr{G} = \mathscr{K} \times \mathscr{L}$$

if only $\mathscr{L}$ is normal in $\mathscr{G}$, the group $\mathscr{G}$ is called the *semidirect product of its subgroups* $\mathscr{K}$ *and* $\mathscr{L}$:

$$\mathscr{G} = \mathscr{K} \wedge \mathscr{L}$$

(The cross $\times$ thus means a "direct product," and the semicross $\wedge$ a "semidirect product".) A subgroup $\mathscr{H}$ of $\mathscr{G}$ is called *normal in* $\mathscr{G}$ or *a normal subgroup of* $\mathscr{G}$ if for each element H of $\mathscr{H}$ and each element G of $\mathscr{G}$ the element GHG^{-1} is an element of $\mathscr{H}$.

Hence:

$$\mathscr{U} = \mathscr{A} \times \mathscr{I} = \mathscr{I} \times \mathscr{J} = \mathscr{J} \times \mathscr{A} \tag{7}$$

The generally accepted postulate concerning the action of the elements of $\mathscr{U}$ is as follows:

E' acting on $\mathbf{E}(z, y, z, t)$ gives $\mathbf{E}(x, y, z, -t)$,
E' acting on $\mathbf{B}(x, y, z, t)$ gives $-\mathbf{B}(x, y, z, -t)$
E' acting on $\mathbf{J}(x, y, z, t)$ gives $-\mathbf{J}(x, y, z, -t)$,
E' acting on $\rho(x, y, z, t)$ gives $\rho(x, y, z, -t)$,
I acting on $\mathbf{E}(x, y, z, t)$ gives $-\mathbf{E}(-x, -y, -z, t)$,
I acting on $\mathbf{B}(x, y, z, t)$ gives $\mathbf{B}(-x, -y, -z, t)$,
I acting on $\mathbf{J}(x, y, z, t)$ gives $-\mathbf{J}(-x, -y, -z, t)$,
I acting on $\rho(x, y, z, t)$ gives $\rho(-x, -y, -z, t)$,

the action of $\mathscr{U}$ on the field **D** being the same as that on **E**; and the action on **H** the same as that on **B**.

The action of $\mathscr{U}$ on the em-field plays a decisive part in a classification of the various EMSG's of the various media as characterized by their CRs. The reason for that is that while a rotation (group $\mathscr{R}_+$), or a translation (group $\mathscr{T}$), acts in exactly the same way on **E**, **B**, **D**, **H**; or, more precisely, gives rise to exactly the same linear combination of the components of **E**, as those of **B**, or **D** or **H**, this is not so in the case of the three discrete space-time transformations. In group-theory language, the fields **E**, **B**, **D**, **H**, and **J**, each generate the same representation of the groups $\mathscr{R}_+$ and $\mathscr{T}$, but different representations of the group $\mathscr{U}$.

Before coming to this problem of classification of the electromagnetic groups, it is useful to characterize the "gross" structure of the group $\mathscr{E}$ by means of a concise formula.

Each element of $\mathscr{E}$ is a triplet: one element of $\mathscr{U}$, one element of $\mathscr{R}_+$, and one element of $\mathscr{T}$. Each element of $\mathscr{U}$ commutes with each element of $\mathscr{R}_+$; but an element of $\mathscr{T}$ need not commute with elements of $\mathscr{U}$ and $\mathscr{R}_+$. All that is summed up in the formula

$$\mathscr{E} = (\mathscr{U} \times R_+) \wedge \mathscr{T} \tag{8}$$

[In passing I may mention that the "gross" structure of the Poincaré group is described by the formula

$$\mathscr{P} = (\mathscr{U} \wedge \mathscr{L}_+{}^{\uparrow}) \wedge \mathscr{T} \tag{9}$$

where $\mathscr{L}_+{}^{\uparrow}$ is the *proper*, *orthochronous Lorentz group*. The elements of $\mathscr{U}$ do not commute with those elements of $\mathscr{L}_+{}^{\uparrow}$ which are not in $\mathscr{R}_+$, that is why $\wedge$ appears in (9) instead of $\times$ when $\mathscr{R}_+$ is replaced by $\mathscr{L}_+{}^{\uparrow}$ in (8).]

Another way of indicating the structure of the group $\mathscr{E}$ is to define the group

$$\mathscr{R} = \mathscr{I} \times \mathscr{R}_+ \tag{10}$$

of all (proper and improper) rotations of space; then, in view of (7), the structure formula (8) can be rewritten as

$$\mathscr{E} = (\mathscr{A} \times \mathscr{R}) \wedge \mathscr{T} \tag{11}$$

Although the group $\mathscr{E}$ is clearly fundamental for the whole structure of physics, and has been used (not always explicitly, it is true) for many years, it has—strangely enough—no generally accepted name. I suggest we call it the *Newton group*. Galileo, Lorentz, Poincaré have "their" groups, why not Newton?

The structure formula (11) for $\mathscr{E}$ is particularly useful when one restricts oneself (as will be done in the remainder of this paper) to considering only media that are static and homogeneous. *Static* means that the functions f and g in the CRs are invariant under the group $\mathscr{T}_t$; *homogeneous* means that they are invariant under the group $\mathscr{T}_s$. Formula (11) then shows that one may simplify the language by redefining the EMSG of such a medium as a subgroup of $\mathscr{A} \times \mathscr{R}$ rather than of $\mathscr{E}$, the invariance of the CRs of the medium under all space- and time-translations always being understood. This simplification of the language will be consistently used in what follows.

The group

$$\mathscr{R}_{st} = \mathscr{A} \times \mathscr{R} \tag{12}$$

could be called the *general space-time rotation group*.

Still another way of looking at the structure of the group $\mathscr{E}$ is to regard it to be a group of pairs, one member of each pair being any element of the three-dimensional Euclidean group $\mathscr{E}_s$, consisting of all translations and all (proper and improper) rotations of (our "physical" three-dimensional Euclidean) space, and the other member of each pair being any element of the one-dimensional Euclidean space group $\mathscr{E}_t$, consisting of all translations and time inversion of the one-dimensional Euclidean space that is called *time* in physics:

$$\mathscr{E} = \mathscr{E}_t \times \mathscr{E}_s \tag{13}$$

where

$$\mathscr{E}_s = \mathscr{R} \wedge \mathscr{T}_s, \quad \mathscr{E}_t = \mathscr{A} \wedge \mathscr{T}_t$$

Proof of (13), starting out from (8), is as follows:

$$\begin{aligned}
\mathscr{E} &= (\mathscr{U} \times \mathscr{R}_+) \wedge \mathscr{T} \\
&= ((\mathscr{A} \times \mathscr{I}) \times \mathscr{R}_+) \wedge (\mathscr{T}_t \times \mathscr{T}_s) \\
&= (\mathscr{A} \times \mathscr{R}) \wedge (\mathscr{T}_t \times \mathscr{T}_s) \\
&= (\mathscr{A} \wedge \mathscr{T}_t) \times (\mathscr{R} \wedge \mathscr{T}_s) \\
&= \mathscr{E}_t \times \mathscr{E}_s
\end{aligned}$$

5 MAGNETIC ROTATION GROUPS

It follows from the preceding discussion that the EMSG (as redefined in Section 4) of a static homogeneous medium is always a subgroup of the group $\mathscr{R}_{st} = \mathscr{A} \times \mathscr{R}$, which is itself a subgroup of the Newton group. It is thus important to have a survey of all such subgroups.

The group $\mathscr{R}_{st}$ can be divided into two disjoint subsets (in group-theory language, the two cosets of $\mathscr{R}$), space rotations, and space-time rotations:

$$\mathscr{R}_{st} = \mathscr{R} + E'\mathscr{R}. \tag{14}$$

"Rotation" means here, as everywhere else in this paper, "proper rotation" or "improper rotation"; space inversion or time inversion are thus rotations in that sense.

Formula (14) implies that each element of a subgroup of $\mathscr{R}_{st}$ is either an element of $\mathscr{R}$, that is, a space rotation R, or an element of the coset $E'\mathscr{R}$, that is, a space rotation combined with time inversion, $E'R$, to be denoted simply by R', and called a *primed rotation* (or *antirotation*, a term which will not be used in this paper).

In particular, if a subgroup of $\mathscr{R}_{st}$ contains some primed rotations, it may or may not contain the primed identity, that is, time inversion E' itself. Therefore all the subgroups of $\mathscr{R}_{st}$ can be divided into three disjoint classes:

1) subgroups containing E', called *non-magnetic rotation groups*;

2) subgroups containing no primed rotations at all, called *trivial magnetic rotation groups*;

3) subgroups containing some primed rotations, but not E', called *non-trivial magnetic rotation groups*.

The reason for this terminology is very simple: because of the way time inversion E' acts on magnetic fields **B** and **H** (and, hence, on the magnetization $\mathbf{M} = (\mathbf{B} - \mathbf{H})/4\pi$), no field **B** or **H** or magnetization can be invariant under E'; in this sense a group that contains E' is "non-magnetic."

Given a group **R** of rotations (that is, a subgroup of $\mathscr{R}$) one obtains a non-trivial magnetic rotation group belonging to the *family of* **R** by appropriately "priming" some of the elements of **R**; "appropriately" means in such a way that the resulting set of rotations and primed rotations forms a group (which then necessarily is a subgroup of $\mathscr{R}_{st}$). For some rotation groups this can be done in more than one way, for some not at all. The *family* of **R** thus consists of **R** and all (if any) the non-trivial magnetic rotation groups obtained from $\mathscr{R}$ in this way. The family of **R** and the non-magnetic group $\mathbf{R}_{st} = \mathscr{A} \times \mathbf{R}$ will be called the *superfamily of* **R**.

For example, if the rotation group **R** is (here I am using the "international notation," in which $E' = 1'$, $I = \bar{1}$ and $I' = \bar{1}'$)

$$2/m = \{1, 2, \bar{1}, m = 2\bar{1}\}$$

then the other members of its family are:

$$2'/m = \{1, 2', \bar{1}', m\}$$

$$2/m' = \{1, 2, \bar{1}', m'\}$$

$$2'/m' = \{1, 2', \bar{1}, m'\}$$

These four groups together with the group

$$2/m1' = \{1, 2, \bar{1}, m, 1', 2', \bar{1}', m'\}$$

constitute the superfamily of $2/m$.

There is a theorem, according to which there is always (just as in the above example) an equal number of primed and unprimed elements in a group if the latter contains primed elements at all. It is that theorem which makes it possible to establish in a simple way a complete list of all magnetic and non-magnetic rotation groups. If one restricts oneself to to the superfamilies of the 32 crystallographic point groups, the list contains 32 (trivial magnetic) + 58 (non-trivial magnetic) + 32 (non-magnetic) = 122 groups altogether. Heesch,[4] in 1930, was the first to define them and to publish the complete list of them, but he did not interpret the element E' as time inversion. The 122 groups are sometimes called the *Heesch (point) groups*, and more often the *Shubnikov (point) groups*. They were rederived (without mentioning Heesch's work) by Tavger and Zaitsev[5] in 1956, who, after Landau and Lifshitz,[6] stressed the importance of primed rotations for the theory of magnetism. A list of the 90 magnetic groups arranged into families can be found in Ref. 7.

6 GEOMETRICAL SYMMETRY GROUPS (GSG) OF A MEDIUM AND NEUMANN'S PRINCIPLE

In what follows only crystalline media will be considered. Traditionally a crystalline medium is defined as a solid medium whose purely geometrical macroscopic properties can be characterized by associating with it a polyhedron whose symmetry group belongs to one of 32 crystallographic point groups. One calls then that symmetry group the *point group of the crystalline medium* in question. Each point group is thus a subgroup of the group $\mathscr{R}$, and therefore also a subgroup of $\mathscr{R}_{st}$. Furthermore if one does not go beyond this purely geometrical characterization of a crystalline medium, the latter can be regarded as trivially invariant under time inversion E'. Hence if a rotation R belongs to the point group **R** of the medium, or, in other words, if the medium is invariant under R, then it is also invariant under $RE' \equiv R'$. This means that, in the absence of any information about some not purely geometrical properties of the medium, the *macroscopic geometrical symmetry group* (GSG) of the medium can be defined equally well as any of the groups constituting the superfamily of **R**.

In other words, from the purely geometrical space-time point of view, a crystalline medium should be characterized by one of the 32 superfamilies of Heesch groups rather than by one of the 32 crystallographic point groups as is usually done.

For example, in the case of a monoclinic crystal whose point group is $2/m$ any of the five groups in its superfamily could equally well be regarded as the GSG of that crystal.

One could try to remove this lack of uniqueness of the assignment of the GSG to a crystal by arbitrarily defining the largest group in a superfamily, that is, the non-magnetic group, as the unique GSG. In fact, this definition was often tacitly adopted in the past, until about twenty years ago (see, for example, Zocher and Török[8]). And one often came to wrong conclusions, because one identified such an arbitrarily defined GSG of a crystalline medium with an invariance group (that is, a subgroup of the EMSG) of the CRs. For example, making this identification one can easily show that the linear magnetoelectric effect is impossible for all crystalline media (see O'Dell's book[2] for the history of the belief in the impossibility of that effect before its discovery in 1960).

If instead of the non-magnetic group belonging to the superfamily of the crystallographic point group of a medium one declares, in an equally arbitrary way, the point group itself to be the unique GSG of the medium, and one again identifies the GSG with an invariance group of the CRs, one often comes again to conclusions contradicted by experiment. For example, with this definition of the GSG crystals with point group $2/m$ should not manifest a linear magnetoelectric effect, contrary to experiment (see D. E. Cox[9]).

It is clear that such arbitrary definitions of GSG of a crystalline medium, and subsequent identification of the GSG with the EMSG does not make much sense. Or I should rather have said, it is now clear, because it was not clear in the past. Why was it not? I think it is instructive to consider this question. It seems that the answer has to be sought in a misinterpretation of a point of view consistently adopted by Franz Neumann in the second half of the nineteenth century, and called by Woldemar Voigt *Neumann's Principle* in his celebrated, almost 1000 pages long, *Lehrbuch der Kristallphysik*,[10] a kind of "summa crystallographica" of classical physical crystallography, that is, of the crystallography preceding von Laue's discovery of x-ray diffraction in 1912. This principle consists, according to Voigt, "in using the symmetry properties of the crystal form to conclude about the symmetries of the physical properties of crystals."

More specifically, Neumann's Principle has been interpreted as requiring that all elements of the point group **R** of a crystalline medium must be a subgroup of EMSG (in particular, **R** may be identical with EMSG). Again according to Voigt, Neumann's Principle can be justified by invoking the experimental result "that the peculiarities and differences of the constitution ('Konstitution') find a more complete expression in the crystal form than in the other physical properties." However, Voigt adds immediately after: "This experimental result is of course as little rigorous as similar experimental laws in other domains of physics. In principle the possibility is not excluded, although is not very probable, that crystal properties will later be discovered which will manifest still more variety than the crystal form; and then the principle which consists in concluding about the laws of constitution of crystalline matter from the laws governing the form would lose its general significance."

In a sense the possibility deemed by Voigt to be not very probable did occur in the meantime, but not in a way he could have forseen. Just about the time Voigt was writing his book, Minkowski was introducing the idea of space-time geometry in physics. Voigt could hardly have forseen that what really matters is—so to speak—the space-time form of a crystal. In any case, it took more than twenty years before the importance of time-inversion in physics became fully appreciated thanks to Kramers'[11] theorem (1930) on the degeneracy of the energy levels of atomic systems with odd number of electrons, and Wigner's[12] interpretation (1932) of that theorem in terms of time-inversion invariance of the Schroedinger equation. And it took another twenty years before Landau and Lifshitz[6] (1951) stressed that the absence of time-inversion symmetry is a necessary condition for the existence of magnetically ordered crystalline media such as ferromagnetic, antiferromagnetic media, etc. Incidentally, the term "magnetic group" was invented by them.

Although time inversion is in space-time a geometrical symmetry, its presence or absence does not manifest itself as a presence or absence of a geometrical symmetry of a crystalline medium. Therefore Neumann's Principle, even if interpreted more generally in terms of space-time, is useless as long as one does not know which among the several groups constituting the superfamily of the point group of a crystalline medium one has to select "to conclude about the symmetries of physical properties" of the medium. However, such knowledge can be obtained only by considering some "physical properties." This vicious circle can be broken in the case of magnetically ordered crystalline media if one knows (for example, from neutron diffraction data) the *magnetic space group of the medium*, that is, the symmetry group of the atomic state of the medium. It is this kind of information that was used by Dyaloshinski[13] to predict the existence of the linear magnetoelectric effect in antiferromagnetic Cr_2O_3.

7 MAGNETIC SPACE GROUPS

Magnetic space groups are defined in a way that is analogous to the way, indicated above, which leads to the definition of magnetic point groups: instead of considering subgroups of $\mathscr{A} \times \mathbf{R}$ one considers subgroups of $\mathscr{A} \times \mathbf{F}$, where **F** is a space group. Thus, a magnetic space group **M** does not contain time inversion E', but it may contain elements of the form $E'F$, where F is any element of a space group different from the unit element. A magnetic space group is called *trivial* if it is identical with a space group; otherwise (that is, if it contains primed elements), it is called *non-trivial*. It is also useful to define a *family* (and *superfamily*) of magnetic space groups (family of **F**) in the same way as a family of magnetic point groups (family of **R**) was defined in Section 5.

It is of importance to distinguish between those non-trivial magnetic space groups (to be denoted by $\mathbf{M_T}$) whose translations are all unprimed, and those (to be denoted by $\mathbf{M_R}$) for which this is not true. For example, consider the two following magnetic space groups belonging to the family of the space group (the symbols between triangular brackets are generators of the whole group)

$$P2 = \langle 1, 2_y, a, b, c \rangle$$

(a, b, c are here the basic primitive translations):

$$P2' = \langle 1, 2'_y, a, b, c \rangle$$

$$P_{2b}2 = \langle 1, 2_y, a, b', c \rangle$$

The group $P2'$ is $\mathbf{M_T}$, the group $P_{2b}2$ is $\mathbf{M_R}$.

More information about magnetic space groups, and a list of all of them ($230 + 1191 = 1421$) arranged into families can be found in Ref. 7. Magnetic space groups were first defined by Heesch[4] in 1930, but, as has already been mentioned, Heesch did not interpret E' as time inversion. Heesch also gave a list of all magnetic space groups of the triclinic and monoclinic systems. A complete list was first published by Belov, Neronova and Smirnova[14] in 1955, who called the 1421 magnetic space groups plus the 230 direct products of space groups and time-inversion groups (1651 groups altogether) *Shubnikov* (*space*) *groups*, following a suggestion of Zamorzaev.

8 POINT GROUP ASSOCIATED WITH A MAGNETIC SPACE GROUP OF A MAGNETICALLY ORDERED CRYSTALLINE MEDIUM

Assuming that one knows the magnetic space group of a magnetically ordered crystalline medium, what is then the macroscopic GSG of that medium in the sense defined in Section 6? If in the spirit of Neumann's Principle one requires that the Constitutive Relations of the medium be invariant under the GSG, quite different theoretical predictions may result depending on the answer to this question.

The GSG of a magnetically ordered crystalline medium is usually taken to be identical with the point group $\mathbf{R_M}$ associated with the magnetic space group **M** of the medium. The definition of $\mathbf{R_M}$ depends on whether the group **M** is $\mathbf{M_T}$ or $\mathbf{M_R}$ or trivial. It seems that there is a consensus to adopt the following, physically natural but mathematically not unique, unambiguous prescription for finding $\mathbf{R_M}$ if **M** is given: disregard all translations in **M** whether they are associated with space-time rotations or not; what then remains of the group **M** is, by definition, the group $\mathbf{R_M}$. According to this prescription, $\mathbf{R_M}$ is always one of the 122 Heesch groups. In particular:

(**M** = **F**) If a magnetic space group is trivial, that is, identical with a space group **F**, then its point group $\mathbf{R_M}$ is of course identical with the point group of **F**; for example, the point group of $P2$ is 2.

($\mathbf{M} = \mathbf{M_T}$) If a non-trivial magnetic space group is a group $\mathbf{M_T}$, then its point group $\mathbf{R_M}$ consists of all primed and unprimed (proper and improper) rotations in $\mathbf{M_T}$. The point group $\mathbf{R_M}$ is thus in this case a non-trivial magnetic point group, and as such does not contain time-inversion among its elements. For example, the point group of $P2'$ is $2'$.

($\mathbf{M} = \mathbf{M_R}$) If a non-trivial magnetic space group is a group $\mathbf{M_R}$, then its point group $\mathbf{R_M}$ consists of all rotations in $\mathbf{M_R}$, each such rotation appearing in $\mathbf{R_M}$ twice: once as an unprimed rotation, and once as a primed one; for example, the point group of $P_{2b}2$ is $21' = \{1, 2_y, 1', 2'_y\}$.

In the case of $\mathbf{M_R}$ the point group is thus always non-magnetic, that is, it contains time inversion, while the group $\mathbf{M_R}$ itself does not contain time inversion. That the (macroscopic) GSG of a medium contains elements that are not present in the symmetry group of the atomic state of the medium is of course nothing to be surprised about. In any case the GSG contains, in addition to the elements of $\mathbf{R_M}$, all space and time translations, according to the convention introduced in Section 4 about static, homogeneous media. Quite generally, the macroscopic symmetry of a medium is never lower than its atomic symmetry because the transition from an atomic description to a macroscopic description of a physical system always involves some kind of averaging or "coarse-graining," and that can only increase the symmetry. This remark can be regarded as a special case of a somewhat vague but more general statement made by Pierre Curie[1]: "les effets produits peuvent être plus symétriques que les causes," macroscopic state being an "effet" whose "cause" is the atomic state of a medium.

Since the GSG of a magnetically ordered crystalline medium is taken to be identical with the point group $\mathbf{R_M}$ of the magnetic space group **M** of such medium, those media whose magnetic space groups are $\mathbf{M_R}$ cannot, according to the above definitions, manifest a linear magnetoelectric effect. For the GSG of those media necessarily contains time inversion, which is incompatible with the non-vanishing of the constants κ_{jk} and λ_{jk} in Eqs. (1) and (2). Only a magnetically ordered medium whose magnetic space group is $\mathbf{M_T}$ is compatible with the presence of a linear magnetoelectric effect. This theoretical prediction has not been contradicted by experiment so far.

The GSG of an antiferromagnetic crystal thus does or does not contain time inversion according as its magnetic space group is $\mathbf{M_R}$ or $\mathbf{M_T}$. The magnetic space group of a ferromagnetic crystal is always $\mathbf{M_T}$, because the magnetization of a ferromagnetic crystal is, by definition, different from zero, and this is incompatible with the presence of time inversion in the GSG of such a crystal.

In Ref. 7 a different definition was proposed for the point group of a magnetic space group $\mathbf{M_R}$, which I now find unacceptable for physical reasons.

9 MAGIC NUMBERS AND THE STRUCTURE OF THE GENERAL SPACE-TIME ROTATION GROUP

Once the GSG of the medium is known (for example, for the case of magnetically ordered media, by adopting the point of view described in Section 8), it is a matter of applying standard group theory techniques to find the form of Constitutive Relations invariant under that particular group, and to determine their EMSG. This has been done for a number of crystalline media such as magnetoelectric media (linear and higher order), ferromagnetic and ferroelectric media, ferromagnetoelectric media, piezoelectric and piezomagnetic media, etc. etc., each of these terms being defined by a specified form of the CRs (see, for example, Birss[15] or Bhagavantam[16]).

Of course, specified CRs have only some among the 122 *a priori* possible GSGs as invariance groups. For example, there are only 31 ferromagnetic Heesch groups, 58 linear magnetoelectric groups, 19 groups admitting the magnetically induced magnetoelectric effect, again 31 ferroelectric groups, 19 groups admitting the electrically induced magnetoelectric effect, etc., etc. If one extends this list, or for example looks at Table II in H. Schmid's paper[17] on classification of magnetoelectric materials, one will be struck by the appearance of certain integers more than once among these numbers. The reason for the existence of such "magic numbers" becomes obvious if one re-examines the structure of the general space-time rotation group of which all these groups are subgroups.

By comparing Eqs. (8) and (11) in Section 4, it is clear that the group $\mathscr{R}_{st}$ could have been defined by

$$\mathscr{R}_{st} = \mathscr{U} \times \mathscr{R}_+ = \{E, E', I, I'\} \times \mathscr{R}_+ \tag{15}$$

rather than by Eq. (12). The latter equation stresses the fact that the group $\mathscr{R}_{st}$ consists of pairs: a time transformation, and a space transformation which may be a proper or improper rotation. On the other hand, Eq. (15) shows that the group $\mathscr{R}_{st}$ can also be regarded as consisting of pairs: a space-time inversion and a proper rotation; and it also shows that the three space-time inversions, I, E' and I' are perfectly interchangeable so far as the mathematical structure of $\mathscr{R}_{st}$ is concerned. This means, in particular, that by replacing E' ($1'$) by I ($\bar{1}$) and vice versa in any subgroup of $\mathscr{R}_{st}$ one will always obtain again a subgroup of $\mathscr{R}_{st}$. The same is true of the pairs E', I' and I, I'. For example, the groups $2/m$, $2'/m$, $2'/m$ $2'/m'$ become $21'$, $2'/m$, $2/m'$, $m1'$ if one interchanges $1'$ and $\bar{1}$.

Hence, instead of dividing the 122 Heesch groups into 90 magnetic groups (32 trivial, and 58 non-trivial) and 32 non-magnetic groups, according as they do not or do contain time-inversion E', one could divide them into 90 *electric groups* (32 trivial and 58 non-trivial) and 32 *non-electric groups*, according as they do not or do contain space-inversion I. (This terminology can be justified by the fact that the action of I on an electric field is analogous to the action of E' on magnetic field.) It follows that there must be, for example, 31 ferroelectric groups if there are 31 ferromagnetic groups, etc. Of course, there is still a third possibility: one could divide the 122 Heesch groups using the absence or presence of total inversion I' in a Heesch group. One would then obtain 31 groups admitting spontaneous current, as emphasized by Ascher.[18]

The existence of "magic numbers" is thus easily understood. However I have not been able to understand why the number of linear magnetoelectric groups is equal to the number of non-trivial magnetic (or electric) groups (58 in both cases). A hint consisting in observing that substituting in a magnetoelectric group $1'$ by $\bar{1}$ and vice versa gives rise to a magnetoelectric group again has been of no help to me.

REFERENCES

1. Pierre Curie, *J. de Physique*, 3e série, **3**, 393 (1894).
2. T. H. O'Dell, *The Electrodynamics of Magneto-electric Media* (North-Holland, Amsterdam, 1970).
3. E. Ascher, *J. Phys. Soc. Japan* **28**, Suppl., 7 (1970).
4. H. Heesch, *Z. Krist.* **73**, 325 (1930).
5. B. A. Tavger and V. M. Zaitsev, *Zh. Eksperim. i Theor. Fiz.* **30**, 564 (1956); *J. Exptl. Theoret. Phys.* (*U.S.S.R.*) **3**, 430 (1956).
6. L. L. Landau and E. M. Lifshitz, *Statisticheskaya Fizika* GITTL (State Tech. Lit. Press, 1951); *Statistical Physics* (Pergamon Press, London, 1958).
7. W. Opechowski and R. Guccione, *Magnetism*, edited by G. T. Rado and H. Suhl (Academic Press, New York, 1965), Vol. IIA Ch. 3: *Magnetic Symmetry*.
8. H. Zocher and C. Török, *Proc. Nat. Acad. Sci.* **30**, 681 (1953).
9. D. E. Cox, to be published.
10. W. Voigt, *Lehrbuch der Kristallphysik* (Teubner, Leipzig, 1910).
11. H. A. Kramers, *Proc. Acad. Amsterdam* **33**, 959 (1930).
12. E. P. Wigner, *Gött. Nachr., Math-Phys.* 546 (1932).
13. I. E. Dyaloshinskii, *Soviet Phys.—JETP* **10**, 628 (1959).
14. N. V. Belov, N. N. Neronova and T. S. Smirnova, *Tr. Inst. Kristallogr. Akad. Nauk. SSSR* **11**, 33 (1955).
15. R. R. Birss, *Symmetry and Magnetism* (North-Holland, Amsterdam, 1964).
16. S. Bhagavantam, *Crystal Symmetry and Physical Properties* (Academic Press, New York, 1966).
17. H. Schmid, to be published.
18. E. Ascher, *Helv. Phys. Acta* **39**, 40 (1966).

PHENOMENOLOGICAL THEORY OF THE HIGH-FREQUENCY PROPERTIES OF FERROMAGNETICS–FERROELECTRICS

V. G. BAR'YAKHTAR and I. E. CHUPIS

Far East University, Vladivostok, USSR

A phenomenological theory of uniaxial ferroelectric–ferromagnetic crystals at temperatures below those of ferroelectric magnet transitions has been developed. Possible equilibrium values for the electric and magnetic moments of a ferroelectric magnet in a constant external magnetic field indicating to a magneto-electric effect and the possibility of phase transitions of the first and second kinds over the magnetic field have been studied. Vibrations of polarization and magnetization vectors have been quantized, and a spectrum for corresponding quasipartial excitations (segnetomagnons) obtained. The tensor for the ferroelectric magnet high-frequency susceptibility has been calculated, and the absorption energy of the electric field in the frequency of a homogeneous ferromagnetic resonance and absorption of the magnetic field in the ferroelectric frequency estimated. Absorption by the crystal of the energy of the variable magnetic field, causing transitions between the branches of the ferromagnetic spectrum, has also been examined. Expressions have been obtained for the life-time of the spin wave (with $K=0$) involved in decay processes resulting in a ferroelectric phonon and magnon and dispersion processes on the ferroelectric phonon.

I INTRODUCTION

The synthesis of compounds possessing simultaneously ferroelectric and ferromagnetic properties is a rather recent development in science.[1–4] The physical properties of such crystals, whose use in engineering[5] appears to be promising, have been investigated very slightly, especially in the experimental plane. This paper expounds certain aspects of a phenomenological theory of uniaxial ferroelectrics–ferromagnetics developed by the authors. Apart from data published previously,[6–9] the article contains some new results. Thus, according to Tomashpolsky,[3] possible instances of the crystal to which our theory refers may be ferroelectric magnet solid solutions of $BaTiO_3$–$Sr_{0.3}La_{0.7}MnO_3$ with 75–100 mole-% concentrations for $BaTiO_3$.

At temperatures lower than those of the electric θ_e and magnetic θ_m transitions, ferroelectric–ferromagnetic crystals may be described phenomenologically by introducing the density vectors of the electric $\mathbf{P}(\mathbf{r})$ and magnetic $\mathbf{M}(\mathbf{r})$ moments. The interaction of the electric and magnetic subsystems, which in itself is of electro- magnetostrictional origin, leads to a connection between the equilibrium values of the moments $\mathbf{M}$ and $\mathbf{P}$ in the presence of external fields. Possible types of dependencies of equilibrium electric and magnetic moments of a uniaxial crystal upon the magnetic field have been obtained.

Slight deviations of vectors $\mathbf{P}$ and $\mathbf{M}$ from their equilibrium states cause weakly attenuating spin and ferroelectric waves. Expressions for the spectrum of these high-frequency vibrations and relationships connecting the operators of the creation and destruction quanta of the bonded ferroelectric–ferromagnetic vibrations (segnetomagnons) with the operators of the system's electric and magnetic moments have been obtained. Green functions have been constructed, and the high-frequency susceptibility tensor that describes the responses of the spin and ferroelectric subsystems to the variable electric and magnetic fields calculated. The tensor helped to compute the electric field energy absorption in a frequency close to that of a homogeneous ferromagnetic resonance, as well as the energy absorption of the variable magnetic field in the frequency of ferroelectric vibrations and in the frequency causing transitions between the branches of the segnetomagnon spectrum.

To estimate the width of the lines of the ferromagnetic resonance, we have examined processes of the third and fourth order in the ferroelectrics–ferromagnetics, both segnetomagnon branches taking part. Expressions have been obtained for the average time of life of the spin wave ($K=0$) on account of the said processes.

II HAMILTONIAN

We examine a uniaxial ferroelectric–ferromagnetic at low temperatures ($T \ll \theta_e, \theta_m$) with an easy magnetic axis along axis z and with a constantly directed electric dipole moment $P(\mathbf{r})$.[6] This model apparently corresponds to crystal ferromagnetics wherein the internal electric fields are strongly anisotropic, requiring an electric field substantially exceeding the breakdown field to make a significant deviation of the polarization vector from its equilibrium direction.[10] The system's Hamiltonian is a functional of the vectors **P**, **M** and their derivatives. It is written as

$$\mathscr{H} = \int \Bigg\{ \tfrac{1}{2}\alpha \left(\frac{\partial \mathbf{M}}{\partial x_i}\right)^2 - \tfrac{1}{2}\beta M_z^2 - \mathbf{MH} - \tfrac{1}{2}\kappa P^2 + \tfrac{1}{4}\delta P^4 + \tfrac{1}{2}\lambda \left(\frac{\partial \mathbf{P}}{\partial x_i}\right)^2 + \frac{1}{2f}\dot{P}^2 + \tfrac{1}{2}\kappa'(P_x^2 + P_y^2) - \mathbf{PE} - \tfrac{1}{2}\gamma P^2 M_z^2 + \tfrac{1}{2}\gamma_0 P^2 \left(\frac{\partial \mathbf{M}}{\partial x_i}\right)^2 + \tfrac{1}{2}\gamma_1 M_z^2 \left(\frac{\partial \mathbf{P}}{\partial x_i}\right)^2 \Bigg\} dV \quad (1)$$

Here, the first three terms correspond to the usual expression for the energy of the ferromagnetic subsystem in the constant external magnetic field **H**; β is the constant of magnetic anisotropy; α is the constant of exchange interaction; and $\alpha \sim (\theta_m a^2/\mu_0 M_0)$, where μ_0 is the Bohr magneton, a is the lattice constant, and M_0 is the magnetic moment per unit volume. The next six terms are the energy of the ferroelectric subsystem in the external electric field **E**. In shift-type ferroelectrics, electric polarization arises on account of the slight shift Δr of the ions in the unit cell, $\Delta r \ll a$, and $P \sim \Delta r$. Hence, in the expression for the ferroelectric energy *per se* we omit anharmonisms whose order is higher than P^4, since they correspond to higher degrees of the relation $\Delta r/a$.

The constant $\kappa \sim \varepsilon_\infty^{1/2}$, where ε_∞ is the optical dielectric constant; constant δ is of the order 10^{-11} cm^2/dyn^{-1},[11] and constant λ of the order of 10^{-16} to 10^{-15} cm^2,[12] respectively. The term $\dot{P}^2/2f$ is the kinetic energy of the ferroelectric subsystem [$f \sim (e^2/mv_0)$, e is the charge of the ions in the unit cell, and v_0 and m are the volume and mass, respectively, of the unit cell]. The term with coefficient κ' describes the energy connected with the deviation of the electric polarization vector from axis z. In accordance with the above-mentioned, $\kappa' \gg \kappa$. The term with coefficient γ_0 describes the interconnection of electric polarization and magnetization by means of exchange interaction, while the terms with coefficients γ and γ_1 describe their connection through spin-orbital and magnetic dipole–dipole interactions. In the order of magnitude, $\gamma P^2 \sim \beta$, $\gamma_1 M_0^2 \sim \lambda$, $\gamma_0 P_0^2 \sim \alpha$, where P_0 is the static polarization. Coefficients α and λ are positive, since we will consider homogeneous ground states.

III GROUND STATE

The equilibrium values for **P** and M_z in one domain may be found through minimizing by **P** and θ (θ is the angle between $\mathbf{M}_0$ and the axis z) the expression (1), wherein the terms containing the moment derivatives are omitted. For simplicity, let us assume the external electric field **E** to be absent. Moreover, due to the relationship $\kappa' \gg \kappa$, one may neglect the perpendicular component of the vector **P**, i.e., one may consider $\mathbf{P} = (0.0, P_z)$.

In orienting the field along spontaneous magnetization (axis z), the moment values naturally do not change in the magnetic field. The dependencies $P(H)$ and $M_z(H)$ arise in the field that rotates the magnetic moment; for simplicity, let us consider this field to be oriented perpendicular to axis z. Omitting the calculations, let us now formulate the results.

The crystal will be in the paraelectric phase ($P = 0$) at all magnetic field values if $\beta > 0$ and $\kappa + \gamma M_0^2 < 0$ ($\gamma > 0$), or $\kappa < 0$ ($\gamma < 0$). For other values of the parameters κ, β and γ, M_z and P depend on the magnetic field (Figure 1). Shown below are the intervals of the values of the parameters corresponding to curves (a1), (a2), (b1), etc. ($\delta > 0$ in all cases).

a1) $\gamma > 0, \quad 0 < \beta < \dfrac{2\gamma^2 M_0^2}{\delta}, \quad \dfrac{\beta\delta}{2\gamma} - \gamma M_0^2 < \kappa < 0$

a2) $\gamma > 0, \quad 0 < \beta < \dfrac{2\gamma^2 M_0^2}{\delta}, \quad -\gamma M_0^2 < \kappa < \dfrac{\beta\delta}{2\gamma} - \gamma M_0^2; \quad \gamma > 0, \quad \beta > \dfrac{2\gamma^2 M_0^2}{\delta}, \quad -\gamma M_0^2 < \kappa < 0$

b1) $\gamma > 0,\quad 0 < \beta < \dfrac{2\gamma^2 M_0^2}{\delta},\quad 0 < \kappa < -\dfrac{\beta\delta}{\gamma} + 2\gamma M_0^2;\quad \gamma > 0,\quad \beta < 0,\quad -\dfrac{\beta\delta}{\gamma} < \kappa < -\dfrac{\beta\delta}{\gamma} + 2\gamma M_0^2$

b2) $\gamma > 0,\quad \beta < \dfrac{2\gamma^2 M_0^2}{\delta},\quad \kappa > -\dfrac{\beta\delta}{\gamma} + 2\gamma M_0^2;\quad \gamma > 0,\quad \beta > \dfrac{2\gamma^2 M_0^2}{\delta},\quad \kappa > 0$

c1) $\gamma < 0,\quad \dfrac{2}{3}\dfrac{\gamma^2 M_0^2}{\delta} < \beta < \dfrac{\gamma^2 M_0^2}{\delta},\quad \dfrac{\beta\delta}{2\gamma} - \gamma M_0^2 < \kappa < -\dfrac{\beta\delta}{\gamma};\quad \gamma < 0,\quad \dfrac{\gamma^2 M_0^2}{\delta} < \beta < \dfrac{2\gamma^2 M_0^2}{\delta};\quad \dfrac{\beta\delta}{2\gamma} - \gamma M_0^2 < \kappa < -\gamma M_0^2;\quad \gamma < 0,\quad \dfrac{2\gamma^2 M_0^2}{\delta} < \beta < \dfrac{3\gamma^2 M_0^2}{\delta},\quad -\dfrac{\beta\delta}{\gamma} + 2\gamma M_0^2 < \kappa < -\gamma M_0^2$

c2) $\gamma < 0,\quad 0 < \beta < \dfrac{2}{3}\dfrac{\gamma^2 M_0^2}{\delta},\quad 0 < \kappa < -\dfrac{\beta\delta}{\gamma};\quad \gamma < 0,\quad \dfrac{2}{3}\dfrac{\gamma^2 M_0^2}{\delta} < \beta < \dfrac{2\gamma^2 M_0^2}{\delta},\quad 0 < \kappa < \dfrac{\beta\delta}{2\gamma} - \gamma M_0^2$

d1) $\gamma < 0,\quad \dfrac{2\gamma^2 M_0^2}{\delta} < \beta < \dfrac{3\gamma^2 M_0^2}{\delta},\quad \tilde{\kappa} < \kappa < -\dfrac{\beta\delta}{\gamma} + 2\gamma M_0^2;\quad \gamma < 0,\quad \beta > \dfrac{3\gamma^2 M_0^2}{\delta},\quad \tilde{\kappa} < \kappa < -\gamma M_0^2$

d2) $\gamma < 0,\quad \beta > \dfrac{2\gamma^2 M_0^2}{\delta},\quad 0 < \kappa < \tilde{\kappa}$

e1) $\gamma < 0,\quad \dfrac{\gamma^2 M_0^2}{\delta} < \beta < \dfrac{3\gamma^2 M_0^2}{\delta},\quad -\gamma M_0^2 < \kappa < -\dfrac{\beta\delta}{\gamma};\quad \gamma < 0,\quad \beta > \dfrac{3\gamma^2 M_0^2}{\delta},\quad -\dfrac{\beta\delta}{\gamma} + 2\gamma M_0^2 < \kappa < -\dfrac{\beta\delta}{\gamma}$

e2) $\gamma < 0,\quad \beta > \dfrac{3\gamma^2 M_0^2}{\delta},\quad -\gamma M_0^2 < \kappa < -\dfrac{\beta\delta}{\gamma} + 2\gamma M_0^2 \qquad (2)$

Here,

$$\tilde{\kappa} = (2\gamma)^{-1}[(\delta^2\beta^2 + 6\delta\beta\gamma^2 M_0^2 + 33\gamma^4 M_0^4)^{1/2} - \delta\beta - 5\gamma^2 M_0^2],$$

and the designations in the figures are as follows

$$H_a = \delta^{-1} M_0(\beta\delta + \gamma\kappa), \quad H_\beta = \beta M_0,$$

$$H_c = \beta\left(\frac{\kappa + \gamma M_0^2}{\gamma}\right)^{1/2}$$

$$H_k = 2(27\gamma^2\delta^2)^{-1/2}|\beta\delta + \gamma\kappa + \gamma^2 M_0^2|^{3/2},$$

$$M_c = \left|\frac{\kappa}{\gamma}\right|^{1/2}$$

$$M_k = (3\gamma^2)^{-1/2}|2\gamma^2 M_0^2 - \kappa\gamma - \delta\beta|^{1/2},$$

$$P_k = |3\delta\gamma|^{-1/2}|2\kappa\gamma + 2\gamma^2 M_0^2 - \delta\beta|^{1/2}$$

$$P_a = \begin{cases} \kappa^{1/2}\delta^{-1/2} & (\gamma > 0) \\ (\kappa + \gamma M_0^2)^{1/2}\delta^{-1/2} & (\gamma < 0) \end{cases}$$

$$P_b = \begin{cases} (\kappa + \gamma M_0^2)^{1/2}\delta^{-1/2} & (\gamma > 0) \\ \kappa^{1/2}\delta^{-1/2} & (\gamma < 0) \end{cases} \qquad (3)$$

From the dependencies shown in Figure 1 it is apparent that the electric polarization of the ferroelectric magnet depends on the magnetic field, i.e., a magnetoelectric effect takes place. The dependency $P(H)$ shall be observed in the fields $H \leqslant H_n$, where H_n is any one of the fields H_c, H_k, or H_a. The highest H_n value in the order of magnitude is the same as the magnetic anisotropy field H_β.

In the fields H_c, H_a and H_β, crystals with dependencies (a2), (b2), (c1), (d1), (d2) and (e2) should have ferroelectric and ferromagnetic phase transitions of the second kind. Near the threshold transition field, H_n, P and $M_z \sim |H - H_n|^{1/2}$. The values of the finite jumps of the magnetic $\chi_{zx}^{m} = \partial M_z/\partial H_x$ and magnetoelectric $\chi_{zx}^{em} = \partial P_z/\partial H_x$ susceptibilities are

$$\Delta\chi_{zx}^{m} = 2\gamma\beta^{-1}|\kappa|^{1/2}|\kappa + \gamma M_0^2|^{3/2} \times (\delta\beta - 2\gamma\kappa - 2\gamma^2 M_0^2)^{-1} \quad (H = H_c)$$

$$\Delta\chi_{zx}^{em} = \gamma M_0 \delta^{1/2}\kappa^{-1/2}(\kappa\gamma + \delta\beta - 2\gamma^2 M_0^2)^{-1} \quad (H = H_a) \qquad (4)$$

Putting here $\gamma P^2 \sim \kappa\gamma\delta^{-1} \sim \beta$, $\gamma M_0^2 \sim \kappa$, we get $\Delta\chi^{m} \sim \beta^{-1}$, $\Delta\chi^{em} \sim |\beta\kappa|^{-1/2}$, i.e., the jumps of susceptibilities in this case may be about the same as the susceptibilities themselves.

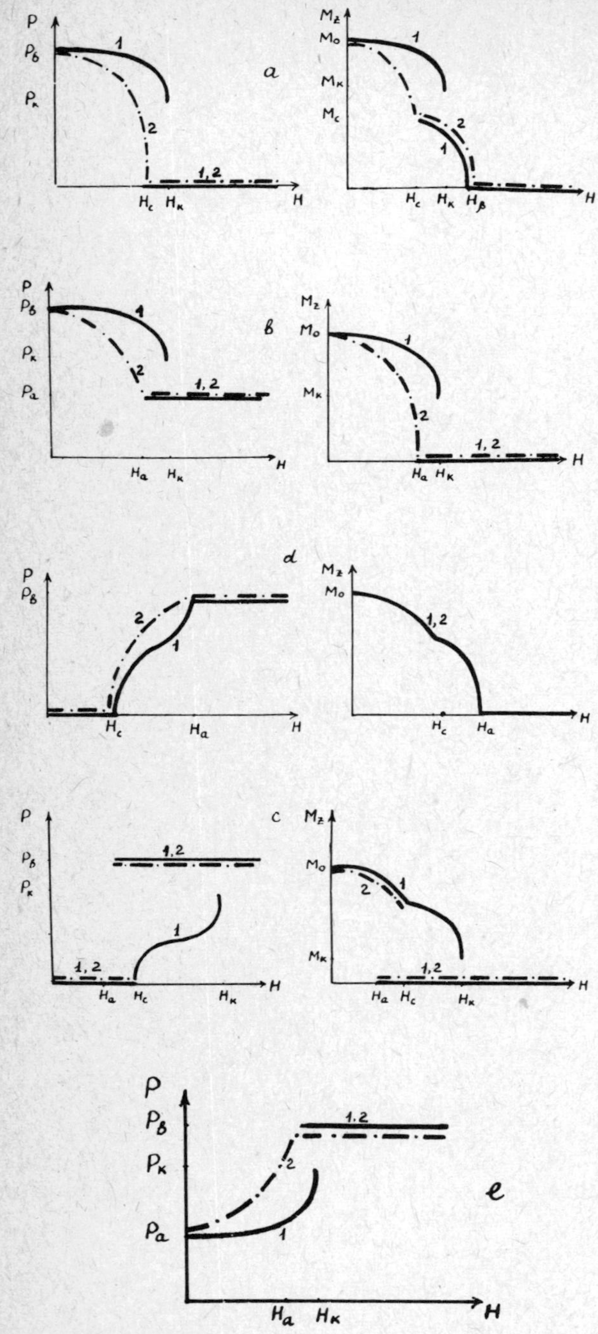

FIGURE 1 Magnetic field-dependencies of electric and magnetic moments of uniaxial ferroelectric magnet. "1" corresponds to continuous curves, and "2" to dashed–dotted curves. Dependency in (e) is the same as in (b). In (d), dotted and continuous curves show same kind of dependency.

The shapes of curves (a1), (b1), (c1), (c2) and (e1) indicate that in a given interval of magnetic field values these crystals have two equilibrium phases with different P and M_z magnitudes, i.e., metastable states are present. The presence of metastable states signifies the possibility of phase transitions of the first kind over the magnetic field and the possibility of a special domain structure in the magnetic field $H < H_n$. Thus, for instance, two equilibrium phases are possible for (c2) in the interval of fields (H_a, H_c), namely one with $P = 0$ and $M_z \neq 0$, and the other with $P \neq 0$ and $M_z = 0$. Now, in the fields (H_a, H') and (H', H_c) the first and second phases have lower energies, respectively. In the field

$$H' = \beta M_0 - \kappa\beta^{1/2}(2\delta)^{-1/2},$$

the phase energies become equal. In such bodies in field H', two types of domains are apparently possible, namely the paraelectric type ($P = 0$) with a predominant orientation of the magnetic moment perpendicular to the magnetic field and the ferroelectric type $P \neq 0$), whose magnetic moment is directed along the magnetic field.

It is also apparent that continuous changes in one of the system's parameters when the other parameters have fixed values may result in a different kind of phase transition. Thus, for instance, for $\gamma > 0$, $0 < \beta < (2\gamma^2 M_0{}^2/\delta)$, a continuous change in κ from the value $-\gamma M_0{}^2$ would lead to the consecutiveness of curves (a2), (a1), (b1) and (b2). A phase transition of the second kind is followed by a transition of the first kind, which is again followed by a transition of the second kind. It is also noteworthy that a phase transition of a given kind should take place simultaneously in the ferroelectric and ferromagnetic subsystems.

IV SEGNETOMAGNON SPECTRUM

The deviation of vectors **P** and **M** from their equilibrium positions in the ground state leads to the existence of connected spin and ferroelectric (segnetoelectric) waves in the ferroelectric magnet. The quanta of these ferromagnetic vibrations are quasiparticles which should naturally be called segnetomagnons.

The segnetomagnon spectrum may be obtained by quantizing the polarization and magnetization vibrations.[7] Let us choose the equilibrium magnetization direction $\mathbf{M}_0$ as the quantization axis ζ, and pass over to a system of coordinates (ξ, η, ζ) by directing the axis ξ in the plane of the vectors $(\mathbf{n}, \mathbf{H})$ ($\mathbf{n}$ is the unit vector along the easy axis z). The electric field is considered directed along axis z. The commutation rules for the components of vector $\mathbf{M}$ are usual, whereas for $\mathbf{P}$, $\dot{\mathbf{P}}$ and $\mathbf{M}$ they are as follows

$$[P_i(\mathbf{r}), \dot{P}_k(\mathbf{r}')] = i\hbar f \delta_{ik}\, \delta(\mathbf{r}-\mathbf{r}'),$$

$$[P_i, M_k] = [\dot{P}_i, M_k] = 0$$

Let us now use the representation M via the Holstein–Primakoff operators, and pass over to the Fourier components of the operators

$$M^+ = M_\xi + iM_\eta = \left(\frac{2\mu_0 M_0}{V}\right)^{1/2} \sum_{\mathbf{k}} a_{\mathbf{k}}^+ \exp(-i\mathbf{kr})$$

$$M^- = M_\xi - iM_\eta = \left(\frac{2\mu_0 M_0}{V}\right)^{1/2} \sum_{\mathbf{k}} a_{\mathbf{k}} \exp(i\mathbf{kr})$$

$$M_\zeta = M_0 - \frac{\mu_0}{V} \sum_{\mathbf{k},\mathbf{k}'} a_{\mathbf{k}}^+ a_{\mathbf{k}'} \exp[i(\mathbf{k}'-\mathbf{k})\mathbf{r}]$$

$$P = P_0 + V^{-1/2} \sum_{\mathbf{k}} b_{\mathbf{k}} \exp(i\mathbf{kr}) \tag{5}$$

Now, since P is Hermitian, we have $b_{\mathbf{k}}^+ = b_{-\mathbf{k}}$. From commutation rules for $\mathbf{P}$, $\mathbf{M}$ and $\dot{\mathbf{P}}$, we find that operators $a_{\mathbf{k}}$ and $a_{\mathbf{k}}^+$ commute with $b_{\mathbf{k}}$, $\dot{b}_{\mathbf{k}}$ and

$$[a_{\mathbf{k}}, a_{\mathbf{k}'}^+] = \delta_{\mathbf{kk}'}, \quad [b_{\mathbf{k}}, \dot{b}_{\mathbf{k}'}^+] = [b_{\mathbf{k}}^+, \dot{b}_{\mathbf{k}'}] = i\hbar f \delta_{\mathbf{kk}'}$$

In finding the normal vibration frequencies in Hamiltonian (Eq. 1), one may omit the energy of the exchange interaction of the subsystem and the last term, since their role is solely to normalize the temperatures of the ferromagnetic and ferroelectric transitions

$$\Theta_{\mathrm{m}} \approx \frac{\alpha\mu_0 M_0}{a^2} = \Theta_{\mathrm{m}}^{0} + \gamma_0 \frac{\mu_0 M_0}{a^2} P_0^2,$$

$$\Theta_{\mathrm{e}}^2 \approx \frac{\lambda f \hbar^2}{a^2} \approx \Theta_{\mathrm{e}}^{(0)2} + \gamma_1 \frac{f\hbar^2 M_0^2}{a^2}$$

We shall also not consider the summand $\frac{1}{2}\kappa' P_\perp^2$, to which vibrations with high activation frequencies $\sim(\kappa')^{1/2}$ correspond. Using Eq. (5), we obtain from (1) the following quadratic part of the Hamiltonian by operators

$$\mathscr{H}_2 = \sum_{\mathbf{k}} \Big\{ A_1 a_{\mathbf{k}}^+ a_{\mathbf{k}} + \tfrac{1}{2} B_1 a_{\mathbf{k}} a_{-\mathbf{k}} + \tfrac{1}{2} B_1^* a_{\mathbf{k}}^+ a_{-\mathbf{k}}^+ + \tfrac{1}{2} A_2 b_{\mathbf{k}}^+ b_{-\mathbf{k}}^+ + \frac{1}{2f} \dot{b}_{\mathbf{k}}^+ \dot{b}_{\mathbf{k}} + C(a_{\mathbf{k}}^+ b_{\mathbf{k}} + a_{\mathbf{k}} b_{\mathbf{k}}^+) \Big\} \tag{6}$$

where

$$A_1 = \mu_0 M_0 \left[\alpha k^2 + \frac{H}{M_0}\cos(\theta-\psi) + (\beta+\gamma P_0^2)(\cos^2\theta - \tfrac{1}{2}\sin^2\theta)\right]$$

$$B_1 = -\tfrac{1}{2}\mu_0 M_0 (\beta + \gamma P_0^2)\sin^2\theta$$

$$A_2 = \lambda k^2 - \kappa + 3\delta P_0^2 - \gamma M_0^2 \cos^2\theta$$

$$C = \left(\frac{\mu_0 M_0}{2}\right)^{1/2} \gamma P_0 M_0 \sin 2\theta \tag{7}$$

In (7), ψ is the angle between the magnetic field and axis z.

To diagonalize the Hamiltonian (6), let us pass over from operators $a_{\mathbf{k}}$ and $b_{\mathbf{k}}$ to the creation and destruction operators of the segnetomagnons $c_{\mathbf{k}}^+$ and $c_{\mathbf{k}}$ in accord with

$$a_{\mathbf{k}} = \sum_{n=1}^{2} (u_{n\mathbf{k}}^1 c_{n\mathbf{k}} + v_{n\mathbf{k}}^{*1} c_{n-\mathbf{k}}^+)$$

$$b_{\mathbf{k}} = \sum_{n=1}^{2} (u_{n\mathbf{k}}^2 c_{n\mathbf{k}} + u_{n\mathbf{k}}^{*2} c_{n-\mathbf{k}}^+) \tag{8}$$

In this case, the Hamiltonian $\mathscr{H}_2$ is reduced to the diagonal form

$$\mathscr{H}_2 = \sum_{\mathbf{k}} (\varepsilon_{1\mathbf{k}} c_{1\mathbf{k}}^+ c_{1\mathbf{k}} + \varepsilon_{2\mathbf{k}} c_{2\mathbf{k}}^+ c_{2\mathbf{k}})$$

if the values of the u–v-transformation amplitudes are

$$u^1_{1\mathbf{k}} = \exp(i\varphi_1) \times \frac{(A_1 - B_1 + \varepsilon_1)}{2[\varepsilon_1(A_1 - B_1)]^{1/2}} \cdot \frac{(\varepsilon_1{}^2 - \varepsilon_e{}^2)}{[(\varepsilon_2{}^2 - \varepsilon_s{}^2)(\varepsilon_2{}^2 - \varepsilon_1{}^2)]^{1/2}}$$

$$v^1_{1\mathbf{k}} = \exp(i\varphi_1) \times \frac{(A_1 - B_1 - \varepsilon_1)}{2[\varepsilon_1(A_1 - B_1)]^{1/2}} \cdot \frac{(\varepsilon_1{}^2 - \varepsilon_e{}^2)}{[(\varepsilon_2{}^2 - \varepsilon_s{}^2)(\varepsilon_2{}^2 - \varepsilon_1{}^2)]^{1/2}}$$

$$u^1_{2\mathbf{k}} = \exp(i\varphi_2) \cdot C \cdot \left(\frac{\hbar^2 f}{2\varepsilon_2}\right)^{1/2} \times \frac{(A_1 - B_1 + \varepsilon_2)}{(\varepsilon_2{}^2 - \varepsilon_s{}^2)} \cdot \left(\frac{\varepsilon_e{}^2 - \varepsilon_1{}^2}{\varepsilon_2{}^2 - \varepsilon_1{}^2}\right)^{1/2}$$

$$v^1_{2\mathbf{k}} = \exp(i\varphi_2) \cdot C \cdot \left(\frac{\hbar^2 f}{2\varepsilon_2}\right)^{1/2} \times \frac{(A_1 - B_1 - \varepsilon_2)}{(\varepsilon_2{}^2 - \varepsilon_s{}^2)} \cdot \left(\frac{\varepsilon_e{}^2 - \varepsilon_1{}^2}{\varepsilon_2{}^2 - \varepsilon_1{}^2}\right)^{1/2}$$

$$u^2_{1\mathbf{k}} = -\exp(i\varphi_1) \cdot \left(\frac{\hbar^2 f}{2\varepsilon_1}\right)^{1/2} \cdot \left(\frac{\varepsilon_2{}^2 - \varepsilon_e{}^2}{\varepsilon_2{}^2 - \varepsilon_1{}^2}\right)^{1/2}$$

$$u^2_{2\mathbf{k}} = \exp(i\varphi_2) \cdot \left(\frac{\hbar^2 f}{2\varepsilon_2}\right)^{1/2} \cdot \left(\frac{\varepsilon_e{}^2 - \varepsilon_1{}^2}{\varepsilon_2{}^2 - \varepsilon_1{}^2}\right)^{1/2} \qquad (9)$$

The energies of the two types of segenetomagnons $\varepsilon_{1\mathbf{k}}$ and $\varepsilon_{2\mathbf{k}}$ are determined by the expressions

$$\varepsilon^2_{1,2} = \tfrac{1}{2}\{\varepsilon_s{}^2 + \varepsilon_e{}^2 \mp [(\varepsilon_e{}^2 - \varepsilon_s{}^2)^2 + 8f\hbar^2 C^2(A_1 - B_1)]^{1/2}\} \qquad (10)$$

where

$$\varepsilon_s{}^2 = A_1{}^2 - B_1{}^2, \quad \varepsilon_e{}^2 = f\hbar^2 A_2$$

ε_s is the spin wave energy, and ε_e the ferroelectric vibration energy with changed values of activation energy due to interaction of the subsystems.

The dispersion curves $\varepsilon_{1\mathbf{k}}$ and $\varepsilon_{2\mathbf{k}}$ are shown in Figure 2, where the dashed lines indicate the branches of the non-interacting ferroelectric and spin waves. In the absence of interaction, i.e., at $\gamma = C = 0$, first-type segnetomagnon operators c^+_{1k} and $c_{1\mathbf{k}}$ become spin wave operators with energy $\varepsilon_{1\mathbf{k}} = \varepsilon_s$, while second-type segnetomagnon operators $c^+_{2\mathbf{k}}$ and $c_{2\mathbf{k}}$ transform into ferroelectric phonon operators with energy $\varepsilon_{2\mathbf{k}} = \varepsilon_e$. In the general case,

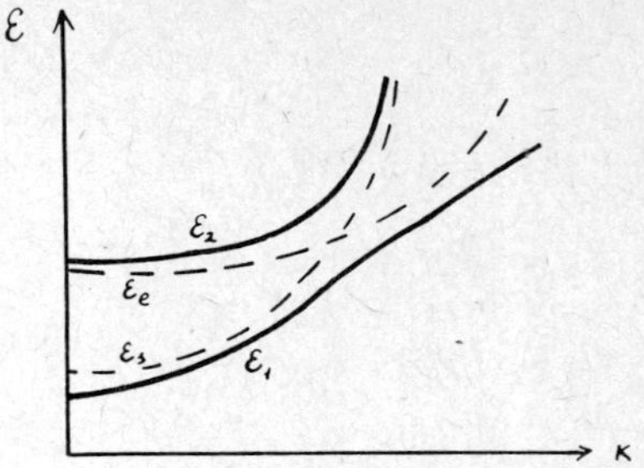

FIGURE 2 Dependency of segnetomagnon energy on wave vector.

however, with interaction between the subsystems, both polarization and magnetization are expressed through the creation operators $c^+_{1\mathbf{k}}$, $c^+_{2\mathbf{k}}$ and destruction operators $c_{1\mathbf{k}}$, $c_{2\mathbf{k}}$ of the two types of segnetomagnons.

V HIGH-FREQUENCY SUSCEPTIBILITY

As is known, the susceptibility tensor for ferromagnetic substances has the form

$$\chi^{(\dots)}_{ik} = \begin{pmatrix} \chi^{m}_{ik}, & \chi^{me}_{ik} \\ \chi^{em}_{ik}, & \chi^{e}_{ik} \end{pmatrix}$$

where χ^m and χ^e are the magnetic and electric susceptibilities, and the "mixed" tensor components χ^{me} and χ^{em} describe the response of the magnetic subsystem to the electric field applied and the response of the ferroelectric subsystem to the magnetic field applied, respectively.

The susceptibility Fourier components and Green functions are connected by the simple relationship

$$\chi^{(\dots)}_{ik}(\mathbf{k}, \omega) = \hbar^{-1} G^{(\dots)}_{ik}(\mathbf{k}, \omega)$$

where $G^{(\dots)}_{ik}(\mathbf{k},\omega)$ is the Fourier component of one of the four Green functions for the ferroelectric magnet. These functions can be introduced in the usual way, e.g.

$$G^{me}_{ik}(\mathbf{r} - \mathbf{r}', t) = -i\Delta(t) < [M_i(\mathbf{r}, t), P_k(\mathbf{r}', 0)] >$$

where

$$\Delta(t) = \begin{cases} 1 & t > 0 \\ 0 & t < 0 \end{cases}$$

Using (5) and taking into account that

$$c_{n\mathbf{k}}(t) = c_{n\mathbf{k}} \exp(-i\hbar^{-1}\varepsilon_{n\mathbf{k}} t), \quad [c_{n'\mathbf{k}'}, c_{n\mathbf{k}}^{+}] = \delta_{nn'}\delta_{\mathbf{k}\mathbf{k}'}$$

the following first-approximation expressions are obtained for χ values not equal to zero

$$\chi_{\xi\xi}^{\mathrm{m}} = \tfrac{1}{2} g M_0 \sum_{n=1}^{2} |u_{n\mathbf{k}}^1 + v_{n\mathbf{k}}^1|^2 \left(\frac{1}{\omega + \omega_{n\mathbf{k}}} - \frac{1}{\omega - \omega_{n\mathbf{k}}}\right)$$

$$\chi_{\xi\eta}^{\mathrm{m}} = \frac{i}{2} g M_0 \sum_{n=1}^{2} \left[\frac{(u_{n\mathbf{k}}^1 - v_{n\mathbf{k}}^1)(u_{n\mathbf{k}}^{*1} + v_{n\mathbf{k}}^{*1})}{\omega + \omega_{n\mathbf{k}}} + \frac{(u_{n\mathbf{k}}^{*1} - v_{n\mathbf{k}}^{*1})(u_{n\mathbf{k}}^1 + v_{n\mathbf{k}}^1)}{\omega - \omega_{n\mathbf{k}}}\right]$$

$$\chi_{\eta\eta}^{\mathrm{m}} = \tfrac{1}{2} g M_0 \sum_{n=1}^{2} |u_{n\mathbf{k}}^1 - v_{n\mathbf{k}}^1|^2 \left(\frac{1}{\omega + \omega_{n\mathbf{k}}} - \frac{1}{\omega - \omega_{n\mathbf{k}}}\right)$$

$$\chi_{zz}^{\mathrm{e}} = \hbar^{-1} \sum_{n=1}^{2} |u_{n\mathbf{k}}^2|^2 \left(\frac{1}{\omega + \omega_{n\mathbf{k}}} - \frac{1}{\omega - \omega_{n\mathbf{k}}}\right)$$

$$\chi_{z\xi}^{\mathrm{em}} = \left(\frac{g M_0}{2\hbar}\right)^{1/2} \sum_{n=1}^{2} \left[\frac{u_{n\mathbf{k}}^{*2}(u_{n\mathbf{k}}^1 + v_{n\mathbf{k}}^1)}{\omega + \omega_{n\mathbf{k}}} - \frac{u_{n\mathbf{k}}^2(u_{n\mathbf{k}}^{*1} + v_{n\mathbf{k}}^{*1})}{\omega - \omega_{n\mathbf{k}}}\right]$$

$$\chi_{z\eta}^{\mathrm{em}} = i \left(\frac{g M_0}{2\hbar}\right)^{1/2} \sum_{n=1}^{2} \left[\frac{u_{n\mathbf{k}}^2(u_{n\mathbf{k}}^{*1} - v_{n\mathbf{k}}^{*1})}{\omega - \omega_{n\mathbf{k}}} + \frac{u_{n\mathbf{k}}^{*2}(u_{n\mathbf{k}}^1 - v_{n\mathbf{k}}^1)}{\omega + \omega_{n\mathbf{k}}}\right]$$

$$\chi_{\zeta\zeta}^{\mathrm{e}} = \chi_{zz}^{\mathrm{e}} \cos^2\theta, \quad \chi_{\xi\xi}^{\mathrm{e}} = \chi_{zz}^{\mathrm{e}} \sin^2\theta, \quad \chi_{\xi\zeta}^{\mathrm{e}} = -\tfrac{1}{2}\chi_{zz}^{\mathrm{e}} \sin 2\theta$$

$$\chi_{\xi\xi}^{\mathrm{em}} = -\chi_{z\xi}^{\mathrm{em}} \sin\theta, \quad \chi_{\zeta\xi}^{\mathrm{em}} = \chi_{z\xi}^{\mathrm{em}} \cos\theta, \quad \chi_{\xi\eta}^{\mathrm{em}} = -\chi_{z\eta}^{\mathrm{em}} \sin\theta$$

$$\chi_{\zeta\eta}^{\mathrm{em}} = \chi_{z\eta}^{\mathrm{em}} \cos\theta, \quad \chi_{ik}^{\mathrm{m}} = \chi_{ki}^{*\mathrm{m}}, \quad \chi_{ik}^{\mathrm{e}} = \chi_{ki}^{*\mathrm{e}}, \quad \chi_{ik}^{\mathrm{em}} = \chi_{ki}^{*\mathrm{me}} \tag{11}$$

VI ABSORPTION ENERGIES OF VARIABLE HOMOGENEOUS EXTERNAL FIELDS

Knowing the high-frequency susceptibility tensor it is easy to calculate the external field energy losses in the ferroelectric magnet. Due to the presence in the expressions of both types of segnetomagnon operators $c_{1\mathbf{k}}$ and $c_{2\mathbf{k}}$ for the moments **M** and P, resonance excitation of the "ferroelectric" frequency branch by the magnetic field and, contrariwise, resonance absorption of electric energy by the "magnetic" branch become possible.[7]

In applying to the crystal a homogeneous magnetic field with the frequency ω, orientated perpendicular to the direction of equilibrium magnetization along axis ξ the absorbed energy equals

$$\dot{Q} = \tfrac{1}{2}\omega g M_0 |h(\omega)|^2 [|u_{10}^1 + v_{10}^1|^2 \delta(\omega - \omega_{10}) + |v_{20}^1 + u_{20}^1|^2 \delta(\omega - \omega_{20})] \tag{12}$$

where $h(\omega)$ is the Fourier component of the magnetic field.

Thus, from (12) we see that resonance absorption can take place either in the ferromagnetic resonance frequency $\omega_{10} = \varepsilon_{10}\hbar^{-1}$, or in the ferroelectric frequency $\omega_{20} = \varepsilon_{20}\hbar^{-1}$. Relative absorption intensities in these frequencies are determined by the relationship

$$\frac{\dot{Q}_{\omega 20}}{\dot{Q}_{\omega 10}} \approx \frac{C^2}{\varepsilon_{s0}} \left(\frac{\varepsilon_{s0}}{\varepsilon_{e0}}\right)^3 \sim \left(\frac{\varepsilon_{s0}}{\varepsilon_{e0}}\right)^3 \ll 1 \tag{13}$$

If a homogeneous electric field is applied to the solid, along axis z, then

$$\dot{Q}_{\omega} = \omega\hbar^{-1} |e(\omega)|^2 [|u_{10}^2|^2 \delta(\omega - \omega_{10}) + |u_{20}^2|^2 \delta(\omega - \omega_{20})] \tag{14}$$

i.e., resonance absorption of the electric field may occur both in the ferroelectric frequency ω_{20} and in the homogeneous ferromagnetic resonance frequency ω_{10}. The relationship of the absorption intensities in this case is

$$\frac{\dot{Q}_{\omega 10}}{\dot{Q}_{\omega 20}} \approx \frac{C^2}{\varepsilon_{e0}} \sim \frac{\varepsilon_{s0}}{\varepsilon_{e0}}$$

This value is also much less than (1), but much greater than (13).

Absorption intensities in the frequencies $\omega_{1\mathbf{k}}$ and $\omega_{2\mathbf{k}}$ become of the same order if heterogeneous magnetic or electric fields with frequencies close to the frequency of noninteracting spin and ferroelectric branches are applied to the ferroelectric magnet; however, we will not consider this case here.

Considerations of segnetomagnon damping leads to the necessity of replacing the function δ in (12) and (14) by $\Gamma[(\varepsilon_0 - \hbar\omega)^2 + \Gamma^2]^{-1}$, where Γ is the damping of vibrations with the frequency $\varepsilon_0 \cdot \hbar^{-1}$.

The homogeneous variable magnetic field, directed along the equilibrium magnetization, may cause transitions between the branches of the segnetomagnetic spectrum if the following condition is fulfilled

$$\hbar\omega = \varepsilon_{2\mathbf{k}} - \varepsilon_{1\mathbf{k}}$$

An analysis of this relationship, assuming that $(C^2/\varepsilon_{e0}) \ll 1$, shows that the absorption is of a

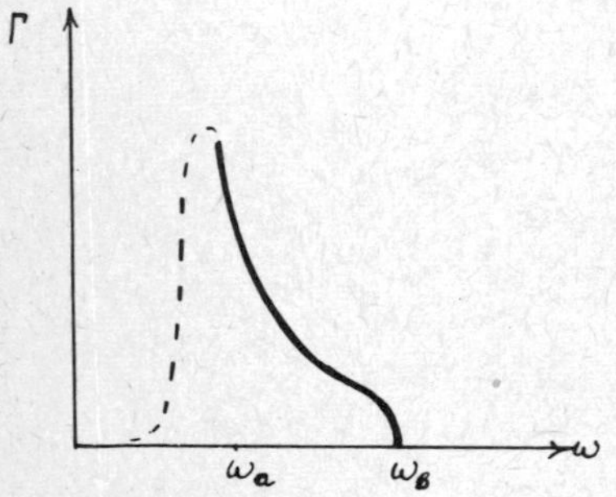

FIGURE 3 Dependency of absorption coefficient on frequency of variable magnetic field causing transitions between segnetomagnetic spectrum branches.

resonance-threshold nature and takes place in the frequency interval (ω_a, ω_b), where

$$\omega_a \approx C\left(\frac{2f}{\varepsilon_{e0}}\right)^{1/2}, \quad \omega_b \approx \omega_{e0} = \varepsilon_{e0} \cdot \hbar^{-1}$$

In the approximation considered, $\omega_a \ll \omega_b$. The frequency dependence of the absorption coefficient in the time unit $\Gamma(\omega)$ is shown in Figure 3. Near the threshold values of the frequencies,

$$\Gamma \sim \begin{cases} (\omega - \omega_a)^{-1/2} & \omega \approx \omega_a \\ (\omega_b - \omega)^{1/2} & \omega \approx \omega_b \end{cases} \tag{15}$$

Consideration of interaction shall, naturally, lead to "erosion" of the absorption curve near the resonance frequency (Figure 3, dotted portion of curve). For numerical evaluations, it is suitable to use the dimensionless magnitude $d\Gamma/d\omega$. Assuming that $T/\theta_m \sim 10^{-1}$, $\omega - \omega_a \sim \tau^{-1}$, $\tau \sim 10^{-8}$ sec, $\omega_a \sim 10^{11}$ sec^{-1}, $\mu_0 M_0 \sim 1°$K, $\varepsilon_e \sim \theta_m \sim 10^2$ °K and $\beta \sim 1$, we get $(d\Gamma/d\omega) \sim 10^{-4}$ near the resonance frequency, i.e., the absorption may be noticeable.

VII RELAXATION PROCESSES

The widths of the absorption lines in the ferroelectric magnet are determined by interaction processes in the segnetomagnon system and their interaction with phonons.

We consider relaxation processes involving three and four segnetomagnons of both types in a constant magnetic field directed along the easy axis of the crystal.[9] Inasmuch as, in this case, the angle $\theta = 0$, the operators $a_\mathbf{k}$ and $a_\mathbf{k}^+$ are expressed only through the operators $c_{1\mathbf{k}}$ and $c_{1\mathbf{k}}^+$, whereas the operator $b_\mathbf{k}$ is expressed via $c_{2\mathbf{k}}$ and $c_{2\mathbf{k}}^+$; the dispersion laws for segnetomagnons of the first and second type take on the form of dispersion laws for magnons and ferroelectric phonons, respectively; however, their activation energies are changed due to the interaction of the electric and magnetic subsystems. In the long-wave approximation $(ak \ll 1)$ they are

$$\varepsilon_{1k} = \varepsilon_{10} + \theta_m(ak)^2, \quad \varepsilon_{2k} = [\varepsilon_{20}^2 + (\theta_e ak)^2]^{1/2}$$

It is readily apparent that if $\varepsilon_{20} \gtrsim \theta_m$, then for processes involving three segnetomagnons the laws of preservation of energy and wave vectors are fulfilled simultaneously only for states close to the boundary of the Brillouin zone $(ak \sim 1)$. In this case, a macroscopic consideration is unacceptable. However, since the relationship $\varepsilon_{20} \gtrsim \theta_m$ is essentially the most probable, one should use the microscopic model to calculate the probabilities of "triple" processes.

The Hamiltonian part (Eq. 1) describing the interaction of the subsystems may also be written as

$$\begin{aligned}\mathscr{H}_{\text{int}} = &-\tfrac{1}{2}(2\mu_0)^{-2}\int f(|\mathbf{r}-\mathbf{r}'|)P(\mathbf{r})P(\mathbf{r}') \\ &\times \mathbf{M}(\mathbf{r})\,\mathbf{M}(\mathbf{r}')\,d\mathbf{r}\,d\mathbf{r}' \\ &-\tfrac{1}{2}(2\mu_0)^{-2}\int \varphi_{ik}(\mathbf{r}-\mathbf{r}')P(\mathbf{r})P(\mathbf{r}') \\ &\times M_i(\mathbf{r})\,M_k(\mathbf{r}')\,d\mathbf{r}\,d\mathbf{r}'\end{aligned} \tag{16}$$

where the functions $f(|\mathbf{r}|)$ and $\varphi(\mathbf{r})$, rapidly decreasing with distance, are connected with the constants γ and γ_0 by the relationships

$$\gamma = (2\mu_0)^{-2}\int \varphi_{zz}(\mathbf{r})\,d\mathbf{r},$$

$$\gamma_0 = \tfrac{1}{6}(2\mu_0)^{-2}\int f(|r|)\,r^2\,dr \tag{17}$$

In (16), we ignore the summand of the Hamiltonian with coefficient γ_1, the summand being much smaller than the exchange interaction energy of the subsystems and contributing to the processes with the participation of four and more segnetomagnons.

In the Hamiltonian (16), let us pass over from the magnetic moment density operators to the spin operators, assuming

$$\mathbf{M}(\mathbf{r}) = 2\mu_0 \sum_l \mathbf{S}_l\, \delta(\mathbf{r} - \mathbf{R}_l)$$

Then, instead of (16), we will have

$$\mathscr{H}_{\text{int}} = -\tfrac{1}{2} \sum_{l \neq l'} f_{ll'} P_l P_{l'} (\mathbf{S}_l \mathbf{S}_{l'}) - \tfrac{1}{2}\gamma \frac{(2\mu_0)^2}{v_0} \sum_l P_l^2 (S_l^z)^2 \tag{18}$$

where index l numbers the crystal unit cells. In this case, the operators P_l and S_l are expressed as in (5) and (8) through operators $a_{\mathbf{k}}$, $b_{\mathbf{k}}$, $c_{\mathbf{k}}$ and $c_{\mathbf{k}}^{+}$. For processes involving three quasiparticles and spin values $S = \frac{1}{2}$, we have

$$\mathscr{H}^{(3)}_{\text{int}} = \sum_{\substack{\lambda,\mu,\nu \\ (\mathbf{k}_\lambda = \mathbf{k}_\mu + \mathbf{k}_\nu)}} \Phi^{1;21}_{\lambda;\mu\nu}\, c^{+}_{1\lambda} c_{2\mu} c_{1\nu} + \text{c.c.}$$

The amplitude, allowed by selection rules, of the magnon decay process into a magnon and ferroelectric phonon equals

$$\Phi^{1;21}_{\lambda;\mu\nu} = \frac{P_0}{2V^{1/2}} \left(\frac{\hbar^2 f}{2\varepsilon_{2\mu}}\right)^{1/2} \left[\frac{2(2\mu_0)^2}{v_0}\gamma - f(-\mathbf{k}_\lambda) - f(\mathbf{k}_\nu) + f(\mathbf{k}_\mu) + f(0)\right] \tag{19}$$

where

$$f(\mathbf{k}) = \sum_l f_{ll'} \exp\left[-i\mathbf{k}(\mathbf{R}_l - \mathbf{R}_{l'})\right]$$

Most probable are the processes involving a spin wave with $K = 0$; these processes determine the line width of the homogeneous ferromagnetic resonance.

For the crystal with a symmetry center, the sum of the last four terms in (19) is zero, i.e., the exchange interaction does not contribute to the processes. In our case, preservation laws make possible for the spin wave with $K \neq 0$ to decay to a spin wave with $K = 0$ and a ferroelectric phonon; reverse processes also become possible. Using (19), we get the following expression for the relaxation time of the spin wave with $K = 0$ on account of the involvement in these processes of

$$\frac{1}{\tau_0^{(3)}} = \frac{2}{\pi^2} \cdot f \cdot \left(\frac{\gamma P_0 \mu_0^2}{v_0}\right)^2 \cdot \int \frac{n_2 - n_1}{|v_e^z - v_m^z|}\, \varepsilon_2^{-1}\, dk_x\, dk_y \tag{20}$$

where v_e^z and v_m^z are the speed z-components of the ferroelectric phonon and magnon, and n_2 and n_1 are the equilibrium functions of phonon and magnon distributions.

Fulfilment of the energy preservation law in the process studied implies crossing of the spin and ferroelectric vibration branches. Now, since, with $K = 0$, $\varepsilon_{20} \gg \varepsilon_{10}$, the crossing is remote from the center of the Brillouin zone. To calculate the integral in (20), let us assume that over the entire Brillouin first zone $\varepsilon_2 = \varepsilon_{20} = \text{const.}$, and crossing of the energy branches takes place on account of the spin frequency, which is determined by the known expression

$$\varepsilon_{1k} = \varepsilon_{10} + 2\mathscr{I}_0 (3 - \cos ak_x - \cos ak_y - \cos ak_z)$$

where $\mathscr{I}_0$ is the exchange integral between the closest neighbors.

Consequently, we find

$$\frac{1}{\tau_0^{(3)}} = \frac{8}{\pi^2} \cdot \frac{\hbar f}{\mathscr{I}_0} \cdot \frac{(\gamma P_0 \mu_0^2)^2}{v_0^3} \cdot \frac{(n_2 - n_1)}{\varepsilon_{20}} \cdot I \tag{21}$$

where

$$n_2 - n_1 = [\exp(\varepsilon_{20} T^{-1}) - 1]^{-1} - [\exp\{(\varepsilon_{10} + \varepsilon_{20}) T^{-1}\} - 1]^{-1}$$

$$I = \iint_R \frac{dx\, dy}{\left\{(1 - x^2)(1 - y^2)\left[1 - \left(3 - \dfrac{\varepsilon_{20}}{2\mathscr{I}_0} - x - y\right)^2\right]\right\}^{1/2}}$$

The integration region R is determined by the inequalities

$$\left|3 - \frac{\varepsilon_{20}}{2\mathscr{I}_0} - x - y\right| < 1, \quad |x| < 1, \quad |y| < 1$$

The approximative value of the integral I in the interval of the values $\mathscr{I}_0 \leqslant \varepsilon_{20} \leqslant 10\mathscr{I}_0$ is equal to $\pi^2/4$.

For the most probable case, when $\varepsilon_{20} \sim \mathscr{I}_0$ (i.e., $\theta_e \sim \theta_m$), we have

$$\frac{1}{\tau_0^{(3)}} = \begin{cases} \dfrac{2\hbar f}{\mathscr{I}_0 \varepsilon_{20}} \cdot \dfrac{(\gamma P_0 \mu_0{}^2)^2}{v_0{}^3} \exp(-\varepsilon_{20} T^{-1}) & T \ll \varepsilon_{10} \\ \dfrac{2\hbar f}{\mathscr{I}_0 \varepsilon_{20}} \cdot \dfrac{(\gamma P_0 \mu_0{}^2)^2}{v_0{}^3} \cdot \dfrac{\varepsilon_{10}}{T} \cdot \exp(-\varepsilon_{20} T^{-1}) & \varepsilon_{10} \ll T \ll \theta_{m,e} \end{cases} \quad (22)$$

In the region of "high" temperatures, $\varepsilon_{10} \ll T \ll \theta_{m,e}$, assuming in (22) for the estimation of $\theta_m \sim \mathscr{I}_0 \sim \varepsilon_{20} \sim 10^{-13}$ erg, $\hbar^2 f \sim \varepsilon_{20}^2$, $v_0 \sim 10^{-24}$ cm^3, $\mu_0{}^2/v_0 \sim 10^{-16}$ erg and $\gamma P_0 M_0 \sim \beta \sim 1$–$10$, we get $\tau_0^{(3)} \sim 10^{-6}$–$10^{-5}$ sec. In other words, at room temperatures the above processes can give a ferromagnetic resonance line width of about 0.1–0.01 Oe. If the ferroelectric magnet has $\varepsilon_{20} \gg \mathscr{I}_0$ or $\varepsilon_{20} \ll \mathscr{I}_0$, then the probabilities of these processes are considerably less than the probabilities determined by (22).

Let us now go over to the processes in the Hamiltonian (18). The corresponding interaction Hamiltonian is

$$\mathscr{H}_{\text{int}}^{(4)} = \sum_{\substack{\kappa, \lambda, \mu, \nu \\ (\mathbf{k}_\lambda + \mathbf{k}_\mu = \mathbf{k}_\kappa + k_\nu)}} \Phi_{\lambda\mu;\kappa\nu}^{12;12} c_{1\lambda}^+ c_{2\mu}^+ c_{1\kappa} c_{2\nu} + \sum_{\substack{\kappa, \lambda, \mu, \nu \\ (\mathbf{k}_\lambda = \mathbf{k}_\kappa + \mathbf{k}_\mu + \mathbf{k}_\nu)}} \Phi_{\lambda;\kappa\mu\nu}^{1;221} c_{1\lambda}^+ c_{2\kappa} c_{2\mu} c_{1\nu} + \text{c.c.} \quad (23)$$

Hence, of the two possible processes, the most probable is the spin wave dispersion process with $K = 0$ on the ferroelectric phonon. The amplitude of this process is determined by the expression

$$\Phi_{0\lambda;\mu\nu}^{12;12} = \frac{\gamma \mu_0 M_0}{V} \cdot \hbar^2 f [\varepsilon_2(\mathbf{k}_\lambda)\, \varepsilon_2(\mathbf{k}_\nu)]^{-1/2}$$

where

$$\varepsilon_2(\mathbf{k}) = (\varepsilon_{20}^2 + \lambda f \hbar^2 k^2)^{1/2}$$

By suitable calculations we find:

$$\frac{1}{\tau_0^{(4)}} = \frac{1}{\pi^3} \cdot \frac{\hbar f}{\lambda \varepsilon_{20}} \cdot \left(\frac{\gamma \mu_0 M_0}{a^2}\right) \left(\frac{T}{\theta_m}\right)^2 \exp(-\varepsilon_{20} T^{-1}) \qquad T \ll \varepsilon_{20} \quad (24)$$

With the parameter values used previously for estimations, we get $\tau_0^{(4)} \sim 10^{-6}$–$10^{-5}$ sec, i.e., the probability of the processes of the fourth order is comparable with that of the processes of the third order.

As is known, a ferromagnetic resonance line width in an ideal uniaxial ferroelectric magnet with a large anisotropy constant is determined by decay processes of the spin wave with $K = 0$ into three other spin waves, the probabilities of such processes giving, at room temperatures, a line width of $\Delta H \sim 0.1$ Oe. As is apparent from (22) and (24), the above spin wave interaction processes with ferroelectric phonons may give a contribution to the width of the ferromagnetic resonance line in the ferroelectric magnet that is comparable to the contribution of the spin system.

CONCLUSIONS

Magnetoelectric effect; phase transitions in a magnetic field; variable electric field energy absorption in a spin frequency and magnetic field energy absorption in a ferroelectric frequency; presence of a lower threshold frequency of magnetic field absorption, causing transitions between the spectrum waves; and other properties described above are proportional to the subsystems' interaction constant γ, i.e., they arise only in crystals having electric and magnetic subsystems.

Thus, interaction between the ferroelectric and magnetic subsystems leads to changes in the values of P in one domain with changed m_z, and vice versa. This change occurs in fields rotating the crystal magnetic moment, i.e., in fields smaller than the magnetic anisotropy field. In cases to which the curves (c1), (c2), (d1) and (d2) correspond in the crystals, which, in the absence of a field, are essentially ferromagnetics and paraelectrics, an induced electric polarization shall arise in a sufficiently high magnetic field. Conversely, in cases to which the curves (a1) and (a2) correspond, a sufficiently high magnetic field $H \geqslant H_n$ shall lead to the disappearance of electric polarization $P_z = P_b$. Again, similar effects are to be observed in such crystals at rotation of field $H \geqslant H_n$ by 90° from the base plane to the easy axis z, or vice versa. In other words, a turn of the magnetic field $H \geqslant H_n$ by 90° in such crystals is tantamount to a "switch-on" or "switch-off" of the ferroelectric subsystem.

Experimental studies of this effect, as well as measurements of magnetoelectric susceptibility, observations of the domain structure in the magnetic field, measurements of energy absorption for variable external fields, etc., would make it possible to verify the correctness of our ideas on ferroelectric magnets. However, such measurements are still pending and, therefore, appear to be highly desirable.

ACKNOWLEDGEMENTS

The authors are very grateful to G. A. Smolensky for his constant interest in the work, and to I. A. Akhiezer and O. V. Kovalev for their fruitful discussions. The translation of the paper from the Russian by Joseph C. Shapiro is also acknowledged.

REFERENCES

1. G. A. Smolensky and A. I. Agranovskaya, *Zhurnal Tekhnicheskoy Fiziki*. **28**, 1491 (1958); G. A. Smolensky, A. I. Agranovskaya, S. N. Popov and V. A. Isupov, *Zh. Tekhn. Fiz.* **28**, 2152 (1958); V. A. Isupov, A. I. Agranovskaya and N. P. Khuchua, *Izvestiya Akad. Nauk SSSR, Ser. Fiz.* **24**, 1271 (1960) (all in Russian).
2. Y. A. Tomashpolsky, Y. N. Venevtsev and V. N. Beznozdrev, *Fizika Tverdogo Tela* **7**, 2763 (1965) (in Russian).
3. Y. A. Tomashpolsky and Y. N. Venevtsev, *Kristallografia* **11**, 731 (1966) (in Russian).
4. K. Leibler, V. A. Isupov and H. Bielska-Lewandowska, *Acta Phys. Pol.* **A40**, 315 (1971) (in Polish).
5. G. A. Smolensky, V. A. Bokov, V. A. Isupov, N. N. Krainik and G. M. Nedlin, *Ferroelectrics* (Rostov University, 1968), p. 129 (in Russian); Y. N. Venevtsev and G. S. Zhdanov, *Ferroelectrics* (Rostov University, 1968), p. 155 (in Russian).
6. V. G. Bar'yakhtar and I. E. Chupis, *Fizika Tverdogo Tela* **10**, 3547 (1968) (in Russian).
7. V. G. Bar'yakhtar and I. E. Chupis, *Fizika Tverdogo Tela* **11**, 3242 (1969) (in Russian).
8. V. G. Bar'yakhtar and I. E. Chupis, *Ukr. Fiz. Zhur.* **17**, 652 (1972) (in Russian).
9. I. E. Chupis and N. Y. Plyushko, *Fizika Tverdogo Tela* **13**, 2252 (1971) (in Russian).
10. V. Kenzig, *Ferroelectrics and antiferroelectrics* (IA, Moscow, 1961) (in Russian).
11. A. F. Devonshire, *Phill. Mag.* **40**, 1040 (1949).
12. V. L. Ginzburg, *UFN* **38**, 490 (1949) (in Russian).

KINETO-ELECTRIC AND KINETOMAGNETIC EFFECTS IN CRYSTALS†

EDGAR ASCHER

Battelle Institute, Advanced Studies Center, CH-1227 Carouge-Genève, Suisse

The kineto-electric, kinetomagnetic and ferrokinetic effects are defined as corresponding to the following terms in the density of stored free enthalpy g of the crystal

$$g = \ldots + \eta_{ik} v_i E_k + \zeta_{ik} v_i cB_k + {}^0p \cdot v$$

and the symmetry conditions for the existence of these effects are discussed. A corresponding relativistic Lagrangian density $\mathscr{L}$ is set up

$$\mathscr{L} = \frac{1}{8c} \chi^{\alpha\beta\sigma\delta} F_{\alpha\beta} F_{\sigma\delta} + \tfrac{1}{2}\xi^{\sigma\alpha\beta} u_\sigma F_{\alpha\beta} + \tfrac{1}{2}\hat{\rho}^{\alpha\beta} u_\alpha u_\beta$$

and from this the electric polarization $\mathbf{P}$, the magnetization $\mathbf{J}$ and the linear momentum $\mathbf{p}$ of a crystal moving in an electromagnetic field are determined as

$$c\mathbf{P}_\perp = \frac{1}{\gamma}(cP + \beta \times J)_\perp, \quad \mathbf{J}_\perp = \frac{1}{\gamma}(J - \beta \times cP)_\perp, \quad c\mathbf{P}_\parallel = P_\parallel, \quad \mathbf{J}_\parallel = J$$

$$\mathbf{p}_\parallel = -\left[\frac{1}{c^2} E \times J + P \times B + ({}^2p^0 - {}^3p^0)\beta\right]_\parallel$$

$$\mathbf{p}^0 = \frac{1}{|\beta|}\left[\frac{1}{c^2} E \times J + P \times B + {}^2p + {}^3p\right]_\parallel$$

The expressions for P, J, ${}^2p$, ${}^3p$ in terms of the applied fields E, B, and v are given in the text.

The expression kineto-electric effect denotes the occurrence of electric polarization P in a moving crystal, proportional to the velocity v and in absence of any applied electromagnetic field: $P \sim v$. Similarly, kinetomagnetic corresponds to $J \sim v$. Besides these we shall consider also bilinear effects such as $P \sim vE$, $J \sim vcB$.

For several reasons, but especially with a view to the relativistic treatment, which we shall undertake, it appears convenient to consider cB as the magnetic field and to measure the magnetization by J. Let us recall the relevant relations:

$$D = \varepsilon_0 E + P$$

$$H = \frac{1}{\mu_0} B - J \quad M = \mu_0 J.$$

To understand the kinetic effects we want to discuss, it is useful to distinguish four types of vector, and not only two; viz.:

type M "axial"
type P "polar"
type j "axio-polar"
type $\dot{M}$

† Lecture given at the "Symposium on Magneto-Electric Interaction Phenomena in Crystals", Battelle Seattle Research Center, May 21–24, 1973.

These four types correspond to the four irreducible representations of the dihedral group $\bar{1}1'$ of order four generated by the space inversion $\bar{1}$ and the time reversal $1'$:

$\bar{1}1'$	1	$\bar{1}$	$1'$	$\bar{1}'$	
	1	1	1	1	$\dot{M}$, grad P
	1	1	−1	−1	M, grad v
	1	−1	1	−1	P, $\dot{v}$
	1	−1	−1	1	j, v, A, $\dot{P}$, grad M

We shall here adopt the physical interpretation of velocity for the "axio-polar" vector.

Ferro-electricity, ferromagnetism and magneto-electricity correspond to the existence, in the density of stored free enthalpy g for a single domain, of terms proportional to the electric field E to the magnetic field B and to the product of both

$$-g(E, B) = \ldots + {}^0P \cdot E + {}^0J \cdot B + \varepsilon_0 \lambda_{ik} E_i cB_k$$

By stored free enthalpy we mean the free enthalpy where the contributions of the vacuum have been subtracted. Here ${}^0P$ and ${}^0J$ denote the spontaneous polarization and magnetization, and $\varepsilon_0 \lambda_{ik}$ the magneto-electric coefficient in the variables and units we have chosen. (λ will be defined below.) These properties are described by polar and axial vectors. If we want to consider the analogous properties corresponding to the axio-polar vector v, we have to introduce the following terms into the density of free enthalpy

$$-g(E, B, v) = \ldots + {}^0p \cdot v + \eta_{ik} v_i E_k + \zeta_{ik} v_i cB_k$$

and we shall denominate the effects described by these terms as ferrokinetic, kineto-electric and kinetomagnetic; we find indeed:

$$P_i = -\frac{\partial g}{\partial E_i} = \ldots + \eta_{ki} v_k$$

$$J_i = -\frac{\partial g}{\partial B_i} = \ldots + c\zeta_{ki} v_k$$

$$p_i = -\frac{\partial g}{\partial v_i} = \ldots + {}^0p_i + \eta_{ik} E_k + \zeta_{ik} cB_k$$

Here p_i of course denotes (linear) momentum. The ferrokinetic effect corresponds to the existence of momentum without applied fields E, B, v.

Formally these new effects are completely analogous to the old ones. There are 31 (out of the 122) Shubnikov groups that are compatible with the ferrokinetic effect as is the case for ferro-electricity and ferromagnetism. These groups are:[1]

> $1, 2, m, 2', m', \bar{1}', mm2, 22'2', 2/m', 2'/m, mm'/2', mmm', 4, \bar{4}', 4/m', 4mm, 42'2', \bar{4}'2'm, 4/m'mm, 3, 6, \bar{6}', \bar{3}', 6/m', 3m, 32', \bar{3}'m, 6mm, 62'2', \bar{6}'m2'$ and $6/m'mm$.

(see also Table I).

Similarly, there are 58 Shubnikov groups compatible with the kineto-electric and kinetomagnetic effect respectively as is the case for the magneto-electric effect. In the three listings of 58 groups, in Table II, 36 are in common, i.e. they are compatible with the three effects. In each listing, the remaining 22 groups permit no other of the three considered effects; in Table II they are underlined. It can be seen

TABLE I

$\bar{1}$	$4/m$	$6/m$	$2m'm'$	$1'$	$41'$	$61'$
$2/m$	$\bar{4}2'm'$	$\bar{3}m'$	$4m'm'$	$21'$	$4'mm'$	$3m1'$
$2'/m'$	$4/mm'm'$	$\bar{6}2'm'$	$3m'$	$m1'$	$4mm1'$	$6'm'm$
$mm'm'$	$\bar{6}$	$6/mm'm'$	$6mm$	$2mm1'$	$6'$	$6mm1'$
$\bar{4}$	$\bar{3}$			$4'$	$31'$	

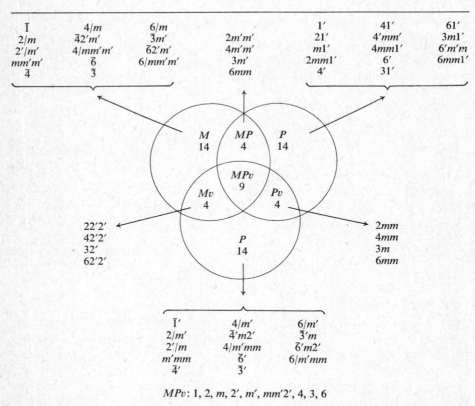

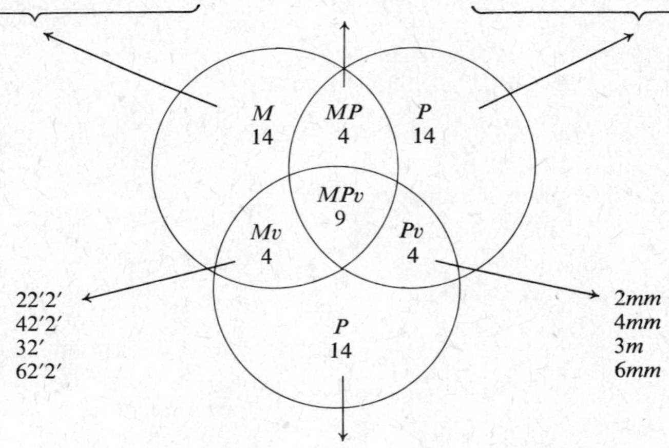

$\bar{1}'$	$4/m'$	$6/m'$
$2/m'$	$\bar{4}'m2'$	$\bar{3}'m$
$2'/m$	$4/m'mm$	$\bar{6}'m2'$
$m'mm$	$\bar{6}'$	$6/m'mm$
$\bar{4}'$	$\bar{3}'$	

MPv: 1, 2, m, 2′, m', $mm'2'$, 4, 3, 6

that for each effect there are altogether 11 types of tensorial representations. The type of the kineto-electric tensor η_{ik} or the kinetomagnetic tensor ζ_{ik} is that of the magneto-electric tensor λ_{ik} for the corresponding group which can be easily found (see e.g. Ref. 2). In Table II these groups are listed on corresponding places.

To conclude this brief excursion into the magic land of symmetry we show in Figure 1 the number of groups that permit a given combination of the 6 effects: ferromagnetism (M), ferro-electricity (P), ferrokinetic effect (p), magneto-electric effect (α), kineto-electric effect (η) and kinetomagnetic effect (ζ). In Table III, the corresponding groups are explicitly listed. Note that, out of the 64 imaginable combinations of 6 effects, only 15 are possible.

When speaking about kineto-electric and kineto-magnetic effects it is always necessary to take into account special relativity, because the Lorentz transformation gives rise to kinetic effects. More precisely it gives rise to second-order kinetic effects, namely to polarizations and magnetizations bilinear in the velocity and the electromagnetic field. This is of course well known, but we shall see it once more later on in our formulae. Furthermore, we know that there are no symmetry conditions for the existence of this effect. However, when certain conditions are fulfilled, viz. if the crystal is ferro-electric or ferromagnetic, or else ferromagento-electric, we observe a polarization, or magnetization, that is proportional to the velocity, and this is precisely a kineto-electric or kineto-magnetic effect. We shall therefore proceed relativistically.

We consider a density of free enthalpy—or a lagrangian density $\mathscr{L}(F.., u.)$, since we are not interested in thermal properties—composed of three parts

$$\mathscr{L}(F.., u.) = {}^1\mathscr{L}(F.., u.) + {}^2\mathscr{L}(F.., u.) + {}^3\mathscr{L}(u.)$$

The first part

$${}^1\mathscr{L}(F.., u.) = \frac{1}{8c}\chi^{\alpha\beta\sigma\delta} F_{\alpha\beta} F_{\sigma\delta}$$

is symmetry independent and well known. We shall however recall the meaning of the symbols and the notations and conventions we use.

The metric tensor is

$$g_{\alpha\beta} = \mathrm{diag}(-1, 1, 1, 1) = g^{\alpha\beta}$$

so that the coordinates of an event are

$$x^\alpha = (ct, x_1, x_2, x_3)$$

The four-velocity is

$$u^\alpha = \frac{1}{\gamma}(c, v_1, v_2, v_3) = \frac{c}{\gamma}(1, \beta_1, \beta_2, \beta_3)$$

TABLE II

	$E_i B_k$	λ	$v_i E_k$	ξ	$v_i B_k$	ξ
1	$1, \underline{\bar{1}'}$		$1, \underline{\bar{1}}$		$1, \underline{1'}$	
2	$2, m', \underline{2'/m}$		$2, m', \underline{2/m}$		$2, m', \underline{21'}$	
3	$m, 2', \underline{2'/m}$		$m, 2', \underline{2'/m'}$		$m, 2', \underline{m1'}$	
4	$222, 2'm'm', \underline{m'm'm'}$		$222, 2m'm', \underline{mmm}$		$222, 2m'm', \underline{2221'}$	
5	$2mm, 22'2', 2'mm', \underline{m'mm}$		$2mm, 22'2', 2'mm', \underline{mm'm'}$		$2mm, 22'2', 2'mm', \underline{mm21'}$	
6	$4, \bar{4}', \underline{4/m'}$ $3, 6, \underline{\bar{6}'}, \underline{\bar{3}'}, \underline{6/m'}$		$4, \bar{4}', \underline{4/m}$ $3, 6, \underline{\bar{6}}, \underline{\bar{3}}, \underline{6/m}$		$4, \bar{4}', \underline{41'}$ $3, 6, \underline{6'}, \underline{31'}, \underline{61'}$	
7	$\bar{4}, 4', \underline{4'/m'}$		$\bar{4}, 4', \underline{4'/m}$		$\bar{4}, 4', \underline{\bar{4}1'}$	
8	$422, \bar{4}'2m', 4m'm'$ $\underline{4/m'm'm'}, 32, 3m', 622$ $\underline{\bar{3}'m'}, \underline{\bar{6}'m'2}, 6m'm', \underline{6/m'm'm'}$		$422, \bar{4}'2m', 4m'm'$ $\underline{4/mmm}, 32, 3m', 622$ $\underline{\bar{3}m}, \underline{\bar{6}m2}, 6m'm', \underline{6/mmm}$		$422, \bar{4}'2m', 4m'm'$ $\underline{4221'}, 32, 3m', 622$ $\underline{321'}, \underline{6'2'2}, 6m'm', \underline{62221'}$	
9	$\bar{4}2m, 4'22', \bar{4}2'm', 4'm'm$ $\underline{4'/m'm'm}$		$\bar{4}2m, 4'22', \bar{4}2'm', 4'm'm$ $\underline{4'/mmm'}$		$\bar{4}2m, 4'22', \bar{4}2'm', 4'm'm$ $\underline{\bar{4}2m1'}$	
10	$4mm, 42'2', \bar{4}'m2'$ $\underline{4/m'mm}, 3m, 32', \underline{\bar{3}'m}$ $6mm, 62'2', \underline{\bar{6}'m2'}, \underline{6/m'mm}$		$4mm, 42'2', \bar{4}\,m2'$ $\underline{4/mm'm}, 3m, 32', \underline{\bar{3}m'}$ $6mm, 62'2', \underline{\bar{6}m'2'}, \underline{6/mm'm'}$		$4mm, 42'2', \bar{4}'m2'$ $\underline{4mm1'}, 3m, 32', \underline{3m1'}$ $6mm, 62'2', \underline{6'mm'}, \underline{6mm1'}$	
11	$23, \underline{m'3}, 432, \underline{\bar{4}'3m'}, \underline{m'3m'}$		$23, \underline{m3}, 432, \underline{\bar{4}3m}, \underline{m3m}$		$23, \underline{231'}, 432, \underline{4'32'}, \underline{4321'}$	

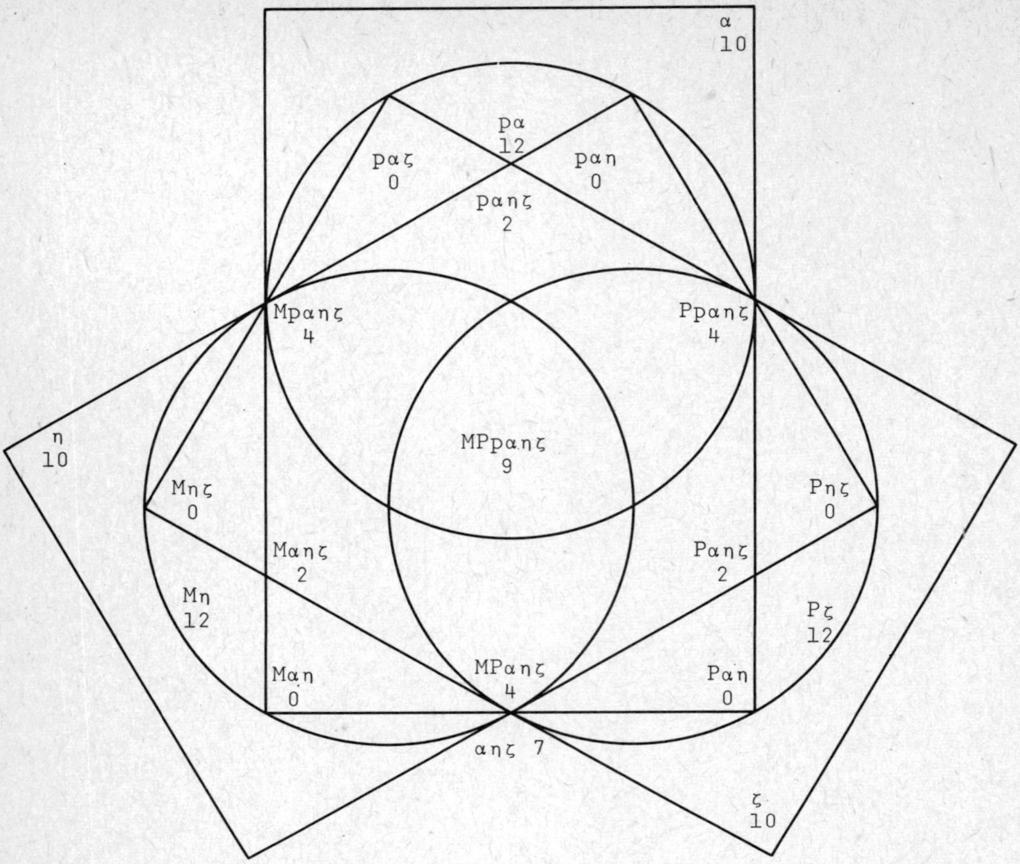

FIGURE 1

with

$$\gamma^2 = 1 - \beta^2, \quad \beta = \frac{v}{c}.$$

The field tensor is

$$F_{\alpha\beta} = \begin{pmatrix} 0 & -E_1 & -E_2 & -E_3 \\ E_1 & 0 & cB_3 & -cB_2 \\ E_2 & -cB_3 & 0 & cB_1 \\ E_3 & cB_2 & -cB_1 & 0 \end{pmatrix}$$

its physical dimension is V/m. The polarization tensor is

$$P^{\alpha\beta} = \begin{pmatrix} 0 & cP_1 & cP_2 & cP_3 \\ -cP_1 & 0 & -J_3 & J_2 \\ -cP_2 & J_3 & 0 & -J_1 \\ -cP_3 & -J_2 & J_1 & 0 \end{pmatrix}$$

with dimension A/m. These tensors are related by the constitutive equation

$$P^{\alpha\beta} = \tfrac{1}{2}\chi^{\alpha\beta\sigma\delta} F_{\sigma\delta}$$

equivalent to the following tensor equations

$$cP = \frac{1}{R_0}[\kappa^B E + \lambda cB]$$

$$J = \frac{1}{R_0}[\tilde{\lambda} E - \varphi^E cB]$$

Here

$$\frac{1}{R_0} = \left(\frac{\varepsilon_0}{\mu_0}\right)^{1/2} = 2.654 \times 10^{-3}\,\text{A/V}$$

and the tilde denotes the transpose.

TABLE III

$MP\rho\alpha\eta\zeta$	9	1, 2, m, 2′, m′, mm′2′, 4, 3, 6
$P\rho\alpha\eta\zeta$	4	mm2, 4mm, 3m, 6mm
$M\rho\alpha\eta\zeta$	4	2′2′2, 42′2′, 32′, 62′2′
$MP\alpha\eta\zeta$	4	m′m′2, 4m′m′, 3m′, 6m′m′
$M\alpha\eta\zeta$	2	$\bar{4}$, $\bar{4}$2′m′
$P\alpha\eta\zeta$	2	4′, 4′m′m
$\rho\alpha\eta\zeta$	2	$\bar{4}$′, $\bar{4}$′m2′
$\alpha\eta\zeta$	9	222, 422, $\bar{4}$2m, 4′22′, $\bar{4}$′2m′, 32, 622, 23, 432
$M\eta$	12	$\bar{1}$, 2/m, 2′/m′, m′m′m, 4/m, 4/mm′m′, $\bar{6}$, $\bar{3}$, 6/m, $\bar{3}$m′, $\bar{6}$m′2′, 6/mm′m′
$P\zeta$	12	1′, 21′, m1′, mm21′, 41′, 4mm1′, 6′, 31′, 61′, 3m1′, 6′mm′, 6mmm1′
$\rho\alpha$	12	$\bar{1}$′, 2/m′, 2′/m, mmm′, 4/m′, 4/m′mm, $\bar{6}$′, $\bar{3}$′, 6/m′, $\bar{3}$′m, $\bar{6}$′m2′, 6/m′mm
η	10	mmm, 4′/m, 4/mmm, 4′/mmm′, $\bar{3}$m, $\bar{6}$m2, 6/mmm; m3, $\bar{4}$3m, m3m
ζ	10	2221′, $\bar{4}$1′, 42221′, $\bar{4}$2m1′, 321′, 6′2′2, 62221′, 231′, 4′32′, 4321′
α	10	mm′m′, 4′/m′, 4/m′m′m′, 4′/m′m′m, $\bar{3}$′m′, $\bar{6}$′m′2, 6/m′m′m, m′3, $\bar{4}$′3m′, m′3m′
ϕ	20	$\bar{1}$1′, 2/m1′, mmm1′, 4/m1′, 4/mmm1′, $\bar{3}$1′, 6/m1′, $\bar{3}$m1′, 6/mmm1′, m31′, m3m1′, 6′/m, 6′/m′mm′, $\bar{6}$m21′, 6′/mmm′, m3m′, $\bar{4}$3m1′, m′3m
Total	122	

The meaning of the dimensionless tensors κ^B, λ and φ can be best understood when they are compared with the tensors occurring in the more familiar constitutive equations

$$P = \varepsilon_0 \kappa^H E + \frac{1}{c} \alpha H$$

$$M = \frac{1}{c} \tilde{\alpha} E + \mu_0 \chi^E H$$

Here χ^E is the magnetic susceptibility at constant E:

$$\chi^E = \mu^E - 1$$

then

$$\varphi^E = (\mu^E)^{-1} - 1 = -\chi^E(\mu^E)^{-1} = -(\mu^E)^{-1}\chi^E.$$

Furthermore κ^B, the electric susceptibility at constant B, is given by

$$\kappa^B = \kappa^H - \alpha\varphi^E \tilde{\alpha} = \kappa^H - \alpha\chi^E(\mu^E)^{-1}\tilde{\alpha}$$

and differs from that at constant H precisely when there is a magneto-electric effect ($\alpha \neq 0$). The magneto-electric coefficient λ for our choice of variables is related to the usual one by

$$\lambda = \alpha(\mu_E)^{-1}$$

After these identifications we find

$$^1\mathscr{L} = \varepsilon_0[-\tfrac{1}{2}\kappa^B_{ik} E_i E_k - \lambda_{ik} E_i cB_k + \varphi^E_{ik} cB_i cB_k]$$

and we can verify that

$$-\frac{\partial^1 \mathscr{L}}{\partial E_i} = \varepsilon_0[\kappa^B_{ik} E_k + \lambda_{ik} cB_k] = {}^1P_i$$

$$-\frac{\partial^1 \mathscr{L}}{\partial B_i} = c\varepsilon_0[\lambda_{ki} E_k - \varphi^E_{ik} cB_k] = {}^1J_1$$

which corresponds indeed to

$$c\frac{\partial^1 \mathscr{L}}{\partial F_{\alpha\beta}} = {}^1P^{\alpha\beta} = \tfrac{1}{2}\chi^{\alpha\beta\sigma\delta} F_{\sigma\delta}$$

We are not yet finished with $^1\mathscr{L}$. We are now interested in moving crystals. Therefore we have to compute

$$c\frac{\partial^1 \mathscr{L}}{\partial F_{\alpha'\beta'}} = \tfrac{1}{2}\Lambda^{\alpha'}_{\alpha} \Lambda^{\beta'}_{\beta} \chi^{\alpha\beta\sigma\delta} F_{\sigma\delta} = {}^1P^{\alpha'\beta'}$$

where $\Lambda^{\alpha'}_{\alpha}$ is the matrix of a Lorentz transformation the inverse of which is denoted by $\Lambda^{\alpha}_{\alpha'}$. The result, in terms of the polarization ^1P and magnetization ^1M induced by the fields, has the well-known form

$$c\,{}^1\mathbf{P}_{\perp} = \frac{1}{\gamma}[c\,{}^1P + \beta \times {}^1J]_{\perp}, \quad c\,{}^1\mathbf{P}_{\parallel} = c\,{}^1P$$

$$^1\mathbf{J}_{\perp} = \frac{1}{\gamma}[{}^1J - \beta \times c\,{}^1P]_{\perp}, \quad {}^1J_{\parallel} = {}^1J_{\parallel}$$

To simplify our writing, we now choose our coordinates so that the velocity of the crystal is parallel to the 2-direction. Thus

$$\Lambda^{\alpha'}_{\alpha} = \begin{pmatrix} \frac{1}{\gamma} & 0 & \frac{\beta}{\gamma} & 0 \\ 0 & 1 & 0 & 0 \\ \frac{\beta}{\gamma} & 0 & \frac{1}{\gamma} & 0 \\ 0 & 0 & 0 & 1 \end{pmatrix}, \quad \Lambda^{\alpha}_{\alpha'} \Lambda^{\alpha'}_{\alpha} = g_{\beta}{}^{\alpha},$$

Then we have, for instance,

$$c^1 \mathbf{P}_1 = \frac{1}{\gamma} \frac{1}{R_0} \left\{ (\kappa^B_{1k} E_k + \lambda_{1k} c B_k) + \frac{v_2}{c} (\lambda_{k3} E_k + \varphi^E_{3k} c B_k) \right\}$$

Up to first order in the velocity, this gives

$$c\,^1\mathbf{P}_1 = c\,^1P_1 + \varepsilon_0(\lambda_{k3} - \tfrac{1}{2}\kappa^B_{1k})\, v_2 E_k$$
$$+ \varepsilon_0 (\varphi^E_{3k} - \lambda_{1k})\, v_2\, c B_k$$

We see, therefore, that the Lorentz transformation gives rise to higher order kinetic effects. The ordinary kinetic effect arises (to first order) when there are spontaneous polarizations. These are taken care of by the second term, $^2\mathscr{L}$, of the Lagrangian density.

For this term we write

$$^2\mathscr{L}(F_{..}, u_.) = \tfrac{1}{2}\xi^{\sigma\alpha\beta} \mu_\sigma F_{\alpha\beta}$$

and we identify the tensor ξ in the following way

$$\xi^{0\alpha\beta} = \frac{\gamma}{c^2}\, {}^0P^{\alpha\beta} = \frac{\gamma}{c^2} \begin{pmatrix} 0 & -c\,^0P_1 & -c\,^0P_2 & -c\,^0P_3 \\ c\,^0P_1 & 0 & ^0J_3 & -^0J_2 \\ c\,^0P_2 & -^0J_3 & 0 & ^0J_1 \\ c\,^0P_3 & ^0J_2 & -^0J_1 & 0 \end{pmatrix}$$

(where $^0P^{\alpha\beta}$ is of course the tensor of spontaneous polarization). Furthermore

$$\xi^{i\alpha\beta} = \gamma \begin{pmatrix} 0 & \eta_{i1} & \eta_{i2} & \eta_{i3} \\ -\eta_{i1} & 0 & -\zeta_{i3} & \zeta_{i2} \\ -\eta_{i2} & \zeta_{i3} & 0 & -\zeta_{i1} \\ -\eta_{i3} & -\zeta_{i2} & \zeta_{i1} & 0 \end{pmatrix}$$

(where the η_{ik} and ζ_{ik} are the kinetic coefficients introduced above). The physical dimension of ξ is As^2/m^3, the same as that of the ratio of the mass m_e to the Bohr magneton b_e of the electron:

$$\frac{m_e}{b_e} = 0.785 \frac{As^2}{m^3}$$

It is interesting to note that the tensor depends on the symmetry of the crystal; for some groups it must vanish. These are the 42 groups listed under $p\alpha$, α and ϕ in Table III.

With this identification, $^2\mathscr{L}$ can be transcribed as

$$^2\mathscr{L} = -\{^0P \cdot E + {}^0J \cdot B + \eta_{ik} v_i E_k + \zeta_{ik} v_i c B_k\}$$

Again we compute the contribution of this term to the electromagnetic polarization of the moving crystal:

$$c \frac{\partial^2 \mathscr{L}}{\partial F_{\alpha'\beta'}} = {}^2\mathbf{P}^{\alpha'\beta'} = c\Lambda^{\alpha'}_{\alpha} \Lambda^{\beta'}_{\beta} \xi^{\sigma\alpha\beta} u_\sigma = \Lambda^{\alpha'}_{\alpha} \Lambda^{\beta'}_{\beta}\, {}^2P^{\alpha\beta}$$

Here we have introduced the abbreviation

$$^2P^{\alpha\beta} = c\xi^{\sigma\alpha\beta} u_\sigma$$

which is equivalent to

$$^2P = {}^0P + \eta v$$

$$^2J = {}^0J + c\zeta v$$

Then we find of course

$$c\,^2\mathbf{P}_\perp = \frac{1}{\gamma} [c\,^2P + \beta \times {}^2J]_\perp, \quad c\,^2\mathbf{P}_\parallel = c\,^2P$$

$$^2\mathbf{J}_\perp = \frac{1}{\gamma} [^2J - \beta \times c\,^2P]_\perp, \quad ^2\mathbf{J} = {}^2J$$

We can now confirm the existence of first-order kinetic effects due to the spontaneous polarizations, by considering the first-order development of the expressions above. For our choice of the velocity in the 2-direction, we find e.g.

$$c\,^2\mathbf{P}_1 = \frac{1}{\gamma} \left\{ c(^0P_1 + \eta_{21} v_2) + \frac{v_2}{c} (^2J_3 + c\zeta_{23} v_2) \right\}$$

$$= c\,^0P_1 + \left[c\eta_{21} + \frac{1}{c} (^0J_3 - \tfrac{1}{2} c\,^0P_1) v_2 \right]$$

This concludes our calculation of the polarization of a moving crystal when the symmetry permits spontaneous polarization and first-order kinetic effects:

$$c\mathbf{P}_\parallel = \left(c\,^0P + \frac{1}{R_0} \kappa^B E + \frac{1}{R_0} \lambda c B + \eta v \right)_\parallel$$

$$\mathbf{J}_\parallel = \left(^0J + \frac{1}{R_0} \tilde{\lambda} E - \frac{1}{R_0} \varphi^E c B + c\zeta v \right)_\parallel$$

$$c\mathbf{P}_\perp = \frac{1}{\gamma}\Big[c\,{}^0P + \frac{1}{R_0}\kappa^B E + \frac{1}{R_0}\lambda cB + \eta v$$

$$\beta \times \left({}^0J + \frac{1}{R_0}\tilde{\lambda}E - \frac{1}{R_0}\varphi^E\, cB + c\zeta v\right)\Big]_\perp$$

$$J_\perp = \frac{1}{\gamma}\Big[\,{}^0J + \frac{1}{R_0}\tilde{\lambda}E - \frac{1}{R_0}\varphi^E\, cB + c\zeta v$$

$$\beta \times \left(c\,{}^0P + \frac{1}{R_0}\kappa^B E + \frac{1}{R_0}\lambda cB + \eta v\right)\Big]_\perp$$

Indeed the third term, ${}^3\mathscr{L}$, of the Lagrangian density does not contribute to the polarization. It does contribute, however, to the momentum. We shall now discuss ${}^3\mathscr{L}$, then calculate the contributions of all three terms to the momentum of the moving crystal.

Now this third term ${}^3\mathscr{L}$ clearly must be related to the kinetic energy which was still missing in our Lagrangian density. It will also contain the ferrokinetic effect. Indeed we put

$${}^3\mathscr{L} = \tfrac{1}{2}\hat{\rho}^{\alpha\beta}\, u_\alpha u_\beta$$

and identify the components of $\hat{\rho}^{\alpha\beta}$ in the following way:

$$\hat{\rho}^{\alpha\beta} = \gamma_2 \begin{pmatrix} -{}^0\rho & \frac{1}{c}\,{}^0p_1 & \frac{1}{c}\,{}^0p_2 & \frac{1}{c}\,{}^0p_3 \\ \frac{1}{c}\,{}^0p_1 & -\rho_{11} & -\rho_{12} & -\rho_{13} \\ \frac{1}{c}\,{}^0p_2 & -\rho_{12} & -\rho_{22} & -\rho_{23} \\ \frac{1}{c}\,{}^0p_3 & -\rho_{13} & -\rho_{23} & -\rho_{33} \end{pmatrix}$$

Then ${}^3\mathscr{L}$ can be transcribed as

$${}^3\mathscr{L} = -(\tfrac{1}{2}\,{}^0\rho c^2 + {}^0p\cdot v + \tfrac{1}{2}\rho_{ik}v_i v_k)$$

The physical dimension of $\hat{\rho}$ is VAs^3/m^5.

We now start determining the linear momentum of a crystal moving in a constant and uniform electromagnetic field.

The first contribution is

$${}^1p^\mu = -\frac{1}{\gamma}\frac{\partial\, {}^1\mathscr{L}}{\partial u_\mu}$$

$$= -\frac{1}{\gamma}\frac{\partial}{\partial u_\mu}\left\{\frac{1}{8c}\Lambda^{\alpha'}_{\alpha}\Lambda^{\beta'}_{\beta}\Lambda^{\sigma'}_{\sigma}\Lambda^{\delta'}_{\delta}\chi^{\alpha\beta\sigma\delta}F_{\alpha'\beta'}F_{\sigma'\delta'}\right\}$$

$$= -\frac{1}{\gamma c}K^{\mu\nu}{}_{\alpha}T_\nu{}^\alpha$$

with

$$K^{\mu\nu}{}_{\alpha} := \frac{\partial \Lambda^{\alpha'}_{\alpha}}{\partial u_\mu}\Lambda^{\nu}_{\alpha'}$$

and

$$T_\nu{}^\alpha := \tfrac{1}{2}\chi^{\alpha\beta\sigma\delta}F_{\nu\beta}F_{\sigma\delta}$$

If you note that the tensor $T_\nu{}^\alpha$ is related to the energy-momentum tensor $\mathscr{T}_\nu{}^\alpha$ in the following way

$$\mathscr{T}_\nu{}^\alpha = \delta_\nu{}^\alpha\mathscr{L} - T_\nu{}^\alpha$$

the final result will not come as a surprise.

For our choice of the velocity in the 2-direction, we find

$${}^1\mathbf{p}^\mu = -\frac{1}{\gamma c}(K^{\mu 0}{}_2 T_0{}^2 + K^{\mu 2}{}_0 T_2{}^0)$$

Furthermore

$$K^{20}{}_2 = K^{02}{}_2 =: K^0 = -\frac{\gamma}{c\beta} = -\frac{1}{\beta}K^2$$

$$K^{00}{}_2 = K^{22}{}_0 =: K^2 = \frac{\gamma}{c} = -\beta K^0$$

and of course

$$K^{10}{}_2 = K^{12}{}_0 =: K^1 = 0$$

$$K^{30}{}_2 = K^{32}{}_0 =: K^3 = 0$$

Thus

$${}^1\mathbf{p}^0 = \frac{1}{\beta c^2}(T_0{}^2 + T_2{}^0) = -\frac{1}{\beta}\,{}^1\mathbf{p}^2$$

$${}^1\mathbf{p}^2 = -\frac{1}{c^2}(T_0{}^2 + T_2{}^0) = -\beta\,{}^1\mathbf{p}^0$$

Now it is easily verified that

$$T_0{}^2 = (E \times {}^1J)_2$$

and

$$T_2{}^0 = c^2({}^1P \times B)_2$$

where ${}^1P$, ${}^1J$ are well-known functions of the applied electromagnetic field, which we have already seen previously. Thus the final result is

$${}^1\mathbf{p}^0 = \frac{1}{\beta}\left(\frac{1}{c^2} E \times {}^1J + {}^1P \times B\right)_{\parallel}$$

$${}^1\mathbf{p}^2 = -\left(\frac{1}{c^2} E \times {}^1J + {}^1P \times B\right)_{\parallel} = -\beta^1 p^0$$

The second contributions is

$${}^2\mathbf{p}^\mu = -\frac{1}{\gamma}\frac{\partial^2 \mathscr{L}}{\partial u_\mu}$$

$$= -\frac{1}{\gamma}\tfrac{1}{2}\xi^{\mu\alpha\beta} F_{\alpha\beta} - \frac{1}{\gamma}\frac{\partial}{\partial u_\mu}\{\tfrac{1}{2}\Lambda_\sigma^{\sigma'}\Lambda_\alpha^{\alpha'}\Lambda_\beta^{\beta'}\xi^{\sigma\alpha\beta} u_{\sigma'} F_{\alpha'\beta'}\}$$

$$= {}^2p^\mu - \frac{1}{\gamma} K^{\mu\nu}_{\ \ \alpha} X_\nu{}^\alpha$$

where we have defined

$${}^2p^\mu = -\frac{1}{\gamma}\tfrac{1}{2}\xi^{\mu\alpha\beta} F_{\alpha\beta} = -\frac{1}{\gamma}\left(\frac{\partial^2 \mathscr{L}}{\partial u_\mu}\right)_{\Lambda=0}$$

and

$$X_\nu{}^\alpha = \tfrac{1}{2}\xi^{\alpha\sigma\beta} u_\nu F_{\sigma\beta} + \xi^{\sigma\alpha\beta} u_\sigma F_{\nu\beta}$$

It is perhaps worthwhile to single out

$${}^2p^0 = -\frac{1}{\gamma}\tfrac{1}{2}\xi^{0\alpha\beta} F_{\alpha\beta} = -\frac{1}{c^2}\tfrac{1}{2}{}^0P^{\alpha\beta} F_{\alpha\beta}$$

$$= \frac{1}{c}({}^0P\cdot E + {}^0J\cdot B)$$

In view of our choice of the velocity

$${}^2\mathbf{p}^\mu = {}^2p^\mu - \frac{1}{\gamma} K^\mu (X_0{}^2 + X_2{}^0)$$

The computation of the tensor X gives

$$X_0{}^2 = \tfrac{1}{2}\xi^{2\sigma\beta} u_0 F_{\sigma\beta} + \xi^{\sigma 2\beta} u_\sigma F_{0\beta}$$

$$= c(\eta_{2k} E_k + \zeta_{2k} cB_k) + \frac{1}{c}[({}^0J_1 + c\zeta_{21} v_2) E_3 - ({}^0J_3 + c\zeta_{23} v_2) E_1]$$

$$= \left(c\,{}^2p + \frac{1}{c} E \times {}^2J\right)_2$$

and

$$X_2{}^0 = \tfrac{1}{2}\xi^{0\sigma\beta} u_2 F_{\sigma\beta} + \xi^{\sigma 0\beta} u_\sigma F_{2\beta}$$

$$= -\frac{v_2}{c^2}\tfrac{1}{2}{}^0P^{\alpha\beta} F_{\sigma\beta} + c[({}^0P_3 + \eta_{23} v_2) B_1 - ({}^0P_1 + \eta_{21} v_2) B_3]$$

$$= \frac{v_2}{c}({}^0P\cdot E + {}^0J\cdot H) + (c^0 P \times B)_2$$

$$= (-v\,{}^2p^0 + c^2 P \times B)_2$$

so that we finally find

$${}^2\mathbf{p}^0 = \frac{1}{\beta}\left({}^2p + \frac{1}{c^2} E \times {}^2J + {}^2P \times B\right)_{\parallel}$$

$${}^2\mathbf{p}^2 = \left(\frac{1}{c}\,{}^2p^0 v + \frac{1}{c^2} E \times {}^2J + {}^2P \times B\right)_{\parallel}$$

$${}^2\mathbf{p}^1 = {}^2p^1 = \eta_{1k} E_k + \zeta_{1k} cB_k$$

$${}^2\mathbf{p}^3 = {}^2p^3 = \eta_{3k} E_k + \zeta_{3k} cB_k$$

The third contribution to the linear momentum is

$${}^3\mathbf{p}^\mu = -\frac{1}{\gamma}\frac{\partial^3 \mathscr{L}}{\partial u_\mu}$$

$$= -\frac{1}{\gamma}\hat{\rho}^{\mu\beta} u_\beta - \frac{1}{\gamma}\frac{\partial}{\partial u_\mu}(\tfrac{1}{2}\Lambda_\alpha^{\alpha'}\Lambda_\beta^{\beta'}\hat{\rho}^{\alpha\beta} u_\alpha u_{\beta'})$$

$$= {}^3p^\mu - \frac{1}{\gamma} K^\mu(\phi_0{}^2 + \phi_2{}^0)$$

where we have defined

$$^3p^\mu = -\frac{1}{\gamma}\left(\frac{\partial^3\mathscr{L}}{\partial u_\mu}\right)_{\Lambda=0} = -\frac{1}{\gamma}\hat{\rho}^{\mu\beta}u_\beta$$

giving

$$^3p^0 = -\left(c\,^0\rho + \frac{1}{c}\,^0p\cdot v\right)$$

$$^3p_i = {}^0p_i + \rho_{ik}v_k$$

and

$$\phi_\nu{}^\alpha = \hat{\rho}^{\alpha\beta}u_\nu u_\beta$$

We find

$$\phi_0{}^2 = c\,^3p_2$$

$$\phi_2{}^0 = -v\,^3p^0$$

so that the final result, after some cancelling of terms, is

$$^3\mathbf{p}^0 = \frac{1}{\beta}\,^3p^2, \quad ^3\mathbf{p}^1 = {}^3p^1$$

$$^3\mathbf{p}^2 = \beta\,^3p^0, \quad ^3\mathbf{p}^3 = {}^3p^3$$

Adding the three contributions, we obtain as final result

$$\mathbf{p}_\parallel \equiv -\left\{\left[{}^0\rho + \frac{1}{c^2}({}^0P\cdot E + {}^0J\cdot B + {}^0p\cdot v)\right]v + \frac{1}{c^2}E\times\left({}^0J + \frac{1}{R_0}\tilde{\lambda}E - \frac{1}{R_0}\varphi^E cB + c\zeta v\right) + \left({}^0P + \frac{1}{R_0}\kappa^B E + \lambda cB + \eta v\right)\times B\right\}_\parallel$$

$$\mathbf{p}_\perp = ({}^0p + \eta E + \zeta cB + \rho v)_\perp$$

$$\mathbf{p}_0 = \frac{1}{|\beta|}\left\{({}^0p + \eta E + \zeta cB + \rho v) + \frac{1}{c^2}E \times\left({}^0J + \frac{1}{R_0}\tilde{\lambda}E - \frac{1}{R_0}\varphi^E cB + c\zeta v\right) + \left({}^0P \times \frac{1}{R_0}\kappa^B E + \frac{1}{R_0}\chi cB + \eta v\right)\times B\right\}_\parallel$$

Concerning the vector products that appear, let us note the following result:

$$\frac{1}{c^2}E\times J + P\times B = D\times B - \frac{1}{c^2}E\times H$$

so that this term corresponds to the Poynting vector of the moving polarized crystal where the contribution of the vacuum has been subtracted (and in vacuum this term vanishes). This corresponds exactly to the fact that we have started with a density of stored free enthalpy, i.e. a free enthalpy where the vacuum contributions have been subtracted.

ACKNOWLEDGEMENT

Fruitful and stimulating discussions with H. Schmid and P. B. Scheurer are gratefully acknowledged.

REFERENCES

1. E. Ascher, *Helv. Phys. Acta* **39**, 40 (1966).
2. S. Bhagavantam, *Crystal Symmetry and Physical Properties* (London and New York, 1966), p. 171.

Discussion

S. IIDA *Comment*: I have succeeded in getting the covariant definition of the magnetic moment from a persistent current. It is:

$$M(\overline{\Pi}, t) = \frac{1}{2c} \int_{t-T/2}^{t+T/2} \iiint \Pi \times \dot{\pi}(\Pi, t)\, dV dT \Bigg/ \int_{t_0-T_0/2}^{t_0+T_0/2} dt_0$$

in which $\{\Pi, ict\}$ is a position four vector, $\{\dot{\pi}(\Pi, t), ic\rho(\Pi, t)\}$ the current density four vector, $\{\overline{\Pi}, ict\}$ the vector for the center of the current distribution, $\{\Pi_0, ict_0\}$ the same vector in the proper frame, and $T = \gamma T_0$ the differential time interval for the average. This definition, as, a tensor identity, automatically defines the magnetic and electric momenta four tensor

$$\begin{bmatrix} 0 & M_3 & -M_2 & +iP_1 \\ -M_3 & & M_1 & +iP_2 \\ M_2 & -M_1 & 0 & +iP_3 \\ -iP_1 & -iP_2 & -iP_3 & 0 \end{bmatrix},$$

in which

$$iP(\overline{\Pi}, t) = \iiiint p(\Pi - \overline{\Pi})\, dV dt \Big/ \int dt_0 = \frac{V_1}{c} \times iM(\overline{\Pi}, t)$$

has been included mathematically.

Part III

Materials and Measurements

A SURVEY OF MAGNETO-ELECTRIC MEASUREMENTS†

T. H. O'DELL

Department of Electrical Engineering, Imperial College, London

The magneto-electric susceptibility is defined and the methods of measurement reviewed. The advantages of the pulsed magnetic field technique, in certain situations, are indicated and a description of a number of experiments on chromium oxide, gallium iron oxide and yttrium iron garnet are discussed. These experiments were directed more towards the domain dependent effects which can be observed through the magneto-electric effect than the microscopic origin of the effect itself and, in particular, measurements of the dynamic behaviour of antiferromagnetic domains in Cr_2O_3 suggest that there is a considerable amount of further work to be done in this field. The paper concludes with a discussion of the higher order magneto-electric effect in YIG and the way in which this can be considered as a magnetic field dependent electric susceptibility.

1 INTRODUCTION

The measurement of magneto-electric susceptibility is useful in studies of magnetic crystal structure, critical behaviour, magnetic transition temperatures and in the detection of antiferromagnetic domains. A recent publication[1] lists twenty compounds which have been studied magneto-electrically, the majority of these measurements being made upon single crystals and the remainder being dealt with by means of the polycrystalline technique worked out by Shtrikman and Treves.[2] The main interest in making these measurements has either been in an attempt to elucidate the spin structure of the material, its magnetic crystal point group, or to provide data on the temperature dependence of the magneto-electric susceptibility which could be used to test a theoretical model describing the atomic scale origin of the magneto-electric effect in the particular material under consideration.

In this paper, a review is given of some measurements on Cr_2O_3, $GaFeO_3$ and the higher order magneto-electric effect in YIG. These measurements were directed more towards the macroscopic properties than the microscopic in that the magneto-electric effect was used as a probe to study the domain dependent effects in these particular crystals. In Cr_2O_3, for example, it proved possible to study the dynamic behaviour of the antiferromagnetic domains.

† Part of this paper was presented at the Symposium on Magneto-Electric Interaction Phenomena in Crystals, May 21, 1973, at Battelle Seattle Research Center, Seattle, Washington.

2 MEASUREMENT TECHNIQUES

The magneto-electric effect is represented by the magneto-electric susceptibility tensors, $\chi_{(\mathrm{me})}^{\alpha\beta}$ and $\chi_{(\mathrm{em})}^{\alpha\beta}$, in the equations,

$$P^{\alpha} = \chi_{(\mathrm{e})}^{\alpha\beta} E_{\beta} + \chi_{(\mathrm{em})}^{\alpha\beta} B_{\beta} \tag{1}$$

$$M^{\alpha} = \chi_{(\mathrm{me})}^{\alpha\beta} E_{\beta} + \chi_{(\mathrm{m})}^{\alpha\beta} B_{\beta} \tag{2}$$

where $\chi_{(\mathrm{e})}^{\alpha\beta}$ and $\chi_{(\mathrm{m})}^{\alpha\beta}$ are susceptibilities which are related to the familiar electric and magnetic susceptibilities and need not concern us here. The magneto-electric susceptibility tensors are the transpose of one another,[3] that is

$$\chi_{(\mathrm{em})}^{\alpha\beta} = \chi_{(\mathrm{me})}^{\beta\alpha} \tag{3}$$

so that a material is characterized either by an experiment which determines the electric polarization, **P**, resulting from the application of a magnetic field **B**, or by an experiment which determines the magnetization, **M**, resulting from an electric field **E**.

Both kinds of experiment may be found in the literature. The technique of applying an electric field and measuring the magnetization was pioneered by Astrov[4] for his work on chromium oxide and developed by Mercier who gave a detailed description of his apparatus.[5] The alternative technique, of applying a magnetic field, was used in some early static measurements,[6,7] at radio frequencies,[8] and in wide-band pulse measurements.[9]

The two experimental techniques offer some advantages over one another, depending on the material and experimental situation. As a rule, however, the applied magnetic field technique is easier because the magneto-electric effect is a weak effect in the sense that the energy density associated with the applied field is very much larger than the energy density of the field produced by the resulting polarization. In fact, the ratio of these energies is the square of the dimensionless magneto-electric susceptibility, $\mu_0 c|\chi^{\alpha\beta}_{(em)}|$ or $\mu_0 c|\chi^{\alpha\beta}_{(me)}|$, and, as this susceptibility is of the order of 10^{-3} in most materials, the electrical energy at the output of the experimental set-up is usually very small. For this reason, it is useful to make the input energy as large as possible and this is easier to do when a magnetic field is used, simply because B^2/μ_0 is a much larger quantity than $\varepsilon_0 E^2$ when typical laboratory values of **B** and **E** are considered.[3] It is for this reason that the applied electric field techniques are used with very narrow bandwidth systems, usually involving synchronous detectors,[5] and the applied magnetic field techniques can be used at much higher frequencies and can also cover wide bandwidths to study situations in which the magneto-electric susceptibility is changing with time.[10]

In sections 3 and 4, a number of experimental measurements will be reviewed in which the pulsed magnetic field technique was used. This pulsed technique makes it possible to input a very large energy into the material under study in a short time and so obtain a reasonably large output signal which gives a measure of the magneto-electric susceptibility.[9]

3 MEASUREMENTS ON CHROMIUM OXIDE

3.1 Magneto-Electric Annealing Experiments

Chromium oxide is antiferromagnetically ordered along the rhombohedral axis and two distinct kinds of ordering are allowed,[4] the relation between them being the time reversal operator, $t \to -t$. A single crystal of chromium oxide will usually be divided into antiferromagnetic domains, each domain belonging to one or the other kind of antiferromagnetic order with equal probability. It follows that the overall magneto-electric susceptibility of a single crystal will usually be very nearly zero because the $t \to -t$ operator changes the sign of the magneto-electric susceptibility so that one domain will cancel out the effect of its neighbour.

Shtrikman and Treves[2] proposed a method of preparing predominantly single domain samples of chromium oxide by cooling the material through the Néel point with both electric and magnetic fields applied along the rhombohedral axis. A glance at Eqs (1) and (2) will show how this technique works. If an electric field is applied just at the Néel point, the domains which nucleate may be magnetized either parallel or anti-parallel to this field through the term $\chi^{33}_{(me)} E_3$ being positive or negative. If a magnetic field is applied at the same time, domains with $\chi^{33}_{(me)}$ positive will be energetically preferred if the two applied fields are parallel and a single crystal sample should end up as one single antiferromagnetic domain with a positive magneto-electric susceptibility.

Brown[11] has made a thorough investigation of this magneto-electric annealing process in high quality single crystal samples of Cr_2O_3, grown by the Verneuil technique and cut into slices a few millimetres in diameter and 1 millimetre thick. The rhombohedral axis was normal to the surface of the slice, the magneto-electric effect being particularly strong in this direction with a maximum value of $\mu_0 c\chi^{33}_{(me)} = 1.1 \times 10^{-3}$ at $\approx 250°$ K. The sample could be saturated into the positive or negative state by cooling through the Néel point in a magnetic field of 0.39 Wb/m^2 and an electric field greater than 35 V/mm. At electric field strengths very much smaller than this an unexpected asymmetry in the results was observed and it was found that the sample showed a definite magneto-electric effect when it was cooled through T_N in a magnetic field alone. Brown's results for these experiments showed that magnetic fields of about 0.4 Wb/m^2 could produce a sample with a magneto-electric susceptibility 60% of that of the single domain sample but there was no really well defined value, this being very dependent upon the past history of the sample. The sign of the magneto-electric susceptibility was very well defined however, and was considered to be determined by the relative orientation of the magnetic field and the stress gradient in the material. A stress gradient would be expected to have a very similar effect to an electric field; both being polar vectors. Magneto-electric annealing in a magnetic field alone was previously reported by Astrov[4] and by Rado and Folen.[12]

Brown[11] also observed a remarkable "temperature memory effect" in Cr_2O_3 which remains unexplained. This effect is observed when a single domain sample is heated above T_N, care being taken to ensure that there is no electric or magnetic field applied, and then cooled again to a temperature below T_N. It was found that the sample remained a single domain, with its previous high value of $\mu_0 c\chi^{33}_{(me)}$ unchanged in magnitude and sign, provided it was not heated above the temperature from which it had been cooled in its original magneto-electric anneal. For example, if the sample had been cooled from 60°C, which is 27°C above T_N, then it could be heated to 58°C and no change in the magneto-electric susceptibility would be observed when it was cooled again to 20°C. Heating to 65°C, however, produced an 80% drop in $\mu_0 c\chi^{33}_{(me)}$. This memory effect was as pronounced as this up to 140°C but at higher temperatures only the sign of the original susceptibility was consistently retained. Heating above 180°C erased all traces of previous annealing.

This memory effect suggests that some magnetic ordering must persist in Cr_2O_3 presumably at impurity centres, at temperatures well above T_N. The application of large electric and magnetic fields was found to have no effect whatsoever unless these were left on as the sample passed through T_N.[11] This effect is still a subject for investigation.

3.2 *Antiferromagnetic Domain Switching in* Cr_2O_3

If a single crystal sample of Cr_2O_3 is heated well above 180°C, and then cooled below T_N, so that it becomes a multi-domain sample, it is possible to then switch it into a very nearly single domain state by applying very large electric and magnetic fields along the rhombohedral axis.[13] This would be expected because the applied electric field would cause the domains to become magnetized either parallel or anti-parallel to the applied magnetic field and there would then be a force on the domain walls which would tend to reduce the volume of domains magnetized against the applied field.

This antiferromagnetic domain switching may be observed by monitoring the magneto-electric susceptibility of the sample and gradually increasing the magnitude of the applied fields. The magneto-electric susceptibility is then seen to change discontinuously as the product of the two applied fields exceeds a number of critical threshold values.[14]

By means of the pulsed magnetic field technique of measuring magneto-electric susceptibility, it has been found possible to observe anti-ferromagnetic domain switching in Cr_2O_3 dynamically and measure the speed at which it occurs.[10, 11] This can be done because the pulse length used in the pulsed field technique is only a few microseconds long and the only limit upon the pulse repetition frequency is the available power supply. By using only a burst of pulses, lasting a few milliseconds, this power supply problem is easily overcome.

The experiment proceeds by applying a large magnetic field to the sample, starting the burst of magnetic field pulses, switching on the applied electric field and observing how the amplitude of the signal pulses varies with time.[10] These signal pulses give a measure of the overall magneto-electric susceptibility of the sample and, consequently, indicate the fraction of the sample volume which is occupied by domains of one polarity or the other, depending upon the sign of the signal pulse. The initial conditions in this experiment were found to be very important because a sample which had been switched very rapidly, under the largest available fields, could not be switched in the reverse direction by simply reversing these fields, presumably because switching had completely saturated the sample so that no small reverse domains were available for nucleation. For this reason each measurement of switching speed was made after the sample had been cooled through T_N with very small fields applied so that it was in, for example, an unsaturated state in which its magneto-electric susceptibility was negative, and then a positive field product was applied, during the burst of magnetic field pulses, to switch it into the positive saturated state.

Under these conditions, the switching time, defined as the 10% to 90% time, τ_s, was found to be given by

$$1/\tau_s \propto [(EB)_A - (EB)_0] \tag{4}$$

where $(EB)_A$ is the product of the fields applied to produce switching in a time τ_s and $(EB)_0$ is a threshold field product below which no switching occurs. This behaviour is remarkably like the well-known switching behaviour of square loop ferrites[15] and Brown[11] was able to show that Cr_2O_3 could be considered as a weak ferromagnet, the weak magnetic moment being proportional to the applied electric field, and that the observed wall velocities were consistant with a wall thickness of between 200 Å and 1000 Å. A great deal needs to be done on this problem of antiferromagnetic domain wall mobility, however, and it appears that any significant step forward depends upon being able to determine the size of the domains more precisely and also being able to see the shape taken by the domain walls during motion. It was with these aims in mind that the magneto-optic technique described in the next section was developed.

3.3 *Magneto-Optic Measurements*

The microscopic theory of the magneto-electric effect in Cr_2O_3 has been the subject of a detailed theoretical study by Hornreich and Shtrikman[16] and by Yatom and Englman.[17,18] This work shows that, at temperatures approaching the Néel point where the magneto-electric susceptibility reaches a maximum value, the magnetization due to an applied electric field may be considered as being due to a difference in the statistical averages associated with the two, very nearly equal, antiferromagnetically coupled sublattices.[3] This means that there is an imbalance between the net magnetizations of the two sublattices and that the resulting magnetization should produce a Faraday rotation of plane polarized light propagating through the crystal, parallel to the rhombohedral axis. It also follows that this Faraday rotation, at constant applied electric field, should be proportional to the magneto-electric susceptibility.

Faraday rotation measurements should, therefore, provide a method of measuring magneto-electric susceptibility and, what is more important, provide a method of actually observing the antiferromagnetic domains in Cr_2O_3 when an electric field is applied along the rhombohedral axis. It would then be possible, in principle, to observe the domain switching process discussed in section 3.2, to find out just how many domain walls are involved in any given switching process and to observe their shape during the switching process.

The only experiments which have been done so far involve a simple measurement of the net Faraday rotation through a thin single crystal of Cr_2O_3 using helium neon laser light.[19] The rotation observed in 4 μm thick hexagonal platelets was of the order of 10°/cm for an applied electric field of the order of 10 V/μm and was very easy to observe, without the use of synchronous detection, when a 3 MHz carrier was used and the detector band-width was limited to 3 KHz. This means that the magneto-optic techniques should prove very useful in magneto-electric studies and are well worth pursuing. Measurements should be made over a range of wavelengths, because the rotary power of the Cr^{3+} ion will be very large at wavelengths close to the absorption bands.[20] With increased sensitivity it should be possible to monitor dynamic changes in the domain structure and to observe the position and shape of the domain walls.

4 MEASUREMENTS ON GALLIUM IRON OXIDE

4.1 *Material Properties and Preparation*

The magneto-electric properties of $Ga_{2-x}Fe_xO_3$ were studied by Rado[21] using both the applied electric and the applied magnetic field techniques to single crystal sample for which $x \approx 1$. The material is not a very easy one to grow in single crystal form but Wood[22] obtained magneto-electric measurements in four good quality crystals grown using a bismuth borate flux[23,24] and having values of x between 1.06 and 1.28. Several other fluxes were tested but the only really successful one was bismuth borate, which yielded crystals of good quality a few millimetres in crossection growing to a few millimetres in length.

4.2 *Magneto-Electric Susceptibility*

The pulsed magnetic field technique was used to measure the magneto-electric susceptibility of $Ga_{2-x}Fe_xO_3$[22] and was particularly useful in overcoming the difficulties introduced by the finite conductivity of this material at temperatures around 250°K. This finite conductivity is a considerable nuisance when low frequency or static measurements are made, due to either the loss and heating produced when the applied electric field technique is used, or the fact that the electric polarization generated by an applied magnetic field can be cancelled out by the finite conductivity of the sample. The pulsed magnetic field technique avoids this latter difficulty because the field pulse need only be a few microseconds long and the magneto-electric polarizatiou can be measured before any leakage has had time to change the result.

The magneto-electric susceptibilities in gallium iron oxide, when it is saturated along its easy direction of magnetization, are $\chi^{23}_{(em)}$, $\chi^{32}_{(em)}$, $\chi^{23}_{(me)}$, and $\chi^{32}_{(me)}$. Only $\chi^{23}_{(em)}$ was measured by Wood,[22] the crystal's easy direction of magnetization, the **c** axis, being the 3 direction. When measurements of magneto-electric susceptibility were taken as a function of the magnetic bias field, which was applied along the c axis, an interesting magneto-electric hysteresis was observed which corresponded to the magnetic hysteresis of the crystals. Values of $\mu_0 c\chi^{23}_{(em)}$ between 10^{-3} and 2×10^{-3} were observed when the temperature was adjusted to give a maxi-num effect but better absolute accuracy was impossible due to the small size of the samples after hey had been cut to the required orientation.

5 MEASUREMENTS ON YTTRIUM IRON GARNET

5.1 *Pulsed Magnetic Field Measurements*

The higher order magneto-electric effect in YIG has been treated as an induced magneto-electric effect and measured by means of the pulsed magnetic field technique.[25] The point of view adopted was that an applied electric field induced a polar axis into the otherwise centrosymmetric YIG structure and allowed the magneto-electric effect to exist from the considerations of symmetry. A more elegant way of looking at this was given by Ascher[26] who considered the induced effect in YIG as an example of a higher order magneto-electric effect which would be allowed in the piezo-electric and piezo-magnetic crystal classes. From this stand point, Ascher was able to predict a number of interesting effects[26] and, for YIG, which is piezomagnetic, one of these was that the electric susceptibility would depend upon the applied magnetic field. This suggestion was taken up by Cardwell[27] whose experiments are discussed below.

5.2 *The Magnetic Field Dependent Electric Susceptibility of YIG*

By considering the higher order magnetic-electric effect in YIG as a magnetic field dependent electric susceptibility, Cardwell[27] was able to observe the frequency shift, due to an applied magnetic field, of an oscillator which had a single crystal (211) plate of YIG as a capacitor in its resonant circuit. This technique proved to be very sensitive and later experiments,[28] which were done using a 100 MHz coaxial line oscillator, made it possible to measure the higher order magneto-electric susceptibility from 100°K to 550°K, the Curie point of YIG. The measured values were found to fit a model which introduced two terms into the spin Hamiltonian of the Fe^{3+} ion in YIG[29]. These were: (1) a perturbation of the axial crystal field parameter by the applied electric field; (2) a perturbation of the g tensor by the applied electric field.

Cardwell's experimental technique also made it possible to observe the very pronounced dielectric anisotropy of YIG which was emphasized by Ascher[26] as a feature of its magnetic crystal class, in contrast to the conventional assumption that cubic crystals, like the garnets, will be isotropic dielectrics. YIG is not cubic below the Curie point, however, but trigonal, belonging to the point group $\bar{3}m$, when magnetized along one of its easy directions [111], of magnetization. In Cardwell's experiments[28] a (110) plate of YIG was used as a capacitor and the change in the dielectric constant could be observed as the magnetization was rotated in (110).

6 CONCLUSIONS

The temperature memory effect in Cr_2O_3, which was described in section 3.1, has some similarity to the kind of behaviour observed in hematite[30, 31] in that the domain pattern in this material, above the magnetic transition temperature, is restored upon subsequent heating. In hematite, however, in contrast to Cr_2O_3, the magnetic transition temperature referred to is one between an antiferromagnetic state and a weak ferromagnetic state so that it is easier to understand why some trace of the initial domain pattern should persist. This problem deserves further experimental study.

The antiferromagnetic domain switching which was considered in section 3.2 had a particularly interesting feature in that some relationship between this dynamic behaviour and the dynamic behaviour of weak ferromagnets seemed to appear. These weak ferromagnets, the orthoferrites and some doped garnets, have become very important recently in the field of "magnetic bubbles"[32] and it would be very interesting to see if the magneto-optic techniques described in section 3.3 could be developed so that the shape of the antiferromagnetic domains in a highly uniaxial magnetic crystal, like Cr_2O_3, could, in fact, take the form of antiferromagnetic bubble domains when the applied electric field was of sufficient strength. The material would then be a potential bubble domain material in which the bubble domains could be manipulated by both magnetic and electric field gradients.

Work on gallium iron oxide is made particularly difficult by the small size of the crystals which can be grown with high perfection. This material is of great interest, however, as it is the only room temperature weak ferromagnet which shows the magneto-electric effect and some measurements[21] indicate that it may have the highest magneto-electric susceptibility at these high temperatures.

The higher order magneto-electric effects suggest

the possibility of making magneto-electric measurements on a very much greater number of magnetic crystals than those limited to the 58 magneto-electric point groups. Ascher's interpretation of the higher order effects[26] is particularly interesting in this connection as it indicates a number of experimental possibilities which have not been fully exploited. These are the magnetic field dependent electric susceptibility considered in section 5.2 and the magnetic field dependent magnetic susceptibility which has been measured by Gorodetsky *et al.*[33] in the material $DyFeO_3$. No work has yet been done upon the electric field dependence of the magnetic or electric susceptibilities.

ACKNOWLEDGEMENTS

The author is grateful to Dr. V. M. Wood, Dr. C. A. Brown and Dr. M. J. Cardwell for many useful discussions and to Dr. E. A. D. White, for his supervision of the crystal growth work associated with these experiments. Thanks are also due to the Science Research Council for the provision of a Research Grant.

REFERENCES

1. R. M. Hornreich, "The Magneto-electric Effect: Materials, Physical Aspects and Applications," *I.E.E.E. Transactions on Magnetics* **8**, 584 (1972).
2. S. Shtrikman and D. Treves, *Phys. Rev.* **130**, 986 (1963).
3. T. H. O'Dell, "The Electrodynamics of Magneto-electric Media" (North Holland, Amsterdam, 1970) chs. 2 and 5.
4. D. N. Astrov, *Soviet Physics, J.E.T.P.* **13**, 729 (1961).
5. M. Mercier, *Revue de Physique Applique*, **2**, 109 (1967).
6. G. T. Rado and V. J. Folen, *Phys. Rev. Letts.* **7**, 310 (1961).
7. E. Ascher, H. Rieder, H. Schmid, and H. Strossel, *J. Appl. Phys.* **37**, 1404 (1966).
8. B. I. Al'shin and D. N. Astrov, *Soviet Physics, J.E.T.P.* **17**, 809 (1963).
9. T. H. O'Dell, *I.E.E.E. Transactions on Magnetics* **MAG2** 449 (1966).
10. C. A. Brown and T. H. O'Dell, *I.E.E.E. Transactions on Magnetics* **MAG5**, 964 (1969).
11. C. A. Brown, "Magneto-electric Domains in Single Crystal Chromium Oxide". Ph.D. Thesis, Imperial College, London (1969).
12. G. Rado and V. J. Folen, *Phys. Rev. Letts.* **7**, 310 (1961).
13. T. J. Martin, *Phys. Letters*, **17**, 83 (1965).
14. T. J. Martin and J. C. Anderson, *I.E.E.E. Transactions on Magnetics* **MAG2**, 446, (1966).
15. E. M. Gyorgy, "Magnetisation reversal in Non-metallic ferromagnets," *Magnetism*, **3**, 525 (1956). Academic Press. Editors G. Rado and H. Suhl.
16. R. Hornreich and S. Shtrikman, *Phys. Rev.* **161**, 506 (1967).
17. H. Yatom and R. Englman, *Phys. Rev.* **188**, 793 (1969).
18. R. Englman and H. Yatom, *Phys. Rev.* **188**, 803 (1969).
19. T. H. O'Dell and E. A. D. White, *Phil. Mag.* **22**, 649 (1970).
20. J. P. Van der Ziel, *Phys. Rev.* **161**, 483 (1967).
21. G. T. Rado, *Phys. Rev. Letts.* **13**, 335 (1964).
22. V. M. Wood, "The Magneto-electric Effect in Gallium Iron Oxide," Ph.D. Thesis, Imperial College, London (1969).
23. J. P. Remeika, *J. Appl. Phys.* **31**, 263s (1960).
24. E. A. Wood, *Acta Cryst.* **13**, 682 (1960).
25. T. H. O'Dell, *Phil. Mag.* **16**, 487 (1967).
26. E. Ascher, *Phil. Mag.* **17**, 149 (1968).
27. M. J. Cardwell, *Phil. Mag.* **20**, 1087 (1969).
28. M. J. Cardwell, "Second order magneto-electric effect in yttrium iron garnet," Ph.D. Thesis, Imperial College, London (1971).
29. M. J. Cardwell, *Phys. stat. sol.* (*b*) **45**, 597 (1971).
30. B. Gustard, *Proc. Roy. Soc.* **297**, 269 (1967).
31. T. E. Gallen, *Proc. Roy. Soc.* **303**, 525 (1968).
32. A. H. Bobeck, *Bell Syst. tech. J.* **46**, 1901 (1967).
33. G. Gorodetsky, B. Sharon, and S. Shtrikman, *Solid State Commun.* **5**, 739 (1967).

Discussion

R. M. HORNREICH Did you obtain an estimate of the antiferromagnetic domain wall widths from your Cr_2O_3 switching studies?

T. H. O'DELL Between 200 Å and 1000 Å. This width was consistent with the observed wall velocity when the material was treated as a weak ferromagnet and modelled by means of a modification of the Landau-Lifshitz-Olbert equation.

G. T. RADO Comment: In regard to your discussion of the theoretical work of M. J. Cardwell, *Phys. Stat. Sol. (b)* **45**, 597 (1971), I wish to point out that since Cardwell neglects to state whether he subtracted a certain self-energy term, it is not clear whether he did avoid double-counting some of the interactions.

RECENT ADVANCES IN MAGNETOELECTRICITY AT THE WEIZMANN INSTITUTE OF SCIENCE†

R. M. HORNREICH

The Department of Electronics, The Weizmann Institute of Science, Rehovot, Israel

A review is given of recent work on magnetoelectricity and magnetoelectric (ME) materials. Areas covered are: (a) The discovery of new ME materials, including $FeSb_2O_4$, Fe_2TeO_6, $A_2M_4O_9$ (A = Ta, Nb; M = Mn, Co), $MGeO_3$ (M = Mn, Co, Fe), $CrTiNdO_5$, ROOH (R = Er, Tb, Dy), and $GdVO_4$. (b) The confirmation of the magnetic space group assignment $I4_1'/a'm'd$ for $GdVO_4$ and a study of the mechanisms underlying the ME effort in this material. It is concluded that the most likely mechanism is an electric-field-induced Dzialoshinsky interaction. (c) Measurements of the critical exponent of the ME susceptibility of Cr_2O_3 ($\beta = 0.35 \pm 0.01$) and $GdVO_4$ ($\beta = 0.50 \pm 0.05$). The latter result is suggestive of significant higher than first neighbor interactions which act to enhance the long-range order of the Gd^{3+} spin system. (d) A study by ME techniques of the variation of the Néel temperature T_N of Cr_2O_3 under conditions of uniaxial stress (along the *a* and *c* crystallographic directions) and hydrostatic pressure. In all cases (dT_N/dp) was found to be positive and $(dT_N/dp)_c$ was approximately 70% greater than $(dT_N/dp)_a$. The implications of these results are discussed.

I INTRODUCTION

A magnetoelectric (ME) medium is one in which there exists a linear relationship between an electric field and the medium's magnetic polarization and between a magnetic field and the medium's electric polarization. The possibility of such an effect occurring in magnetically ordered materials was first pointed out by Landau and Lifshitz.[1] Subsequently, Dzialoshinsky[2] predicted that the ME effect should occur in Cr_2O_3, and the effect was first observed[3] in 1960 in this material.

The discovery of the ME effect in Cr_2O_3 led to a search for additional ME materials. It was quickly recognized that the existence of the effect depended on the symmetry of a particular compound in its magnetically ordered phase or, in other words, on its magnetic crystal class.[4] Unfortunately, however, the great majority of magnetically ordered materials belong to those classes in which the ME effect is forbidden. Thus the discovery of a second ME material was not reported[5] until 1964 and, in the entire decade 1960–1970, only 13 ME materials were discovered.[6]

For the past four years, the Department of Electronics at the Weizmann Institute of Science has been engaged in a program of research on various aspects of the ME effect. These activities have included: (a) a search for new ME materials, (b) the determination of the magnetic structures of ME materials, (c) measurements of the critical behavior of the ME susceptibility, and (d) applications of the ME effect to materials research. These studies have resulted in the discovery of 14 new ME materials and have demonstrated the usefulness of magnetoelectricity as a technique for determining the magnetic crystal class to which a given material belongs and also for sensing changes in magnetic symmetry caused by temperature and/or external fields.

II NEW MAGNETOELECTRIC MATERIALS

Since almost all potentially ME materials are antiferromagnetic and are generally difficult to obtain in the form of single crystals, the technique known as ME annealing[7,8] has been employed to study polycrystalline specimens. By this means a nonzero ME effect in polycrystalline powder specimens of antiferromagnetic materials was obtained.

† This work was performed in part under the sponsorship of the U.S. National Bureau of Standards and was supported in part by the Commission for Basic Research of the Israel Academy of Sciences and Humanities. Part of the paper was presented at the Magnetoelectric Interaction Phenomena in Crystals Symposium, Battelle Seattle Research Center, Seattle Washington, May 1973.

TABLE I
New magnetoelectric materials

Material	T_c	Magnetic class	α_{ij}	$\alpha_{max} \times 10^4$	Reference
1. $FeSb_2O_4$	46	$mm2$	(a)	?	12
2. Fe_2TeO_6	210	$4/m'm'm'$	(b)	0.3^b	16
3. $Nb_2Co_4O_9$	27	$\bar{3}'m'$	(b)	0.2^a	23
4. $Nb_2Mn_4O_9$	108	$\bar{3}'m'$	(b)	0.02^b	23
5. $Ta_2Co_4O_9$	21	$\bar{3}'m'$	(b)	1.1^a	23
6. $Ta_2Mn_4O_9$	104	$\bar{3}'m'$	(b)	0.09^b	23
7. $MnGeO_3$	11	$m'mm$	(a)	0.02^b	27
8. $CoGeO_3$	34	$m'mm$	(a)	1.0^a	29
9. $FeGeO_3$	14	?		0.01^a	29
10. $CrTiNdO_5$	20.5	$m'mm$	(a)	0.1^a	29
11. ErOOH	4.1	$2'/m$	(c)	4.5^a	38
12. TbOOH	10.0	$2/m'$	(d)	3.8^a	38
13. DyOOH	7.2	$2/m'$	(d)	0.98^a	38
14. $GdVO_4$	2.5	$4'/m'mm'$	(e)	2.8	42

Non-zero elements: (a) α_{12}, α_{21}; (b) $\alpha_{11} = \alpha_{22}, \alpha_{33}$; (c) $\alpha_{13}, \alpha_{23}, \alpha_{31}, \alpha_{32}$; (d) $\alpha_{11}, \alpha_{12}, \alpha_{21}, \alpha_{22}, \alpha_{33}$; (e) $\alpha_{11} = -\alpha_{22}$.

T_c—maximum temperature of ME effect; α_{ij}—form of ME tensor; α_{max}—maximum value of α_{ij} (^{a}powder measurement; ^{b}single crystal α_{max} as derived from powder measurement).

In this section we discuss the new ME materials found by us. A list of these materials and a summary of their properties is given in Table I.

1) $FeSb_2O_4$

The first material studied was the antiferromagnet $FeSb_2O_4$, which belongs to the crystallographic space group $P4_2/mbc$. An analysis[9] of neutron diffraction[10] and Mössbauer[11] data indicated that $Pmc2$ was the appropriate magnetic space group for this material and that the ME effect could therefore be present. This prediction was confirmed[12] by the observation of the electrically induced ME effect in $FeSb_2O_4$ below $T_N = 46°K$ following the annealing of a powder specimen in parallel electric and magnetic fields. $FeSb_2O_4$ is interesting in another respect. Since the magnetic group $Pmc2$ does not contain the space inversion operation, ferroelectricity is not forbidden in the antiferromagnetic state and can therefore occur below the Néel point.[13]

2) Fe_2TeO_6

The next material studied was Fe_2TeO_6, which had also been identified on the basis of neutron diffraction studies,[14,15] as potentially ME. The results obtained[16] for both perpendicular and parallel ME anneals are shown in Figure 1. As is well known,[17,18] the magnitude of the measured ME

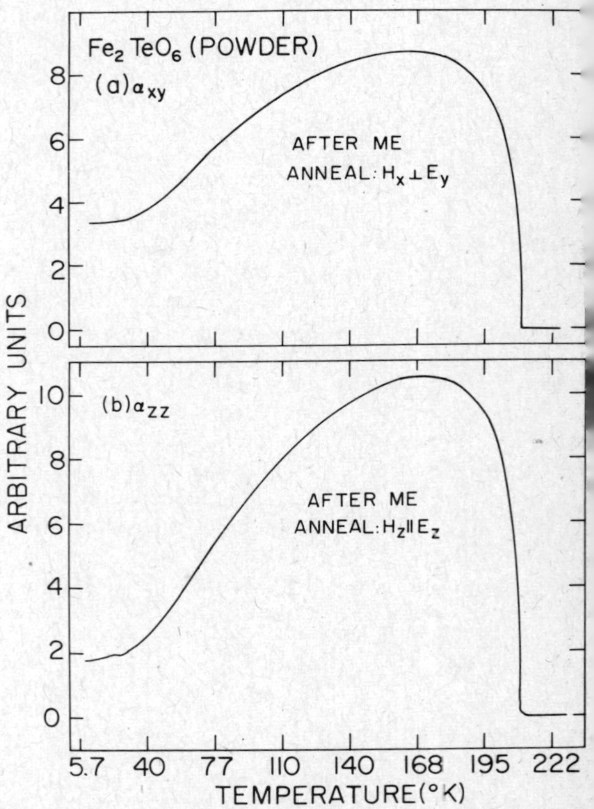

FIGURE 1 Temperature dependence of the powder ME susceptibilities of Fe_2TeO_6 measured following (a) annealing in perpendicular electric and magnetic fields and (b) annealing in parallel fields.

susceptibilities is dependent upon the magnitude of the product EH of the annealing fields. By measuring the susceptibilities as a function of EH, we found that 10^8 V-G/cm (11 kV/cm × 9 kG) was sufficient to achieve better than 90% of saturation for either annealing treatment.

Figure 1(a) shows the ME susceptibility $\alpha_{xy} = M_x/E_y$, measured in the x direction following an $H_x \perp E_y$ annealing treatment and Figure 1(b) shows the ME susceptibility, $\alpha_{zz} = M_z/E_z$, measured in the z direction following an $H_z \| E_z$ annealing treatment. For both cases, the ME susceptibility vanishes above $T = 209 \pm 1$°K, which we therefore identify as the Néel temperature T_N of the compound. From the neutron diffraction studies[14,15] it was concluded that the magnetic space group of Fe_2TeO_6 is $P4_2/m'n'm'$, and this assignment was confirmed[16] by the ME measurements.

For the case of a Fe_2TeO_6 crystal, the only non-zero elements of the ME susceptibility tensor are then[4] $\alpha_{11} = \alpha_{22}$ and α_{33}. For a randomly distributed polycrystalline powder, the powder susceptibilities are given by[7,8]

$$\alpha_{zz} = \tfrac{1}{3}(\alpha_{33} + 2\alpha_{11})(1 - 2\cos^3\gamma) - 2\alpha_{11}\cos\gamma\sin^2\gamma, \tag{1a}$$

$$\alpha_{xy} = (2/3\pi)(\alpha_{33} - \alpha_{11}), \tag{1b}$$

where

$$\tan^2\gamma = -(\alpha_{33}/\alpha_{11})_0, \tag{2}$$

and the subscript indicates that the ratio is evaluated in the region just below T_N. If $(\alpha_{33}/\alpha_{11})_0 > 0$, we set $\gamma = \pi/2$.

Using (1) and (2), the single crystal ME susceptibility tensor elements were derived from the results shown in Figure 1. In the range $199°\text{K} \leqslant T \leqslant 207°\text{K}$, we found $(\alpha_{zz}/\alpha_{xy})_0 = 1.21 \pm 0.01$. Upon assuming that $(\alpha_{33}/\alpha_{11})_0 > 0$, no solution of (1) consistent with the assumption could be found. Thus $(\alpha_{33}/\alpha_{11})_0 < 0$ and, in the region just below T_N, substituting (2) into (1) yields

$$(\alpha_{zz})_0 = \pm(\alpha_{11})_0\,(1 - 3\cos^2\gamma + 4\cos^3\gamma)/3\cos^2\gamma, \tag{3a}$$

$$(\alpha_{xy})_0 = \pm(\alpha_{11})_0\,(2/3\pi\cos^2\gamma). \tag{3b}$$

The choice of sign in each case is such that $(\alpha_{zz})_0$ and $(\alpha_{xy})_0$ are positive quantities. Thus

$$(\alpha_{zz}/\alpha_{xy})_0 = \pm(\pi/2)(1 - 3\cos^2\gamma + 4\cos^3\gamma), \tag{4}$$

and, choosing the upper sign, we obtained $\gamma = 65.2 \pm 0.5°$. For the lower sign there was no solution in the range $0 \leqslant \gamma \leqslant \pi/2$.

Using this value of γ, (1) was inverted to obtain α_{11} and α_{33} in terms of α_{zz} and α_{xy}. The resulting single crystal ME susceptibilities are shown in Figure 2. The maximum value of α_{33} is (in Gaussian units) $(\alpha_{33})_{max} = 3 \times 10^{-5}$.

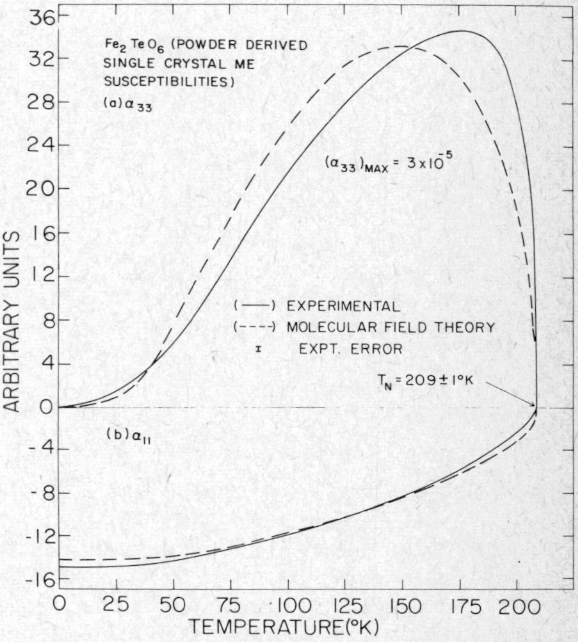

FIGURE 2 Single crystal ME susceptibilities of Fe_2TeO_6 as derived from the powder results (a) parallel to and (b) perpendicular to the four-fold crystal axis. Also shown (dashed curves) are best-fit theoretical susceptibilities derived using a molecular field approximation.

In order to compare the results of Figure 2 with theory, the ME susceptibilities were calculated in the molecular field approximation. Following Rado,[19] expressions for α_{33} and α_{11} were calculated on the assumption that either "single-ion" or "two-ion" mechanisms could underlie the ME susceptibilities. The paramagnetic Néel temperature was estimated from the susceptibility results of Dehn *et al.*[20] to be 455°K. The theoretical expressions were then fitted to the experimental results using a least squares technique with the only free parameter for each case being the overall magnitude of the susceptibility. The best results are shown by the dashed curves in Figure 2. For α_{33}, the best fit was obtained with a two-ion mechanism, wherein the ME susceptibility is proportional to the product of the sublattice magnetization and the c axis susceptibility. For this case, the least squares deviation obtained using a single-ion mechanism was twice as large. Turning to α_{11}, the best fit was obtained with a single-ion mechanism, wherein the ME susceptibility is proportional to the product of the sublattice magnetization and a temperature-dependent factor defined by Rado.[19] For this case, a two-ion mechanism resulted in a least squares deviation that was twice as large.

We emphasize, however, that the identification of α_{33} and α_{11} with two-ion and single-ion mechanisms respectively can only be regarded as tentative. The molecular field approximation is not sufficiently accurate to enable an unambiguous selection of the mechanisms underlying the ME susceptibilities to be made. This has been well illustrated by the case of Cr_2O_3.[19,21] Interestingly, however, the results for Cr_2O_3[21] indicate that the ME susceptibilities parallel and perpendicular to the antiferromagnetic axis are due to two-ion and single-ion effects respectively, i.e., the same as our tentative results for Fe_2TeO_6.

3) $A_2M_4O_9$ *(A = Ta, Nb; M = Mn, Co)*

The transition metal niobates and tantalates having the composition $A_2M_4O_9$ were first prepared and studied by Bertaut *et al.*[22] Their neutron diffraction and powder susceptibility studies showed that $Nb_2Mn_4O_9$ and $Nb_2Co_4O_9$ order antiferromagnetically and belong to $P3'c'1$. Since this magnetic group allows magnetoelectricity,[4] the entire group of transition metal niobates and tantalates were suitable candidates for ME studies.

Our study[23] was restricted to the compounds $Ta_2Mn_4O_9$, $Ta_2Co_4O_9$, $Nb_2Mn_4O_9$, and $Nb_2Co_4O_9$. Magnetoelectricity was found after both parallel and perpendicular annealing treatments in all of these materials. For the case of the $A_2Mn_4O_9$ samples, better than 90% of saturation was achieved for either annealing treatment. For the $A_2Co_4O_9$ samples, we encountered breakdowns at high electrical fields and saturation was not achieved. The $\alpha_{zz}(T)$ and $\alpha_{xy}(T)$ curves obtained[23] for both of the $A_2Mn_4O_9$ compounds resembled those of Fe_2TeO_6 (see Figure 1). The Néel temperatures were 110 ± 1 and 103 ± 1°K for $Nb_2Mn_4O_9$ and $Ta_2Mn_4O_9$ respectively.

Using (1) and (2), the single crystal ME susceptibility tensor elements of Nb_2MnO_4 were derived from the experimental results. In the range 107–109°K, we found $(\alpha_{zz}/\alpha_{xy})_0 = 1.72 \pm 0.03$. The only solution of (1) consistent with this value was with $\gamma = \pi/2$ and $(\alpha_{33}/\alpha_{11})_0 \gg 1$. Inverting (1) then gave α_{11} and α_{33} in terms of α_{zz} and α_{xy}. Since the element α_{11} is much smaller than α_{33} it is much more susceptible to errors caused by small misalignments of the annealing fields from an exactly parallel or perpendicular condition. For this reason we show, in Figure 3, only α_{33}. The maximum value of α_{33} is 1.7×10^{-6} at $T = 89$°K.

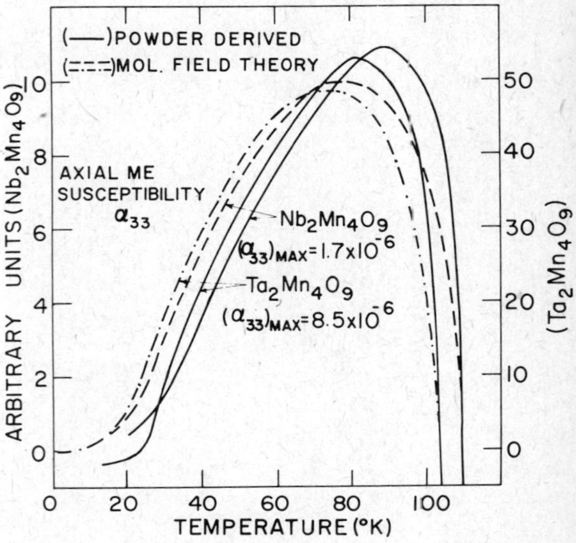

FIGURE 3 Single crystal axial ME susceptibilities of $Nb_2Mn_4O_9$ and $Ta_2Mn_4O_9$ as derived from the powder results. Also shown are best-fit theoretical susceptibilities derived using a molecular field approximation.

Turning to $Ta_2Mn_4O_9$, results similar to those found for $Nb_2Mn_4O_9$ were obtained. In the range 99 – 102°K, $(\alpha_{zz}/\alpha_{xy})_0 = 1.65 \pm 0.04$, so that again $\gamma = \pi/2$ with $(\alpha_{33}/\alpha_{11})_0 \gg 1$. As before, α_{11} is much smaller than α_{33} and we therefore show, in Figure 3, only the element α_{33}. The maximum value of α_{33} for $Ta_2Mn_4O_9$ is 8.5×10^{-6} at $T = 81$°K. The transition of α_{33} from positive to slightly negative values at low temperatures is believed to be due to an annealing misalignment of the type described above.

To compare the results of Figure 3 with theory, the axial ME susceptibilities of $Nb_2Mn_4O_9$ and $Ta_2Mn_4O_9$ were calculated in the molecular field approximation as described previously for Fe_2TeO_6. The paramagnetic Néel temperature of Nb_2MnO_4 was taken from the susceptibility results of Bertaut *et al.*[22] The same temperature was taken for

Ta_2MnO_4, since the unit cell parameters and the Néel points of the two compounds are essentially the same. The best fits obtained are shown as the broken curves in Figure 3. For both materials the best fit was obtained with a two-ion mechanism. The best least squares fit for a single-ion mechanism was approximately three and a half times as large for both compounds. Again we stress that the identification of α_{33} with a two-ion mechanism must be regarded as tentative.

The measured ME susceptibilities of the $A_2Co_4O_9$ compounds following parallel and perpendicular annealing treatments are shown in Figures 4 and 5. The Néel temperatures are 27.0 ± 0.5 and 20.6 ± 0.5°K for $Nb_2Co_4O_9$ and $Ta_2Co_4O_9$ respectively. For these materials we did not achieve saturation during the annealing process. Thus $(\alpha_{zz}/\alpha_{xy})_0$ could not be determined from the experimental results. Several points can be noted however. First, the temperature dependence of the ME susceptibilities of $Nb_2Co_4O_9$ is considerably different from those of the other compounds. Its powder susceptibilities increase monotonically as the temperature is decreased without exhibiting the peak in the region below T_N typical of the other materials studied.

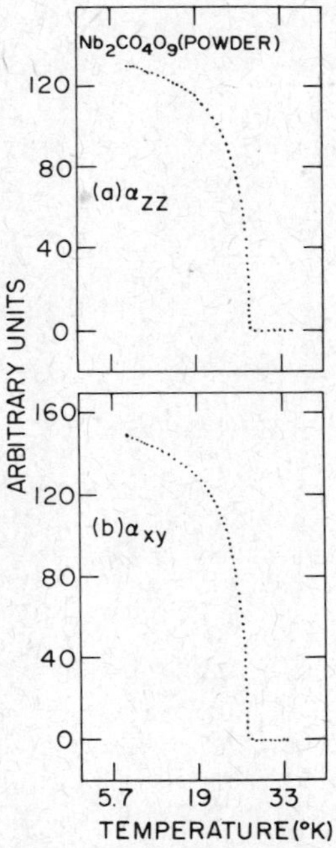

FIGURE 4 Temperature dependence of the powder ME susceptibilities of $Nb_2Co_4O_9$ measured following (a) annealing in parallel electric and magnetic fields and (b) annealing in perpendicular fields. Saturation was not achieved.

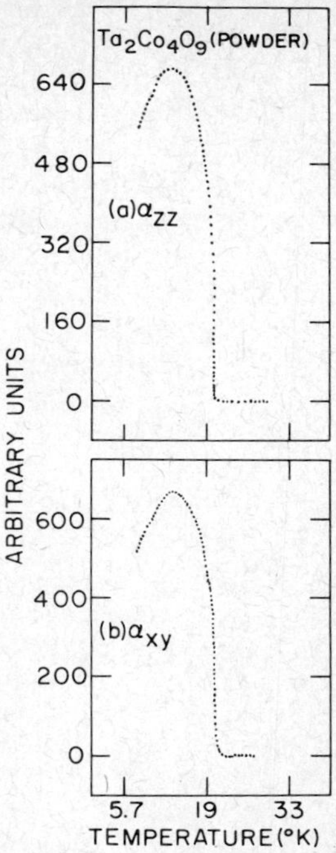

FIGURE 5 Temperature dependence of the powder ME susceptibilities of $Ta_2Co_4O_9$ measured following (a) annealing in parallel electric and magnetic fields and (b) annealing in perpendicular fields. Saturation was not achieved.

Second, the powder susceptibilities of each of the $A_2Co_4O_9$ compounds are at least 30 times greater than those of the corresponding $A_2Mn_4O_9$ compound. This is probably connected with the much larger spin-orbit coupling energy in Co^{2+} ions ($L = 3$) as compared with Mn^{2+} ($L = 0$). Similar results are found in $LiMPO_4$ where the maximum ME susceptibility of $LiCoPO_4$ is 32 times greater than that of $LiMnPO_4$.[24] Finally, the maximum values of α_{zz} measured for $Ta_2Co_4O_9$ and $Nb_2Co_4O_9$ were 11×10^{-5} and 2.1×10^{-5} respectively. These may be compared with the value $(\alpha_{zz})_{max} = 6.4 \times 10^{-5}$ obtained for Cr_2O_3 powder with better than 90% of saturation achieved.[17]

4) $MGeO_3$ ($M = Mn, Co, Fe$)

The magnetic properties of the metagermanates $GeMO_3$ (M = Mn, Co, Fe) have been studied by Sawaoka *et al.*[25] For $MnGeO_3$, they found a peak in the powder susceptibility at 10°K, which they identified as the Néel temperature of the compound. Herpin *et al.*,[26] on the other hand, found no such maximum. Their neutron diffraction studies showed, however, that $MnGeO_3$ orders antiferromagnetically in a colinear structure with magnetic space group *Pb'ca* and $T_N = 16 \pm 1°K$. These results indicated that $MnGeO_3$ should exhibit magnetoelectricity.

To confirm this, a cylindrical sample of sintered $MnGeO_3$ powder was cooled to 4.2°K in the presence of either parallel or perpendicular electric and magnetic fields in order to induce a ME remanent state. Better than 95% of saturation was achieved for either annealing treatment.

The measured ME susceptibilities[27] are shown in Figure 6. We see that α_{xy} vanishes above $11.1 \pm 0.3°K$, which we therefore identify as T_N. Further,

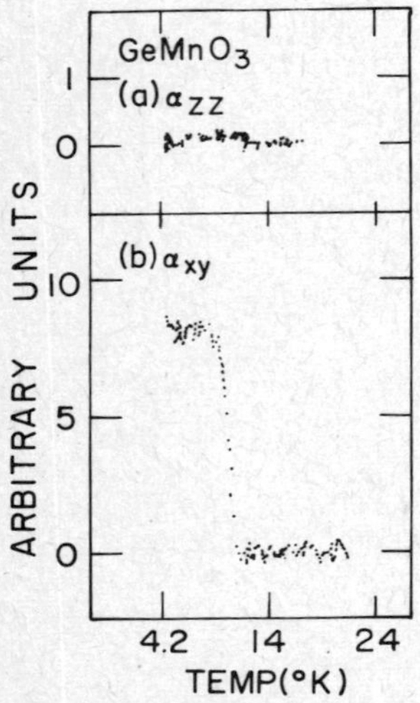

FIGURE 6 Temperature dependence of the powder ME susceptibilities of $MnGeO_3$ measured following (a) annealing in parallel electric and magnetic fields and (b) annealing in perpendicular fields.

α_{zz}, to within our experimental error, is zero. For a single crystal of $MnGeO_3$ the nonzero elements of the ME tensor are[4] α_{12} and α_{21} and α_{zz} will vanish only if $\alpha_{12} = -\alpha_{21}$.[8] Thus our results indicate that the tensor of $MnGeO_3$ is essentially antisymmetric. It then follows[8] that α_{xy} is equal to $|\alpha_{12}|/2$ and we find $\alpha_{12} = (15 \pm 1) \times 10^{-7}$ at 4.2°K. Similar results were obtained by Holmes and Van Uitert[28] in their ME study of the spin flop behavior of $MnGeO_3$.

To further study the discrepancy in T_N between the ME and the neutron diffraction results, specific heat measurements were carried out. A sharp anomaly characteristic of magnetic ordering was found[27] at $10.8 \pm 0.2°K$ in excellent agreement with the ME results. After subtracting off the lattice contribution, the total magnetic entropy per mole ΔS was found to be 1.66, only 7% short of the ln 6 value expected. Upon dividing ΔS into two parts, one for $0 \leqslant T \leqslant T_N$ and the second for $T > T_N$, it was found that 52.5% of the magnetic entropy is removed above T_N. Further, a broad, low maximum appeared in the magnetic specific heat above T_N. It thus seems possible that the considerable short-range order that exists in $MnGeO_3$ in the region immediately above T_N is responsible for the $T_N = 16°K$ transition temperature cited in the neutron diffraction study.[26]

Turning to $CoGeO_3$, this compound also has the orthopyroxine orthorhombic crystallographic structure and at 41°K exhibits a susceptibility peak characteristic of antiferromagnetic ordering.[25] The results of our ME study[29] are shown in Figure 7. As in the case of $MnGeO_3$, better than 95% of saturation was

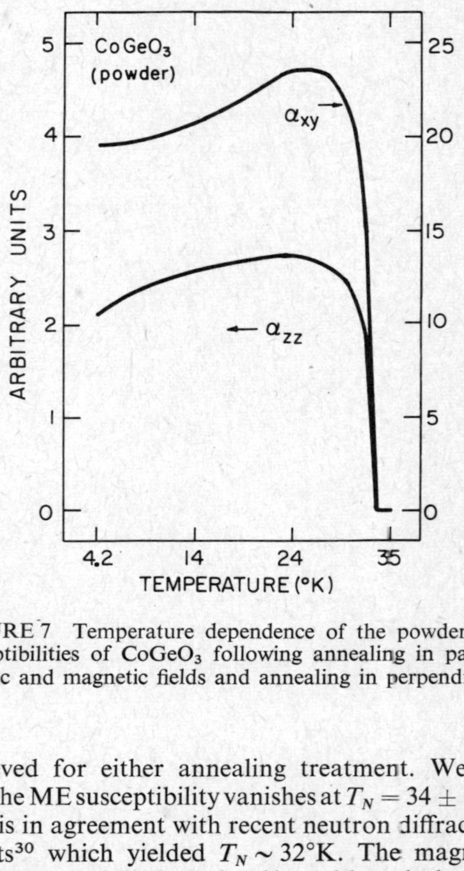

FIGURE 7 Temperature dependence of the powder ME susceptibilities of $CoGeO_3$ following annealing in parallel electric and magnetic fields and annealing in perpendicular fields.

achieved for either annealing treatment. We see that the ME susceptibility vanishes at $T_N = 34 \pm 1°K$. This is in agreement with recent neutron diffraction results[30] which yielded $T_N \sim 32°K$. The magnetic space group of $CoGeO_3$ is *Pb'ca*, although the spin structure is quite different from that of $MnGeO_3$.[30,31]

Finally, we have also studied the ME effect in $FeGeO_3$. Unlike $MnGeO_3$ and $CoGeO_3$, $FeGeO_3$ has the clinopyroxene monoclinic crystalline structure with space group $C2/c$. The magnetic suscepti-bility of this material exhibits peaks suggestive of antiferromagnetic ordering at both 51 and 14°K.[25] The results of our ME study[29] are shown in Figure 8. Here again, better than 95% of saturation was achieved by the ME annealing treatments. We see

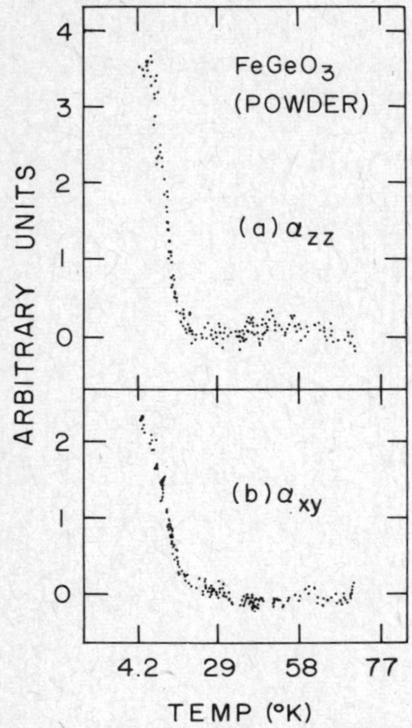

FIGURE 8 Temperature dependence of the powder ME susceptibilities of $FeGeO_3$ following (a) annealing in parallel electric and magnetic fields and (b) annealing in perpendicular fields.

that a nonzero ME susceptibility is found only below the 14°K phase transition point. To confirm that it is nevertheless ordered antiferromagnetically between 14 and 51°K, a Mössbauer effect study of this material was carried out. Hyperfine spectra characteristic of magnetic ordering were found in this temperature interval. We can thus conclude that $FeGeO_3$ does indeed order at 51°K in an antiferro-magnetic but non-ME state and undergoes a second transition to an antiferromagnetic ME state at 14°K. Both of these transitions appear to be second order in nature. Assuming, as is usual, that the group of the lower temperature phase at each phase boundary is a subgroup of that characterizing the higher temperature phase, the possible magnetic groups of $FeGeO_3$ may be determined. Upon eliminating those groups that are capable of ex-hibiting a ferromagnetic moment[4] we find that the magnetic point group of $FeGeO_3$ below 14°K must be $2'/m$, $2/m'$, or $\bar{1}'$. In the range $14°K < T < 51°K$, the only magnetic point group consistent with our assumption is $2/m1'$. ME measurements on a single crystal specimen would permit a definite deter-mination to be made.

5) *CrTiNdO₅*

The compounds $MTiRO_5$ (M = Cr, Mn, Fe; R = Pr, Nd, Sm, Eu, Gd) have been studied by Boisson[32] using X-ray and neutron diffraction techniques. These compounds have the crystallo-graphic space group *Pbam* with the M^{3+} and Ti^{4+} ions distributed over (4*f*) and (4*h*) sites and the R^{3+} ions in (4*g*) sites. For $CrTiNdO_5$, 95% Cr–5% Ti are in (4*f*) sites and 5% Cr–95% Ti are in (4*h*) sites. From neutron diffraction and magnetic measurements, Boisson[32] has reported that $CrTiNdO_5$ is antiferromagnetic below $T_N = 13°K$ with both the Cr^{3+} and Nd^{3+} spin systems ordering cooperatively at this temperature. The spins of the Nd^{3+} and the Cr^{3+} ions in the (4*h*) sites order in a G_xA_y mode[33] and those of the Cr^{3+} ions in (4*f*) sites order in an A_z mode. The magnetic space group

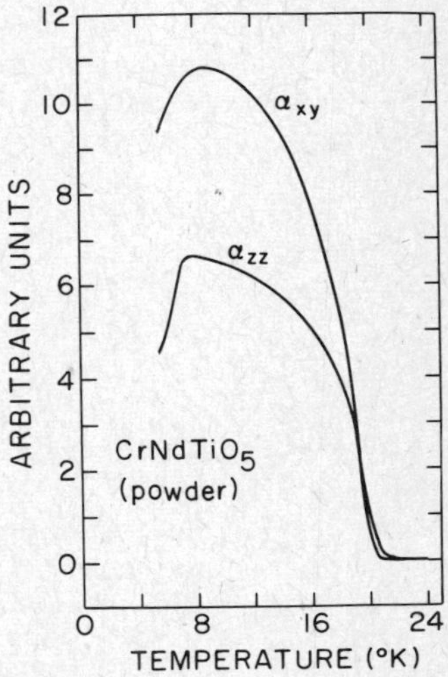

FIGURE 9 Temperature dependence of the powder ME susceptibilities of $CrTiNdO_5$ following annealing in parallel electric and magnetic fields and annealing in perpendicular fields.

assignment is *Pbam'*, indicating that $CrTiNdO_5$ is likely to be ME.

The results of our ME study[29] on a powder speci-men are shown in Figure 9. For both annealing treatments better than 95% of saturation was achieved. From the ME measurements we find that the ordering temperature is $T_N = 20.5 \pm 0.5°K$, considerably higher than the neutron diffraction result. From the ME results, we cannot determine whether the Cr^{3+} and Nd^{3+} spin systems both order cooperatively at 20.5°K or only one of them. Thus the nature of the ordering process in this material is as yet unclear and will require further study.

6) ROOH (R = Er, Tb, Dy)

The monoclinic form (space group $P2_1/m$) rare earth oxide hydroxides ROOH (R = Er, Tb, Dy, Ho, and Yb) were first prepared and their magnetic properties studied by Christensen.[34] The compounds ErOOH, TbOOH, and DyOOH exhibited susceptibility peaks characteristic of antiferromagnetic ordering at 7.2, 10, and 9°K respectively. The other two compounds show no signs of magnetic ordering down to 2.4°K. An independent study of TmOOH by Mössbauer spectroscopy[35] indicated that this compound also does not order down to 0.04°K.

Neutron diffraction studies of DyOOH[36] and ErOOH[37] show that the magnetic space groups of these compounds are $P2_1/m'$ and $P2_1'/m$ respectively. Since these groups allow magnetoelectricity, a study of the ME properties of ROOH (R = Er, Tb, Dy) was undertaken. All three of these compounds exhibited magnetoelectricity after both parallel and perpendicular annealing treatments although saturation was not achieved for either case.[38] The Néel temperatures, as determined from the ME results, were 4.1 ± 0.1, 10.0 ± 0.2, and $7.2 \pm 1°K$ for R = Er, Tb, and Dy respectively. The following conclusions follow from the ME data: (1) For all three materials $\alpha_{zz} \neq 0$. From this it follows that their single crystal ME tensors are not antisymmetric.[8] (2) Since TbOOH exhibits magnetoelectricity and is antiferromagnetic, its magnetic point group must be either $2/m'$, $2'/m$, or, possibly, $\bar{1}'$. (3) The maximum measured values of the powder ME susceptibilities were 4.5×10^{-4}, 3.8×10^{-4}, and 9.8×10^{-5} for R = Er, Tb, and Dy respectively. The single crystal susceptibilities may in fact be considerably higher as powder values are generally 2 to 5 times smaller than those of single crystals[8] and also because saturation was not achieved.

7) $GdVO_4$

Gadolinium vanadate crystallizes in a tetragonal zircon structure with space group $I4_1/amd$.[40] Magnetic measurements have shown that this compound orders antiferromagnetically at 2.5°K with the Gd^{3+} moments aligned along the *c* crystallographic axis.[41] These measurements can be understood[4] in terms of a Heisenberg model of the simplest two sublattice type, wherein antiferromagnetic exchange between a given Gd^{3+} moment and its four nearest neighbors is the predominant interaction. This proposed structure, belonging to the magnetic space group $I4_1'/a'm'd$, was confirmed by the results[42] of our ME study on a single crystal of $GdVO_4$. For this crystal class $\alpha_{11} = -\alpha_{22} = \alpha$ are the only nonzero elements of the ME susceptibility tensor.[4]

The measured temperature dependence of α is shown in Figure 10. In the region studied, α decreased monotonically as the temperature was raised and vanished at $T_N = 2.4°K$.

To compare the results of Figure 10 with theory,

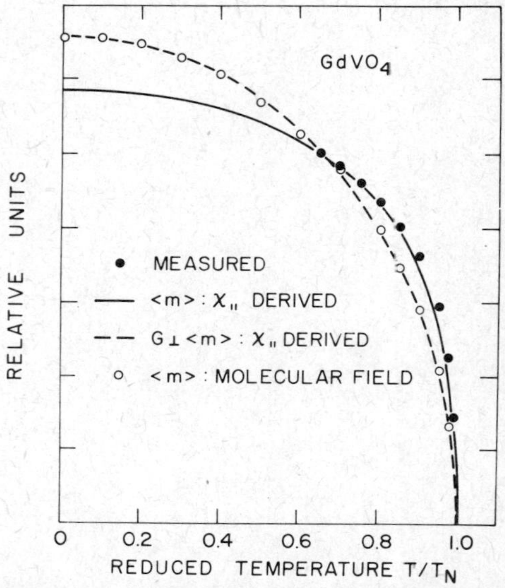

FIGURE 10 Temperature dependence of the ME susceptibility of $GdVO_4$. Also shown are three theoretical fits to the observed data ($\langle m \rangle$: sublattice magnetization. $\chi_{\parallel}$: susceptibility parallel to the antiferromagnetic axis, $G_{\perp}$: temperature-dependent factor, see. Ref. 19).

the perpendicular (to the antiferromagnetic axis) ME susceptibility was calculated on the assumption that either an electric-field-induced shift in the anisotropy energy,[43] an electric-field-induced Dzialoshinsky-type term,[21] or an electric-field-induced *g* shift[44] underlies the observed effect. For both of the latter two mechanisms α has been shown[21] to be simply proportional to the sublattice magnetization. For the case of the first mechanism, however, α is proportional to the product of the sublattice magnetization and a temperature-dependent factor defined by Rado.[19]

Using molecular field theory, our attempts to fit the experimental data did not lead to satisfactory results for either case. This is believed to be due to the limitations of the molecular field theory itself.[21] The better of the two results, obtained using the anisotropy mechanism, is shown in Figure 10.

An alternate approach, using a modified field theory in which the sublattice magnetization is derived from parallel magnetic susceptibility measurements, was more successful. The best fits obtained using this susceptibility-derived technique are also shown in Figure 10. An inspection of the results indicates that, in $GdVO_4$, the electric-field-induced Dzialoshinsky mechanism and/or the electric-field-induced g shift are/is the mechanism(s) responsible for the perpendicular ME effect. Additional arguments, based on the magnitude of the observed effect, indicate that the Dzialoshinsky mechanism is the more likely of these two possibilities.[42]

The maximum measured value of the ME susceptibility α (at $T/T_N = 0.65$) was 2.8×10^{-4}. This is essentially the saturation value of α at this temperature as the annealing fields employed were sufficient to achieve better than 95% of saturation.

III CRITICAL BEHAVIOR

1) Cr_2O_3

A study of the critical behavior of the parallel (to the antiferromagnetic axis) ME susceptibility of Cr_2O_3 has been carried out[45] in the range $0.970 \leqslant T/T_N \leqslant 0.9997$. In analogy with the critical behavior of other physical properties[46] $\alpha(T)$ was assumed, in the critical region, to be given by

$$\alpha = D(1 - T/T_N)^{\beta}. \tag{5}$$

The best fit to (5), obtained by choosing that value of T_N for which a log–log plot showed the best linearity for small values of $(1 - T/T_N)$, yielded $\beta = 0.35 \pm 0.01$. This result was in agreement with the critical behavior of the sublattice magnetization of Cr_2O_3 as obtained from a neutron diffraction study.[47]

2) $GdVO_4$

As a part of our study[42] of $GdVO_4$, the critical behavior of the ME susceptibility was measured. $GdVO_4$ is a particularly suitable candidate for such studies as it is, to a good approximation, a simple three-dimensional two-sublattice Heisenberg antiferromagnet on a diamond lattice. Further, as noted earlier, the magnetic measurements can be understood in terms of a model in which only nearest neighbor interactions are considered. It was thus of interest to compare the critical behavior of $GdVO_4$ with that of isostructural $DyPO_4$,[48] which is an excellent approximation to an ideal three-dimensional two-sublattice Ising system on a diamond lattice with nearest neighbor interactions predominant.[48,49]

For the critical behavior study α was measured in the range $2.370 \leqslant T \leqslant 2.424°K$. The results are shown in Figure 11. As in the case of Cr_2O_3, $\alpha(T)$ was assumed to be given by (5) in the critical region.

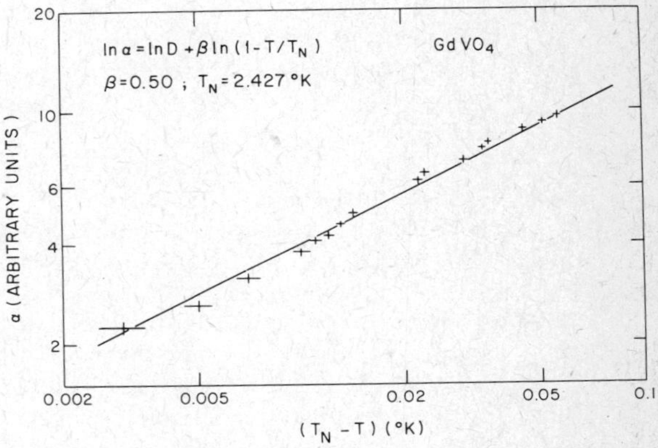

FIGURE 11 Critical behavior of the ME susceptibility of $GdVO_4$.

The best least squares fit over the range $0.977 \leqslant T/T_N \leqslant 0.999$ was obtained for $T_N = 2.427 \pm 0.006°K$ and $\beta = 0.50 \pm 0.05$ and is shown in Figure 11.

For the case of $DyPO_4$, a critical exponent of $\beta = 0.314$ has been measured,[48] in excellent agreement with the theoretical prediction[50] for the critical behavior of the sublattice magnetization of a three-dimensional two-sublattice Ising lattice on a diamond lattice. Now, if in fact α can be expanded in terms of the sublattice magnetization M near T_N, the critical exponents for these two qualities would be expected to be equal. Thus, as found previously for $DyPO_4$ and Cr_2O_3, we expect that the same critical exponent describes the behavior of α and M near the critical point for the case of $GdVO_4$ also. This then implies that the critical behavior of the sublattice magnetization of $GdVO_4$ is also described by $\beta = 0.50$, a result at variance with those found in other antiferromagnets, but in agreement with the prediction of the classical mean field-Landau theory.[46]

As is well known,[46] the mean field theory is appropriate to the case of long range forces, in which spin interactions at distances much greater than those between nearest neighbors play a significant role. Since the long-range dipole–dipole interaction between the Gd^{3+} moments in $GdVO_4$ is relatively small,[41] it appears that the exchange interactions in this compound must extend to higher-order neighbors. This conclusion is also supported by the specific heat data of Cashion *et al.*[41] whose results differed considerably from those expected for a Heisenberg model on a diamond lattice with four nearest neighbor interactions only, but were in fair agreement with those expected for a fcc lattice with twelve nearest neighbor exchange interactions. The effect of higher-neighbor interactions on critical behavior in a diamond lattice is as yet unknown, but our results indicate that, in $GdVO_4$ there exist higher-neighbor interactions whose effect is to enhance the long-range magnetic order. These interactions could have a significant effect on the critical behavior of the system due to the small number of nearest neighbors in the diamond lattice.

Essentially the same conclusion has also been reached by Colwell *et al.*[51] who studied the isomorphous compound $GdAsO_4$ and compared its properties with those of $GdVO_4$. Their specific heat data were in much better agreement with the predictions for a Heisenberg model with nearest neighbor interactions only than was the case for $GdVO_4$. Assuming that the magnetic structure of $GdAsO_4$ is the same as that of $GdVO_4$, a comparison of its critical ME behavior with that of $GdVO_4$ could be an excellent method to determine the effect of higher-neighbor interactions on critical exponents for the diamond lattice.

IV PRESSURE DEPENDENCE OF THE MAGNETOELECTRIC EFFECT

The magnetic interactions in Cr_2O_3 have been studied in detail by Samuelsen *et al.*[52] using inelastic neutron scattering. These studies have shown that the dominant exchange interactions in this material are between the spins of nearest (J_1) and next nearest (J_2) neighbor Cr^{3+} ions. The ratio J_2/J_1 is approximately 0.43 and all other exchange interactions are one-to-two orders of magnitude smaller than these two.

The exchange interactions J_1 and J_2 will of course be sensitive to changes in the interatomic distances. It thus follows that the magnetic properties of Cr_2O_3 should be affected differently by uniaxial stresses applied parallel or perpendicular to the rhombohedral axis of the crystal.

Using the ME effect, we have studied this phenomenon by measuring the variation of the Néel temperature T_N as a function of the uniaxial pressure p applied along each of the principal axes of Cr_2O_3. Some typical results[53] are shown in Figure 12.

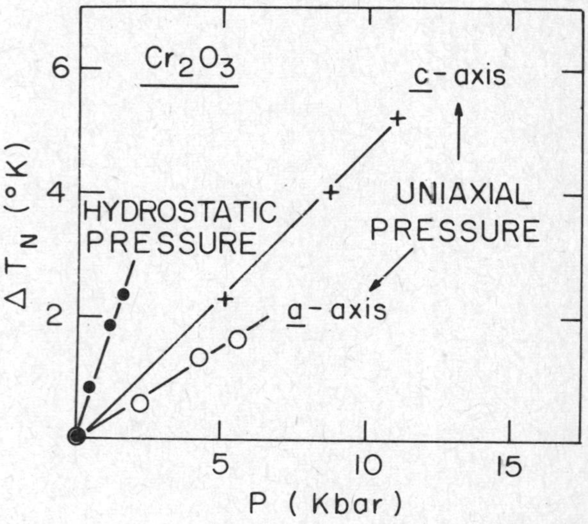

FIGURE 12 Shift in the Néel point of Cr_2O_3 as a function of uniaxial (*a* and *c* axis) and hydrostatic pressure.

For the *c* and *a* crystallographic directions respectively, we find that

$$(dT_N/dp)_c = 0.5 \pm 0.05°\text{K/kbar},$$
$$(dT_N/dp)_c = 0.3 \pm 0.05°\text{K/kbar}. \quad (6)$$

Thus for both crystallographic directions, we find that dT_N/dp is positive and that $(dT_N/dp)_c$ is approximately 60–70 % greater than $(dT_N/dp)_a$.

Under conditions of hydrostatic pressure we would expect that

$$(dT_N/dp)_h = (dT_N/dp)_c + 2(dT_N/dp)_a. \quad (7)$$

From (6) and (7), it follows that $(dT_N/dp)_h$ should also be positive and of magnitude $1.1 \pm 0.15°$K/kbar. This conclusion is in sharp disagreement with the results of Worlton *et al.*[54] who have measured T_N in Cr_2O_3 as a function of hydrostatic pressure by neutron diffraction techniques. They found that $(dT_N/dp)_h = -1.6 \pm 0.3°K/kbar$.

To resolve this question, we measured $(dT_N/dp)_h$ directly on a Cr_2O_3 crystal using our ME techniques. The results of these measurements are also shown graphically in Figure 12. We find

$$(dT_N/dp)_h = 1.5 \pm 0.1°K/k\,bar, \tag{8}$$

in good agreement with (6) and (7), but in disagreement with the neutron diffraction study. The reasons for this disagreement are not clear to us.

It is known that both the c and a unit cell dimensions of Cr_2O_3 decrease when a hydrostatic pressure is applied and our results indicate that both J_1 and J_2 increase under these conditions. Since Samuelson *et al.*[52] have shown that the strength of the exchange interactions increases with decreasing Cr^{3+}–Cr^{3+} distances, we can conclude that these distances must decrease as the unit cell dimensions are decreased.

The anisotropy in the Cr_2O_3 uniaxial pressure data is also indicative of a significant variation in the overall strength of exchange bonds between the spins of Cr^{3+} ions connected by lines lying parallel and perpendicular to the body diagonal of the rhombohedral unit cell. For the particular case of Cr_2O_3 this was of course known previously from other measurements. Our results do, however, illustrate that ME techniques provide a simple and effective means of measuring variations in the ordering temperature of antiferromagnetic materials produced by an applied uniaxial stress or hydrostatic pressure.

ACKNOWLEDGEMENTS

The contributions of many of my colleagues to the work reported here and, in particular, the assistance of Mr. Y. Chopin in the design and fabrication of much of the equipment used in this work and also in the preparation of materials are gratefully acknowledged.

REFERENCES

1. L. D. Landau and E. M. Lifshitz, *Electrodynamics of Continuous Media* (Addison-Wesley Publishing Co., Inc., Reading Mass., 1960) (English transl. of a 1958 Russian edition).
2. I. E. Dzialoshinsky, *Zh. Eksp. Teor. Fiz.* **37**, 881 (1959) [English transl.: *Soviet Phys.—JETP* **10**, 628 (1960)].
3. D. N. Astrov, *Zh. Eksp. Teor. Fiz.* **38**, 984 (1960); **40**, 1035 (1961) [English transl.: *Soviet. Phys.—JETP* **11**, 708 (1960); **13**, 729 (1961)].
4. R. R. Birss, *Rept. Progr. Phys.* **26**, 307 (1963).
5. G. T. Rado, *Phys. Rev. Lett.* **13**, 335 (1964).
6. For a recent review, see R. M. Hornreich, *IEEE Trans. Magnetics* **8**, 584 (1972).
7. S. Shtrikman and D. Treves, *Phys. Rev.* **130**, 986 (1963).
8. R. M. Hornreich, *J. Appl. Phys.* **41**, 950 (1970).
9. R. M. Hornreich, *Sol. State Commun.* **7**, 1081 (1969).
10. J. A. Gonzalo, D. E. Cox, and G. Shirane, *Phys. Rev.* **147**, 415 (1966).
11. F. Varret, P. Imbert, A. Gerald, and F. Hartmann-Boutron, *Sol. State Commun.* **6**, 889 (1968).
12. G. Gorodetsky, M. Sayar, and S. Shtrikman, *Mat. Res. Bull.* **5**, 253 (1970).
13. S. Goshen, D. Mukamel, H. Shaked, and S. Shtrikman, *J. Appl. Phys.* **40**, 1590 (1969).
14. W. Kunnmann, S. La Placa, L. M. Corliss, J. M. Hastings, and E. Banks, *J. Phys. Chem. Solids* **29**, 1359 (1968).
15. M. C. Montmory, M. Belankhovsky, R. Chevalier, and R. Newnham, *Sol. State Commun.* **6**, 317 (1968).
16. S. Bukshpan, G. Gorodetsky, and R. M. Hornreich, *Sol. State Commun.* **10**, 657 (1972).
17. T. J. Martin and J. C. Anderson, *Phys. Lett.* **11**, 109 (1964).
18. T. H. O'Dell, *Phil. Mag.* **13**, 921 (1966).
19. G. T. Rado, *Phys. Rev.* **128**, 2546 (1962).
20. J. T. Dehn, R. E. Newnham, and L. N. Mulay, *J. Chem. Phys.* **49**, 3201 (1968).
21. R. M. Hornreich and S. Shtrikman, *Phys. Rev.* **161**, 506 (1967).
22. E. F. Bertaut, L. Corliss, F. Forrat, R. Aleonard, and R. Pauthenet, *J. Chem. Solids* **21**, 234 (1961).
23. E. Fischer, G. Gorodetsky, and R. M. Hornreich, *Sol. State Commun.* **10**, 1127 (1972).
24. M. Mercier, J. Gareyte, and E. F. Bertaut, *Compt. Rend.* **264B**, 979 (1967).
25. A. Sawaoka, S. Miyahara, and S. Akimoto, *J. Phys. Soc. Japan* **25**, 1253 (1968).
26. P. Herpin, A. Whuler, B. Boncher, and M. Songi, *Phys. Stat. Sol.* (b) **44**, 71 (1971).
27. G. Gorodetsky, R. M. Hornreich, and B. Sharon, *Phys. Lett.* **39A**, 155 (1972).
28. L. M. Holmes and L. G. Van Uitert, *Sol. State Commun.* **10**, 853 (1972).
29. G. Gorodetsky, M. Greenblatt, R. M. Hornreich, and B. Sharon, *Bull. Israel Phys. Soc.* 1973, p. 45.
30. N. Shamir and H. Shaked, *Bull. Israel Phys. Soc.* 1973, p. 47.
31. H. Shaked, private communication.
32. G. Buisson, *J. Phys. Chem. Solids* **31**, 1171 (1970).
33. E. F. Bertaut, in *Magnetism*, ed. G. T. Rado and H. Suhl (Academic Press, Inc., New York, 1963), Vol. III, p. 149.
34. A. Nørland Christensen, *J. Sol. State Chem.* **4**, 46 (1972).
35. T. E. Katda, E. R. Seidel, G. Wortmann, and R. L. Mössbauer, *Sol. State Commun.* **8**, 1025 (1970).
36. A. Nørland Christensen, S. Quézel, and M. Belakhovsky, *Sol. State Commun.* **9**, 925 (1971).
37. A. Nørland Christensen and S. Quézel, *Sol. State Commun.* **10**, 765 (1972).
38. A. Nørland Christensen, R. M. Hornreich, and B. Sharon, to be published.
39. A recent neutron diffraction study of TbOOH has shown that its magnetic space (point) group is $P2_1/a'$ $(2/m')$. A. Nørland Christensen and S. Quézel, submitted to *J. Sol. State Chem.*
40. R. W. G. Wyckoff, *Crystal Structures* (Interscience Publishers, New York), Vol. 3.

41. J. D. Cashion, A. H. Cooke, L. A. Hoel, D. M. Martin, and M. R. Wells, Colloques Internationaux du CNRS, No. 180, *Les Elements de Terres Rares* (Paris–Grenoble, 1969), Vol. II, p. 417.
42. G. Gorodetsky, R. M. Hornreich, and B. M. Wanklyn, *Phys. Rev.* B (in press).
43. G. T. Rado, *Phys. Rev. Lett.* **6**, 609 (1961).
44. S. Alexander and S. Shtrikman, *Sol. State Commun.* **4**, 115 (1966).
45. E. Fischer, G. Gorodetsky, and S. Shtrikman, *J. de Phys.* **32**, C1, 499 (1971).
46. H. E. Stanley, *Introduction to Phase Transitions and Critical Phenomena* (Clarendon Press, Oxford, 1971).
47. H. Shaked and S. Shtrikman, *Sol. State Commun.* **6**, 425 (1968).
48. G. T. Rado, *Sol. State Commun.* **8**, 1349 (1970).
49. J. C. Wright, H. W. Moos, J. H. Colwell, B. M. Mangum, and D. D. Thornton, *Phys. Rev.* **B3**, 843 (1971).
50. G. A. Baker and D. S. Daunt, *Phys. Rev.* **155**, 545 (1967).
51. J. H. Colwell, B. W. Mangum, and D. D. Thornton, *Phys. Rev.* **B3**, 3855 (1971).
52. E. J. Samuelsen, M. T. Hutchings, and G. Shirane, *Physica* **48**, 13 (1970).
53. G. Gorodetsky, R. M. Hornreich, and S. Shtrikman, *Phys. Rev. Lett.* (in press).
54. T. G. Worlton, R. M. Brugger, and R. B. Bermion, *J. Phys. Chem. Solids*, **29**, 435 (1968).

Discussion

D. B. LITVIN In the determination of the magnetic space group of $GdVO_4$ you do not consider antiferromagnetic structures with a doubling of the unit cell. But an antiferromagnetic arrangement with a doubling of the unit cell, could allow a non-vanishing ME-Tensor. Could you comment?

R. M. HORNREICH In principle, this is correct although structures of this type are quite rare. Further, I believe such an ordering would be accompanied by a crystallographic distortion and the reported specific heat results show no evidence for such a distortion.

L. M. HOLMES In studying critical behavior in the absence of applied static biasing fields, it seems to me that the antiferromagnetic domain structure may be changing through the critical region.

R. M. HORNREICH This is possible but, since we obtained essentially the same result upon measuring the critical behavior of several $GdVO_4$ crystals of varying sizes and shapes, we do not think it is likely.

R. ENGLMAN With reference to the critical exponent of about 0.5 and the possibility of its connection to the existence of long-range (dipole–dipole) forces for which the molecular field approximation is thought to be approximately valid, I would like to call attention to a recent result by Mr. Zir Friedman that the molecular field theory is not good for anisotropic long range forces but only for isotropic ones.

R. M. HORNREICH The anisotropic energy of $GdVO_4$ is only about 5% of the total magnetic energy, thus the anisotropic part of the long range dipole–dipole interaction does not seem to play a dominant role in this particular system.

A. KIEL In the case of spin-fluctuations at the paramagnetic-antiferromagnetic phase transition, theory (Huber) predicts $\gamma = 1.3–1.6$. In fact recent measurements show that near T_N, γ is close to the values you measure, 0.3–0.5. Why is there a difference of the predicted γ for the two cases?

R. M. HORNREICH I believe Huber's work is concerned with the critical behavior of the spin–spin correlation function while the critical behavior of the magnetoelectric susceptibility is expected to be similar to that of the sublattice magnetization. The critical exponents of these two quantities are quite different in general.

MAGNETOELECTRIC BEHAVIOUR IN GARNETS†

M. MERCIER

Institut Universitaire de Technologie, Av. A. Briand, 03107 Montlucon, France

O'Dell has made first magnetoelectric measurements on a single crystal of yttrium garnet, at ambient temperature. Next, Ascher studied theoretically that effect, and Cardwell made measurements of magnetic field dependent electric susceptibility.

He measured a magnetoelectric coefficient of the yttrium garnet versus temperature and explained very satisfactorily the curve with a statistical model incorporating the perturbations on the single ion anisotropy and Lande *g*-factor. With a phenomenological theory Lee found only a qualitative agreement.

Lee studied very accurately the junction between the magnetization process and the magnetoelectric effect at ambient and high temperatures.

The results are well interpretated with the assumption that the magnetoelectric effect exists when the spins can rotate.

The studies were extended in low temperatures in connection with Mercier and Bauer.

Some experiments on GdIG and DyIG are also described and commented on.

1 INTRODUCTION

The crystallographic structure of rare earth iron garnets was found by Bertaut and Forrat in 1956[1] and the explanation of the magnetization was made by Neel.[2] These garnets, whose formula is $R_3Fe_5O_{12}$ (R = rare earth), are cubic with *I* a 3*d* symmetry. The chemical cell contains eight formulae and Fe^{+++} ions belong to 16*a* octahedral sites with $\bar{3}$ symmetry and to 24*d* tetrahedral sites with $\bar{4}$ symmetry while R^{+++} ions belong to 24*c* irregular hexahedral sites with 222 symmetry (Figure 1). Oxygen ions are on 96*h* sites.

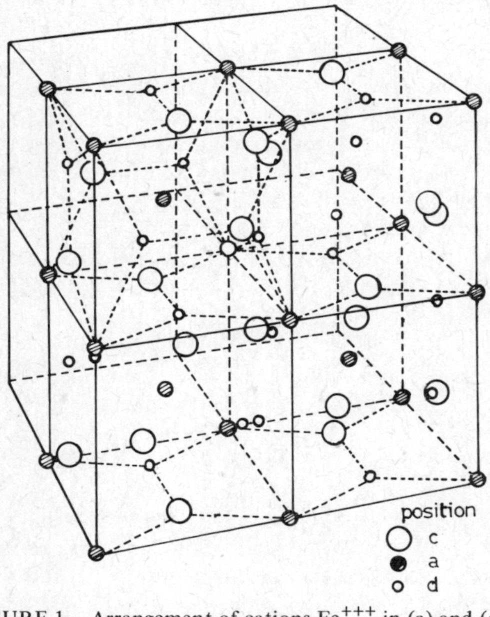

FIGURE 1 Arrangement of cations Fe^{+++} in (a) and (d) sites, and R^{+++} in (c) sites for four octants of the garnet unit cell (3).

Garnets order ferrimagnetically at $T_c \simeq 550$ K and the spontaneous magnetization curves versus temperature, from 4.2 K to T_c, were obtained by Pauthenet (Figure 2). Due to the fact that the Fe (16a) sublattice is antiparallel to the Fe (24d) sublattice and the latter is antiparallel to the R (24c) sublattice, a compensation temperature is observed (Figure 3). Other investigations for magnetic properties are summarized by Von Aulock[6] and Shieber[7].

2 MAGNETIZATION PROCESS AND SYMMETRY

The elementary cell of the garnets has the symmetry of the cube, that is to say:

—four ternary axes ⟨111⟩ that join, by twos, the opposite tops of the cube relative to the center of it,

—three quaternary axes ⟨100⟩ which are perpendicular to the faces of the cube in their centers,

—six binary axes ⟨110⟩ which are diagonals of the faces.

The magnetoelectric effect is not sensitive to the translations. This is the reason why we only consider

† Presented at the Symposium on Magnetoelectric Interaction Phenomena in Crystals, Seattle, Washington, U.S.A., May 21–24, 1973.

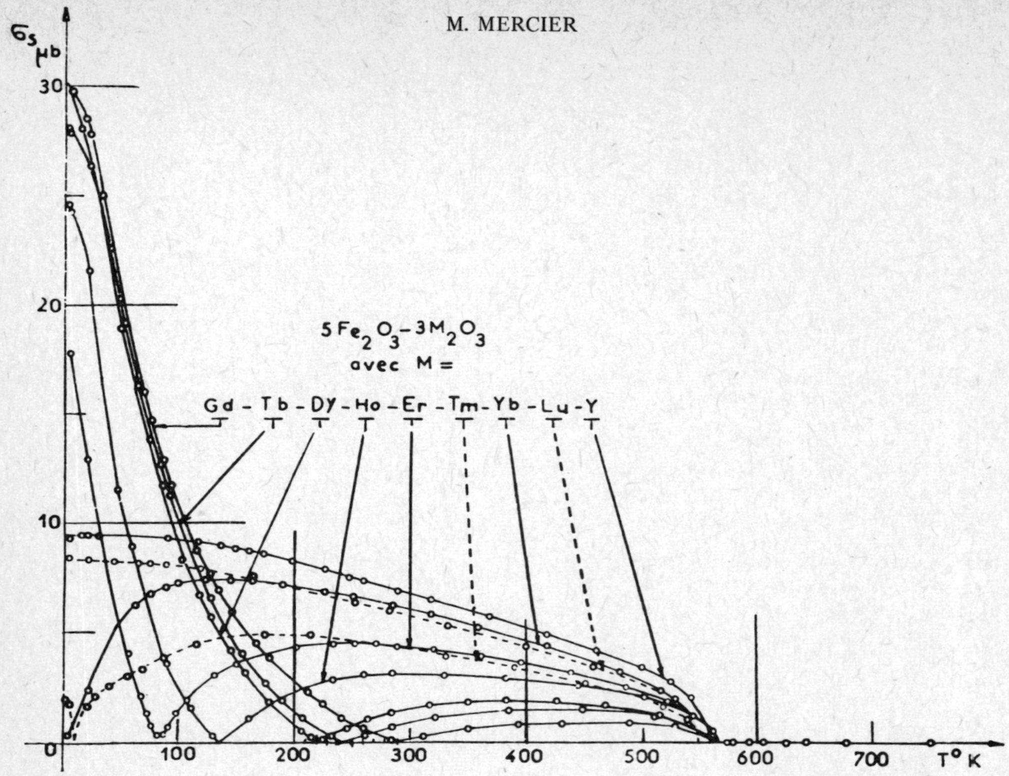

FIGURE 2 Spontaneous magnetization curves versus temperature for several iron garnets (4).

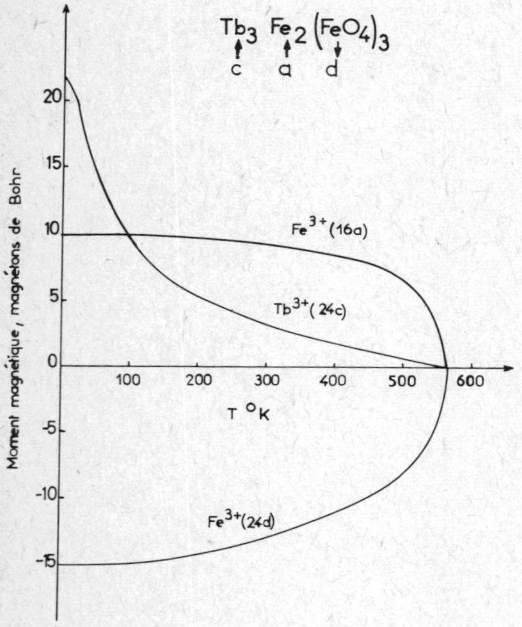

FIGURE 3 Explanation of compensation temperature, for instance in $Tb_3Fe_5O_{12}$ (5).

the point group to which the garnets belong, i.e., $m3m\underline{1}$ in non-ordered state. The moments of the magnetic atoms belonging to the three sublattices of YIG, GdIG and DyIG are, spontaneously, in the directions of easy magnetization i.e., the four ternary axes that we will refer to as ⟨111⟩.[8] This is true for all temperatures in the case of YIG and GdIG but down to about 20 K only for DyIG because, below 20 K, no principal direction of symmetry is of easy magnetization.[9]

The description of the magnetization process of the garnets is discussed by Lee[10] after the works of Neel[11] about iron and Birss and Hegarty[12] about nickel, in terms of modes. During mode I the size of the domains changes through domain wall motion between the eight possible easy ways of the four ⟨111⟩ axes when the applied magnetic field H_0 increases from zero upwards. Then the magnetic domains rotate towards the direction of the field, during mode II if there are three or four fractions or during mode III if there are only two. Mode IV indicates saturation along the field, i.e., approximately one axis.

H_0 *Along a ⟨100⟩ Direction* (Figure 4)

Calling H the internal field, J_s the saturated magnetization, θ the angle between ⟨100⟩ and the magnetization,

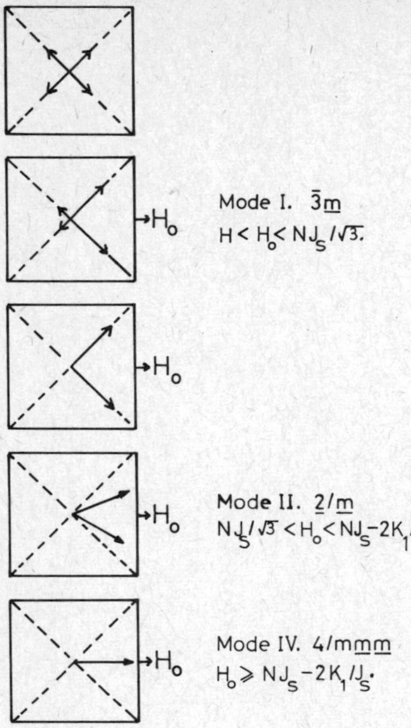

FIGURE 4 Direction of domain magnetization during the magnetization process with H_0 along a ⟨100⟩ direction. Easy directions ⟨111⟩ are projected in dashed lines on a (100) plane.

N the demagnetizing factor and K_1 the first order anisotropy constant, mode I is obtained when $H < H_0 < NJ_s/\sqrt{3} = H_4$ and the symmetry of domains is $3\underline{m}$, sub-group of the $m3m\underline{1}$. Mode II occurs when $H_4 < H_0 < NJ_s - 2K_1/J_s = H_1$ and the symmetry is $\underline{2}/\underline{m}$. When the domains are in-line with ⟨100⟩, we have mode IV with $H_0 > H_1$ and the symmetry is $4/m\underline{mm}$.

H_0 Along a ⟨110⟩ Direction

At the end of mode I, only two of the eight ⟨111⟩ directions are occupied by domains and rotation towards H_0 occurs in mode III whose symmetry is $\underline{2}/\underline{m}$ with $H_2 = \sqrt{2}NJ_s/\sqrt{3} < H_0 < H_1$. Then we have mode IV whose symmetry is $m\underline{mm}$. $H_1 = NJ_s - K_1/J_s$.

H_0 Along a ⟨111⟩ Direction

At the end of mode I mode IV occurs whose symmetry is $\bar{3}\underline{m}$. There is no rotation mode. $H_1 = NJ_s$ is the characteristic value of H_0 at the beginning of mode IV.

3 SYMMETRY AND MAGNETOELECTRIC EFFECT

If we write the free energy of a magnetic system it is possible to have terms of the second, third and fourth order in electric and magnetic fields and the expression is then:

$$F = F_0 - \sum_{ij} \frac{\alpha_{ij}}{4\pi} H_i E_j - \sum_{jkl} \frac{\beta_{jkl}}{4\pi} H_j E_k E_l - \sum_{jkl} \frac{\gamma_{jkl}}{4\pi} H_j H_k E_l - \sum_{ijkl} \frac{\partial_{ijkl}}{4\pi} H_i H_j E_k E_l$$

The term in H.E. gives rise to first order effect and the terms in EH^2 and HE^2 give rise to second order effect. The coefficient of magnetoelectric susceptibility, interesting for us, is:

$$\beta_{jkl} = -4\pi \frac{\partial^3 F}{\partial H_j \partial E_k \partial E_l}$$

In the Gaussian system the units for β_{jkl} are the same as for E^{-1}, i.e. (e.s.u. C.G.S. of E)$^{-1}$. Using the relations $J_j = -(\partial F/\partial H_j)$ and $P_k = -(\partial F/\partial E_k)$ one deduces $\beta_{jkl} = 4\pi J_j/E_k E_l = 4\pi P_k/H_j E_l$.

The tensor β connects a polar tensor E to an axial tensor H. E is an i tensor because it is not modified by time reversal whereas H is a c tensor because it is. So β is a c axial tensor of rank three like piezomagnetic tensor.[13]

Kleiner[14] and Lee[10] have shown that it is possible to classify the magnetic groups according to three types:

I without $\underline{1}$ { I(a) with $\bar{1}$ / I(b) without $\bar{1}$ }

II with $\underline{1}$ explicitly present { II(a) with $\bar{1}$ / II(b) without $\bar{1}$ }

III with $\underline{1}$ in combination { III(a) with $\bar{1}$ explicitly / III(b) without $\bar{1}$ or $\bar{\underline{1}}$ explicitly / III(c) with $\bar{\underline{1}}$ explicitly }

The magnetoelectric groups of first and second order are given in Table I in connection with the magnetic classification used. All the magnetic groups of the different types of domains in YIG allow E^2H effect and exclude EH or H^2E effects, during the magnetization process. The tensor elements have been enumerated by Ascher[15], Sivardiere[16] and Birss[13]. There are four independent elements for $\bar{3}\underline{m}$ symmetry. In a trirectangular system built with one ternary axis ⟨111⟩, as z, and one of the binary axes perpendicular to the first, as y, the different magnetizations induced by

TABLE I
Magnetoelectric groups of first or second order according to Kleiner (14) and Lee (10)'s magnetic classification (exceptions: *EH forbidden, **E^2H forbidden, †EH^2 forbidden).

I(a)	I(b)	II(a)	II(b)	III(a)	III(b)		III(c)
$\bar{1}$	1	$\bar{1}\underline{1}$	$\underline{1}$				$\underline{\bar{1}}$
$2/m$	2	$2/\bar{m}\underline{1}$	$2\underline{1}$	$\underline{2}/\underline{m}$	$\underline{2}$	$\underline{m}$	$\underline{2}/\underline{m}$
	m		$m\underline{1}$	$\underline{mm}m$			$\underline{2}/m$
mmm	222	$mm\underline{1}$	$222\underline{1}$		$\underline{22}2$	$\underline{mm}2$	$\underline{mmm}$
	$mm2$		$mm2\underline{1}$			$\underline{2m}m$	$mm\underline{m}$
$4/m$	4	$4/m\underline{1}$	$4\underline{1}$	$\underline{4}/m$	$\underline{4}$	$\underline{\bar{4}}$	$4/\underline{m}$
$4/mmm$	$\bar{4}$	$4/mmm\underline{1}$	$\bar{4}\underline{1}$	$\underline{4}/mm\underline{m}$	$\underline{4}2\underline{2}$	$\underline{4}m\underline{m}$	$\underline{4}/\underline{m}$
	422		$422\underline{1}$	$4/m\underline{mm}$	$4\underline{22}$	$\underline{\bar{4}}2\underline{m}$	$4/\underline{mmm}$
	$4mm$		$4mm\underline{1}$			$\underline{\bar{4}}m\underline{2}$	$4/\underline{m}mm$
	$\bar{4}2m$		$\bar{4}2m\underline{1}$		$\bar{4}\underline{2m}$	$4\underline{mm}$	$\underline{4}/\underline{mm}m$
$\bar{3}$	3	$\bar{3}\underline{1}$	$3\underline{1}$	$\bar{3}\underline{m}$	$3\underline{2}$	$3\underline{m}$	$\underline{\bar{3}}$
	32		$32\underline{1}$				$\underline{\bar{3}m}$
$\bar{3}m$	$3m$	$3m\underline{1}$	$3m\underline{1}$				$\underline{\bar{3}}m$
$6/m$	6	$6/m\underline{1}$	$6\underline{1}$	$\underline{6}/\underline{m}$	$\underline{6}$	$\underline{\bar{6}}$	$6/\underline{m}$
$6/mmm$	$\bar{6}$*	$6/mmm\underline{1}$	$\bar{6}\underline{1}$	$\underline{6}/\underline{m}m\underline{m}$	$\underline{6}2\underline{2}$*	$\underline{6}m\underline{m}$	$\underline{6}/m$*
	622		$622\underline{1}$	$6/m\underline{mm}$	$6\underline{22}$*	$\underline{\bar{6}}2\underline{m}$*	$6/\underline{mmm}$
	$6mm$		$6mm\underline{1}$			$\underline{\bar{6}}m\underline{2}$	$6/\underline{m}mm$
	$\bar{6}m2$*		$\bar{6}m2\underline{1}$		$\bar{6}\underline{m2}$*	$6\underline{mm}$	$\underline{6}/mm\underline{m}$*
$m3$	23	$m3\underline{1}$	$23\underline{1}$	$m3\underline{m}$	$\underline{4}3\underline{2}$*†	$\underline{\bar{4}}3\underline{m}$	$\underline{m}3$
$m3m$**	432*†*	$m3m\underline{1}$	$432\underline{1}$†				$\underline{m}3\underline{m}$
	$\bar{4}3m$***		$\bar{4}3m\underline{1}$				$\underline{m}3m$*
Explicit appearance of symmetry elements $\underline{1}$, $\bar{1}$ or $\underline{\bar{1}}$:							
$\bar{1}$		$\bar{1}$, $\underline{1}$	$\underline{1}$	$\bar{1}$			$\underline{\bar{1}}$
Total number of groups:							
11	21	11	21	10	27		21
32		32			58		
m.e. effects allowed in each class type (exceptions starred):							
E^2H	EH, E^2H, EH^2	none	EH^2	E^2H	EH, E^2H, EH^2		EH

different components of electric and magnetic fields, are:

$$J_x = \beta_1 E_x^2 - \beta_1 E_y^2 + \beta_3 E_x E_z$$
$$J_y = -2\beta_1 E_x E_y + \beta_3 E_y E_z$$
$$J_z = \beta_4 E_x^2 + \beta_4 E_y^2 + \beta_2 E_z^2$$

For $\underline{mm}m$ symmetry, five independent tensor elements are allowed, while three are allowed for $4/m\underline{mm}$ and ten for $\underline{2}/\underline{m}$.

4 APPARATUS FOR MAGNETOELECTRIC MEASUREMENTS

The first magnetoelectric investigation on the garnets was performed by O'Dell[19] in 1967 by measuring the polarization induced by a pulsed magnetic field in presence of static electric field E_0. The work of Lee, Mercier and Bauer[10,18] on the garnets was carried out by means of magnetoelectric measurements obtained with an a.c. technique that we describe summarily below.

The static (E_0) and alternating ($E_1 \sin \omega t$) electric fields induce an alternating magnetization $J_{m.e.} = \beta_m(E_0 + E_1 \sin \omega t)^2$ which is measured by a detection coil whose e.m.f. is proportional to $dJ_{m.e.}/dt$ that is to say to $2\beta_m E_0 E_1 \cos \omega t + \beta_m E_1^2 \sin 2\omega t$. β_m is the measured value of β.

1 *Measurement Cell*

Disk shaped crystals are held in place on the ceramic mount by Durafix cement. The electrodes are painted with silver paste DAG. The detection coil is about 15 mm long with an internal diameter of about 15 mm. The resistance of the 4,000 turns of aluminium wire is 582Ω at ambient temperature. To avoid vibrations the H.T. line is fused into glass and surrounded by ceramic coating.

2 *Applied Fields* (Figure 5)

A.C. voltages are produced by an Advance HIB oscillator, a Mullard 20 Watt power amplifier and a transformer, so that the distortion of the signal is less than 1% in the 370–1925 Hz band used.

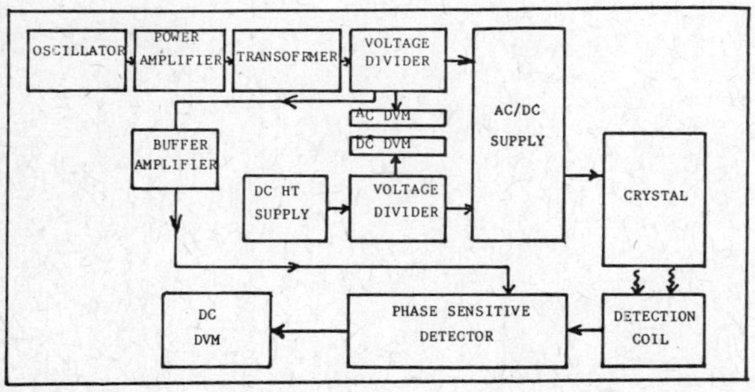

FIGURE 5 Block diagram of the whole device for magnetoelectric measurements (19).

D.C. voltages are produced by a Brandenburg generator and a device to mix A.C. and D.C. voltages was built. Magnetic biasing field is provided by a Newport electromagnet and measured by an R.F.L. Gaussmeter.

3 *Variable Temperature*

High temperatures up to 300°C are obtained, which do not lead to a danger of breakdown if the field applied to the crystals is lower than 1,000 V/mm. Low temperatures were obtained with the use of a cryostat,[19] down to 1.8 K using pumped helium and intermediate temperatures are stabilized by regulation.

4 *Choice of Frequency and Detector*

It has been necessary to use a frequency high enough to avoid the parasites of 50 Hz, but low enough to reduce capacity coupling. The choice was 1 KHz. A phase sensitive detector P.A.R.–H.R. 8 has been used for the measurements of the e.m.f. induced in the detection coil.

5 *Pick-up and Shielding*

Because of the smallness of the magnetoelectric signal it has been necessary to have no more than 0.1 μV of pick-up signal.

The electrostatic pick-up is eliminated by a good shielding between the detection coil and the H.T. line. Loops were avoided in order to eliminate magnetic pick-up. Vibration pick-up is reduced by fusing the H.T. line in glass.

6 *Calibration and Error*

Calibration was made by a small coil in which a known current flows, that coil being inside the detection coil, in place of the crystal. The accuracy of all the apparatus leads to the global error of ±10 to ±15% on measurements with $\pm 0.1K$ for $1.8K < T < 10K$, $\pm 1K$ for $10K < T < 300K$ and $\pm 3K$ for $T > 300K$. Magnetic fields can be measured to ±3%.

7 *Single Crystals*

The crystals are easily obtained by flux method and the experiments were performed on well polished disks whose faces were perpendicular to some principal axis of the cell. That form allows one to apply an electric field under good conditions and makes the demagnetizing field not too inhomogeneous. The misorientation of the crystals with respect to the direction of the applied fields is ±1°.

5 WORK OF O'DELL, CARDWELL AND ASCHER

O. Dell[17] was the first to investigate the behaviour of the magnetoelectric effect versus one applied magnetic field (Figure 6). He observed the enhancement of β during mode I, the maximum for mode III and the disappearance of the effect at saturation. The displacement of the curve from the origin indicates that the

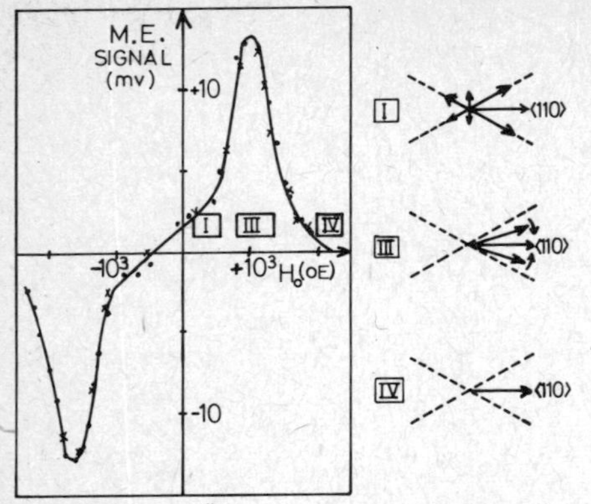

FIGURE 6 O'Dell's results for the variation of the magnetoelectric signal in YIG with H_0 applied along a ⟨110⟩.

pulsed applied field H changes the state of magnetization as H_0 does. The interpretation of the phenomenon was not given in terms of HE^2 effect but in terms of "induced" effect. E_0 was assumed to induce a change in the symmetry of the crystal to a symmetry which allows a first order effect.

Ascher interpreted, in 1968, the results of O'Dell in terms of second order effect HE^2, listed the magnetic groups that allow the EH^2 and HE^2 effect, and described the magnetoelectric tensor in the symmetry $\bar{3}\underline{m}$ when all the magnetic moments are in the easy ⟨111⟩ directions of the cube.[15] He suggested the study of the phase diagrams (H_0, T) of the garnets according to the different directions of H_0.

In 1969, Cardwell showed the magnetic field dependence of the electric susceptibility of yttrium iron garnet.[20] He measured the change of the capacitance of the crystal, as a function of H_0, by making it part of the resonant circuit of a tunnel diode oscillator.

6 CORRELATION BETWEEN MAGNETOELECTRIC EFFECT AND MAGNETIZATION PROCESS

a *The Magnetoelectric Effect is of the Second Order*

To test if the magnetoelectric effect may be interpreted as second order effect, measurements of β were made at ω and 2ω according to $e = K(2\beta_m E_0 E_1 \cos \omega t + \beta_m E_1{}^2 \sin 2\omega t)$ with e = e.m.f. induced in the coil and β_m the value of the global magnetoelectric coefficient measured in a direction. For a crystal whose ⟨110⟩ is perpendicular to the surface of the disk, with $H_0 = 900$ *Oe*, both values obtained for β_m are $\beta_m = 1.08 \times 10^{-6}$ at ω and $\beta_m = 1.03 \times 10^{-6}$ at 2ω.

The good agreement between both values and the linearity of $J_{m.e.}$ with E_1 at 2ω and with E_0 at ω, shows that the interpretation is good (Fig. 7).

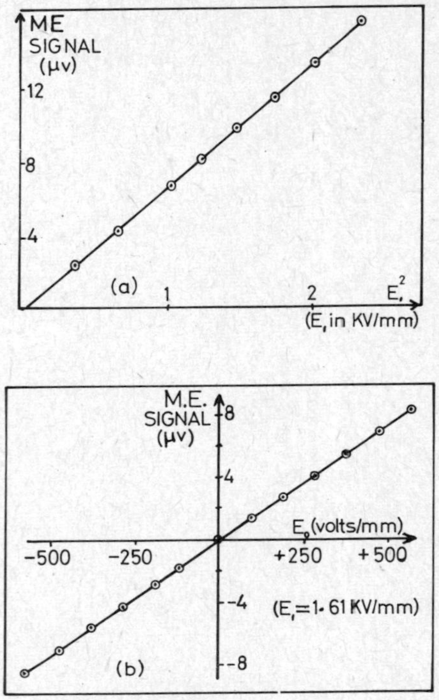

FIGURE 7 YIG (a) $J_{m.e.}$ measured along ⟨110⟩, at 2ω, versus E_1, with $H_0 = 900Oe$ along the same ⟨110⟩ (10). (b) $J_{m.e.}$ measured along ⟨110⟩, at ω, versus E_0 with $H_0 = 900Oe$ along the same ⟨110⟩.

b *Hypothesis 1*

Following Ascher[15] and O'Dell[19], Lee[10] enunciates hypothesis 1, that is to say: the magnetoelectric effect originates from domains with magnetization along ⟨111⟩ directions for which the magnetic point symmetry is $\bar{3}\underline{m}$.

c *Hypothesis 2*

Another explanation, considered by Lee,[10] is that the magnetoelectric effect occurs only when the spins can rotate i.e., during the phase whose symmetry is $2/\underline{m}$. Thus the effect occurs during modes II and III but not during I or IV.

An effect is possible during IV because, for evident experimental reasons, H_0 is not exactly applied along the direction of the chosen axis and the moments can rotate a little. Another reason is that the demagnetizing factor is not exactly uniform in disk shaped crystals. Thus a small number of moments can rotate during the early stages of magnetization.

d *Correlation Between Magnetoelectric Measurements and Magnetization*

i) *Room temperature measurements*

$J(m.e.)$ is always measured along the same direction as H_0.

For H_0 along ⟨110⟩ in the plane of the disk perpendicular to ⟨100⟩ the magnetization process is shown with theoretical and experimental values on (Figure 8). $4\pi J_s$ is 1790*Oe* as quoted in literature.[6] The demagnetizing factor $N/4\pi$ is 0.79 for axial measurements and

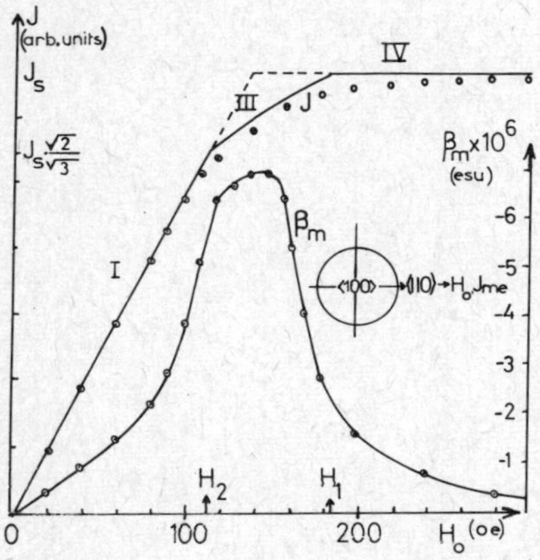

FIGURE 8 YIG. Variation of β_m versus H_0 along ⟨110⟩ for disk perpendicular to ⟨100⟩ and magnetization curves (theoretical and experimental) (10).

0.079 for in-plane measurements. The value of K_1 is taken -6.2×10^3 erg/cm.[36] So one calculates $H_2 = NJ_s(\sqrt{2}/\sqrt{3}) = 118Oe$ and $H_1 = NJ_s - K_1/J_s = 186Oe$. Good agreement is obtained between experimental and theoretical curves of magnetization. A good correlation with the magnetoelectric measurements is evident and the effect is maximum when it is mode III, i.e., when the spins can rotate.

Hypothesis 2 is indicated by $\beta_m = f(H_0)$. In agreement with Craik[21] the non-zero value of β_m, during mode I, is attributed to the lack of uniformity of the demagnetizing field which leads to rotation of some moments.

The importance of that effect is very much bound to crystal shape, temperature and strain configuration. For saturation the non-uniformity of the demagnetizing field and the non-exact alignment of H_0 with ⟨110⟩ correctly explain the non zero value of β_m at the beginning of mode IV.

When H_0 is high enough the saturation is complete and β_m becomes null. It is evident that no modification of the magnetization is possible by means of the applied electric field when the saturation is obtained. The non linearity of β_m for low values of H_0 and the cancellation of β_m for high values of H_0 seems to exclude hypothesis 1. A little effect is nevertheless observed for H_0 along ⟨111⟩ whereas the effect is to be zero in the hypothesis 2. The explanation is perhaps the non uniformity of the demagnetizing field and the non exact alignment of H_0 with ⟨111⟩ (Figure 9).

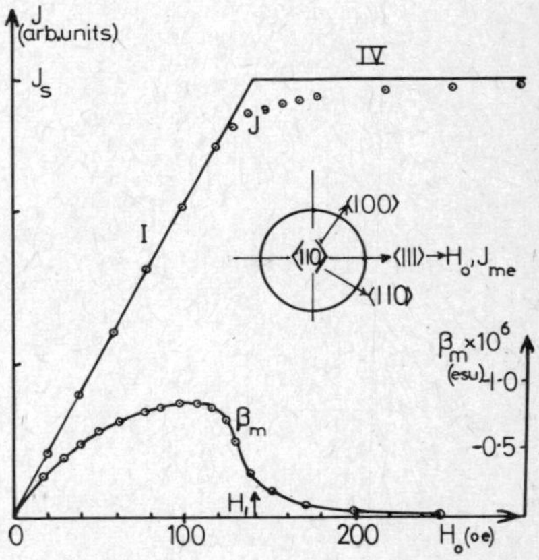

FIGURE 9 YIG. Variation of β_m versus H_0 along ⟨111⟩ for disk perpendicular to ⟨110⟩ and magnetization curves (theoretical and experimental) (10).

ii) *Low and high temperature measurements*

The magnetoelectric effect disappears at T_C when the ordered state is destroyed by thermal motion, according to the disappearance of the symmetry which allows HE^2 effect. The experimental curves of H_1, H_2, H_4, $(H_4 - H_1)$ and $(H_2 - H_1)$ versus temperature, deduced from the magnetoelectric measurements of $\beta_m = f(H_0)$ at different values of T, increase when the temperature decreases in good agreement with the theoretical curves of these fields obtained with the expressions of chapter II and the $J_s(T)$ and $K_1(T)$ given by Von Allock[6] (Figure 10).

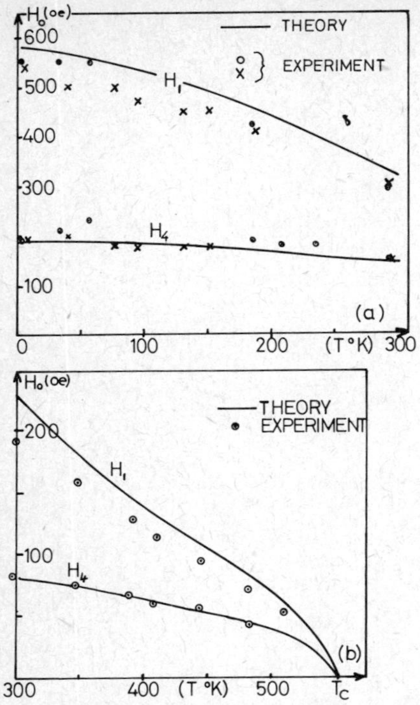

FIGURE 10 YIG. Variation of H_1 and H_4 versus temperature with $J_{m.e.}$ and H_0 along the same in plane ⟨100⟩ (10): (a) for 4.4 $K < T <$ 300 K; (b) for $T >$ 300 K (on another crystal with N different).

7 THE MAGNETOELECTRIC EFFECT IN YIG

1 *Interpretation of the Magnetoelectric Effect According to Lee (10)*

a) *Method*

The spin Hamiltonian is deduced from the statistical method stated by White and Philips[22] for piezomagnetism and magnetoelastic constants. The expression for the magnetoelectric susceptibility is similar to that of Hornreich, Shtrikman[23] and Rado[24] for Cr_2O_3.

For Fe^{+++} the largest splitting is due to the basically octahedral or tetrahedral coordination of the two sites. The perturbation of the ground state gives rise to terms quadratic in E as considered by Kiel,[25] Weber and Fehel,[26] and Clogston.[29] These terms may be considered as possible explanations of the magnetoelectric effect and give rise to the three following contributions.

$$\beta_{jkl} = -4\pi \,.\sum_{\sigma}\sum_{ij} D'^{\sigma}(i,j)\frac{\delta}{\delta H_j}\langle S_i S_j\rangle$$

is due to the spin orbit perturbation.

The indices i and j in the summations indicate the different components of the spin and σ the different sites, for the calculation of the free energy of the crystal. The indices j, k, l of β refer to the crystallographic axes along which the different fields are applied.

$$\beta_{jkl} = -4\pi \,.\sum_{\sigma}\sum_{ij} J'^{\sigma}(i,j)\frac{\delta}{\delta H_j}\langle S_i'' S_j\rangle$$

is due to the exchange field perturbation (S_i'' indicates a component of a spin on a site different from that of S_j).

$$\beta_{jkl} = -4\pi \sum_{\sigma}\sum_{i} g'^{\sigma}(i)\langle S_i\rangle$$

is due to the Landé factor perturbation.

If the D shift term, for instance, is taken Lee deduces from the expression of β_{jkl} that $\beta(T) = D' \,.\, J_s(T) \,.\, \chi(T)$ with averages over the sublattices and over the spin Hamiltonian constants. $\chi(T)$ is the susceptibility with respect to internal field H.

b *Corrections due to the demagnetizing field*

i) *At ambient temperature*

With H_0 along ⟨100⟩ the correction on $J_{m.e.}$, due to the demagnetizing field, leads, in mode II, to the expression:

$$\beta_m = \beta \,.\, \frac{K_1(1 - 9\cos^2\theta)}{NJ_s^2 + K_1(1 - 9\cos^2\theta)}$$

For H_0 along ⟨110⟩:

$$\beta_m = \beta \,.\, \frac{K_1(2 - 9\sin^2\theta)}{NJ_s^2 + K_1(2 - 9\sin^2\theta)} \qquad (10)$$

If the g shift term is dominant the magnetoelectric effect will have the same temperature dependence as the magnetization, which is not the case. If the other terms are dominant β_{jkl} is non-zero only when $\langle S_i S_j\rangle$ and $\langle S_i'' S_j\rangle$ vary with H_j. So, at saturation, β_{jkl} will be zero but not during the rotation mode. That behaviour is in agreement with hypothesis 2 and it is very likely that g shift is not important because Fe^{+++} is in ^6S ground state. Thus g has a value close to 2 which does not vary noticeably under external influences.

Measurements with different shapes of crystals and in different directions, corrected by the demagnetizing factor, lead to the same value of β in the same conditions for H_0. The comparison between $\beta_m = f(H_0)$ and $\beta = f(H_0)$ is made on (Figure 11).

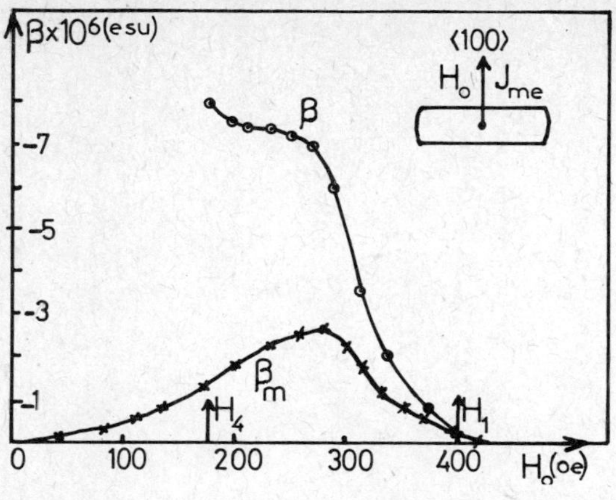

FIGURE 11 YIG. Effect of the demagnetizing field factor on β during mode II for β_m measured in the same in plane ⟨100⟩ direction as H_0, at ambient temperature (thick crystal) (10).

ii) *At high or low temperatures*
In the hypothesis of D shift term, the variation of $J_s(T)\,.\,\chi(T)$ is given, for the value of β at H_4, by $J_s^3(T)/-2K_1(T)$. A qualitative agreement is obtained between values of β deduced from experimental measurements and theoretical values.

2 *Explanation of Cardwell*[28]

Cardwell use the Hamiltonians

$$\mathcal{H}_{s/d} = -g\mu_B S_z B_z^{\text{eff}} + g\mu_B A_d S_z^2 E_1^2 + g\mu_B G_d S_z B_3 E_1^2$$
$$\mathcal{H}_{s/a} = -g\mu_B S_z B_z^{\text{eff}} - g\mu_B A_a S_z^2 E_1^2 + g\mu_B G_a S_z B_3 E_1^2$$

where a and d indicate each sublattice, B_z^{eff} is the effective magnetic field, E_1 and B_3 are the electric and magnetic fields applied along $\{110\}$ and $\{1\bar{1}\bar{1}\}$. The first term is the Zeeman energy, the second an electric field dependent perturbation of the single ion anisotropy energy and the third a perturbation of the Landé g-factor.

Using a statistical theory in the approximation of the molecular field he obtains the expression for the α^{311} magnetoelectric coefficient (that is to say the β_4 of Lee):

$$\alpha^{311} = \frac{\{2Ng^2\mu_B^2|KT\}\{3A_d(\langle m_d^3\rangle - \langle m_d^2\rangle\langle m_d\rangle) + 2A_a(\langle m_a^3\rangle - \langle m_a^2\rangle\langle m_a\rangle)\}}{1 - \{2Ng^2\mu_B^2/KT\}\mu_0\{3C_d(\langle m_d^2\rangle - \langle m_d\rangle^2) - 2C_a(\langle m_a^2\rangle - \langle m_a\rangle^2)\}}$$
$$+ \frac{2Ng\mu_B\{3G_d\langle m_d\rangle + 2G_a\langle m_a\rangle\}}{1 - \{2Ng^2\mu_B^2/KT\}\mu_0\{3C_d(\langle m_d^2\rangle - \langle m_a\rangle^2) - 2C_a(\langle m_a^2\rangle - \langle m_a\rangle^2)\}}$$

where N is the number of formulae per unit volume, m_a and m_d the spin quantum numbers of the sublattices. $C_d = \gamma n_{dd} - (1-\gamma)n_{ad}$ with γ a temperature independant constant. The values of n_{dd} and n_{ad}, intra and intersublattice molecular field coefficients, and the different averaged quantities are deduced from the data of Pauthenet.[29] $C_a = \gamma n_{aa} - (1-\gamma)n_{da}$.

The values of the unknown parameters were computed by fitting the expression of α^{311} with the experimental curve obtained between 100 K and the Curie point.

The quantitative agreement between the theoretical and experimental curves is good (Figure 12) except below 150 K. According to Cardwell this is probably due to the assumption that γ is not a function of temperature and to the approximations of the molecular field theory.

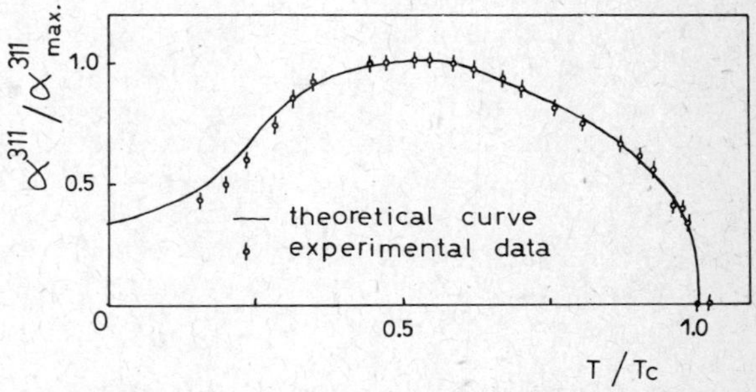

FIGURE 12 YIG. Temperature variation of α^{311} second order magnetoelectric coefficient (28).

3 *Magnetoelectric Tensor Elements*

During the rotation modes, the domain symmetry is $\underline{2/m}$, but when the moments begin to rotate, in mode I with $\bar{3}m$ symmetry, there are four tensor elements for β while there are ten for $\underline{2/m}$.

Lee has considered $\bar{3}m$ symmetry to calculate the tensor elements. For given directions of E_1, H_0 and corrected $J_{m.e.}$ the expression of β is calculated in terms of the β_i elements by summing over the relevant domains. Between these elements β_i each domain gives rise to equations given in Chapter III and an equal distribution of domains is assumed.

The solutions of the possible linear systems of equations have been found, with the aid of a computer program in order to obtain the minimum least square total residual. Taking into account the errors on the experimental values of β_m the limits for the β_i are:

$-7 \times 10^{-6} \leqslant \beta_1 \leqslant -2 \times 10^{-6}$,

$-22 \times 10^{-6} \leqslant \beta_2 \leqslant +13 \times 10^{-6}$

$+22 \times 10^{-6} \leqslant \beta_3 \leqslant +54 \times 10^{-6}$,

$-1.4 \times 10^{-6} \leqslant \beta_4 \leqslant +3 \times 10^{-6}$

The conclusion is that the values are imprecise. The largest tensor element is $\beta_3(H_xE_xE_z$ or $H_yE_yE_z)$ with a positive value. β_1 is negative and small. The values for β_2 and β_4 are undetermined for the sign. The non-ellipsoidal nature of the disks used in experiments explains the non-accuracy of the results.

Lee has investigated the possible explanation of the magnetoelectric effect as a magnetostrictive and electrostrictive mechanism. His calculations show an insignificant contribution to the measured magnetoelectric effect in the light of the present knowledge of the theory of electrostrictive effect. However such a mechanism can qualitatively explain the results. Are the electrostrictive constants in YIG unusually large?

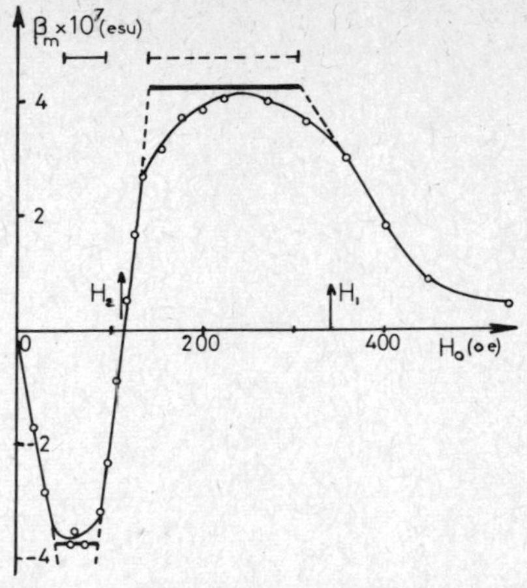

FIGURE 13 GdIG. $T = 203$ K. $\beta m = f(H_0)$ for H_0 and $J_{m.e.}$ along the same in plane ⟨110⟩ direction.

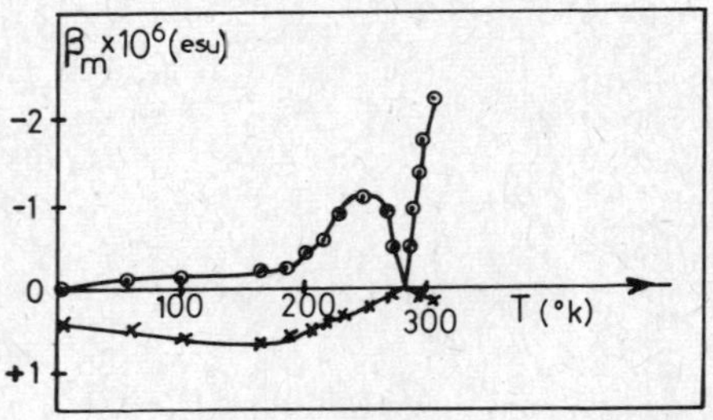

FIGURE 14 GdIG. The extrema of $\beta m = f(T)$.

8 MAGNETOELECTRIC EFFECT IN GdIG and Dy IG[10,18]

1 *Gadolinium Iron Garnet*

The correlation between the magnetoelectric effect and the magnetization process is illustrated by (Figure 13), at 203 K, with H_0 and $J_{m.e.}$ along the same in-plane ⟨110⟩ direction. The values of H_2 and H_1 were calculated by Lee[10] with the equations used above, in which the J_s values are given by Anderson[30] and the K_1 values by Von Aulock[6], Pearson and Lacklison[31]. The sign of β_m changes near H_2 and two extreme values are obtained because, as for YIG, β_m vanishes in mode IV. The change of the sign of β_m occurs in the region of the end of mode I. No satisfactory explanation is yet given to that phenomenon.

The curves of the extrema of β_m versus T are given (Figure 14).

The effect falls smoothly to zero at compensation temperature 285 K. A simple explanation of this fact is that the contribution of the magnetoelectric effect of each sublattice is proportional to the spontaneous magnetization near the compensation temperature. This is not true at low temperatures because the effect is not very high. Correction factor of β_m (due to the demagnetizing field) are 1 near the compensation temperature, and reach 2.6 at 50 K, for H_2. The values of β_m for YIG and GdIG are of the same order of magnitude

Anderson[30] has shown that the introduction of Gd^{+++} instead of Y^{+++} in YIG increases the intra F^{+++} ion sublattice interactions. So an effect on the exchange energy (i.e., a modification of the second expression of β above in VII, 1 *a*) seems quite plausible. Concerning the Gd^{+++} alone a modification of the spin orbit coupling may bring changes in the first expression for β, with respect to the case of YIG. The *g*-shift term seems to have not a very significant effect because, as above (VII, 1, *ba*) the magnetoelectric effect decreases to zero for magnetic saturation.

2 *Dysprosium Iron Garnet*

H_4 and H_1 were calculated from the values of J_s obtained by Harrison *et al.*[32] and these of K_1 obtained by Pearson and Lacklison.[31] The curves of $\beta_m = f(H_0)$, as for YIG, do not show change of the sign of β_m (Figure 15).

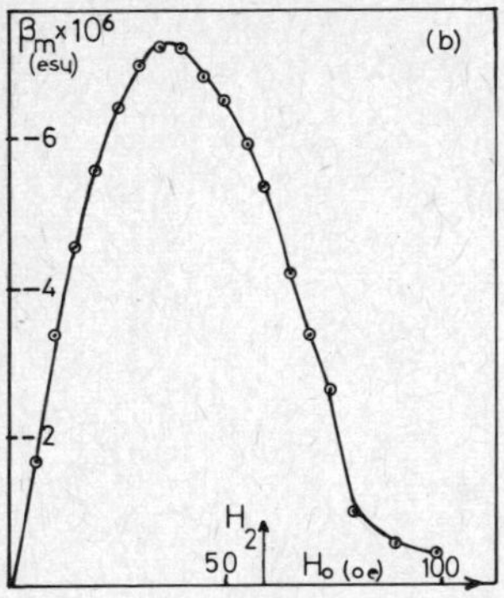

FIGURE 15 DyIG. $T = 296\ K.\ \beta m = f(H_0)$ for H_0 and $Jm.e.$ along the same in plane ⟨110⟩.

Correction factors of β_m are always close to 1 because of the large values of anisotropy constant K_1. The curves $\beta_m = f(T)$ (Figure 16) shows the null value

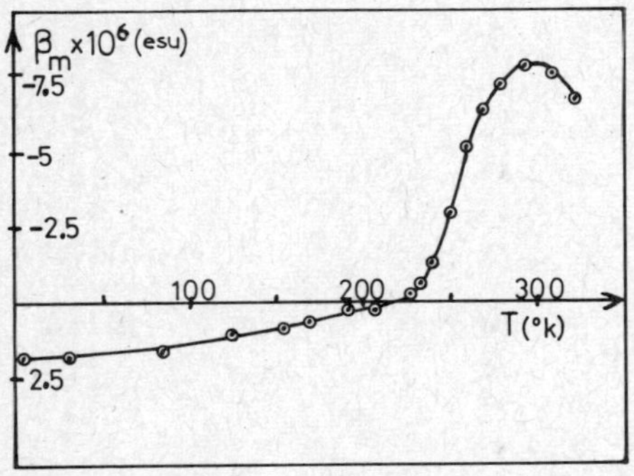

FIGURE 16 DyIG. Maximum values of $\beta m = f(H_0)$ versus temperature, with H_0 and $Jm.e.$ along the same in plane ⟨110⟩.

of β_m at compensation temperature 220 K and the change of the sign of β_m. The values of β_m are of the same order of magnitude as for YIG.

IX CONCLUSION

Important contribution to the study of the magnetoelectric effect in garnets have been made for the last five years, initiated by the work of O'Dell. The magnetoelectric effect is of the second order E^2H in all the point symmetry groups $\bar{3}m$, $\underline{2}/\underline{m}$, $4/m\underline{mm}$ and $m\underline{m}m$ (Table I) of the different modes I, II, III, IV obtained during the magnetization process.

The correlation between the magnetoelectric effect and magnetization is important for the three garnets studied (YIG, GdIG and DyIG). However, due to the non ellipsoidal form of the crystals, there is some uncertainty concerning the very beginning of the modes. There are more important arguments to explain the magnetoelectric behaviour by hypothesis 2, i.e., that magnetoelectric effect occurs when the spins can rotate, rather than by hypothesis 1, i.e., the magnetoelectric effect occurs when the spins are in-line with ⟨111⟩.

However this explanation is perhaps not perfectly satisfactory and further investigations are necessary in order to determine that important point with greater certainty.

ACKNOWLEDGEMENTS

Interesting discussions were held with Dr. Lee who has read the manuscript.

REFERENCES

1. F. Bertaut and F. Forrat, *C.R. Acad. Sci. Paris,* **242**, 382 (1956).
2. L. Neel, *C.R. Acad. Sci. Paris,* **239**, 8 (1954).
3. S. Geller, H. J. Williams, R. C. Sherwood, G. P. Espinoza and E. A. Nesbitt, *J. Appl. Phys.*, **35**, 520 (1964).
4. R. Pauthenet, *C.R. Acad. Sci. Paris,* **243**, 1499 (1956).
5. F. Tcheou, Thèse 3e cycle. University of Grenoble p. 14 (1966).
6. W. H. Von Aulock. Handbook of microwaves ferrites materials. Academic Press (1965).
7. M. M. Schieber. Experimental magnetochemistry p. 327–378. North Holland Publishing Company (1967).
8. K. J. Standley. Oxide magnetic materials, p. 107, and 117, Oxford University Press (1962).
9. A. E. Clark and E. Callen, *J. Appl. Phys.*, **39**, 5972 (1968).
10. G. R. Lee, Philosophical Doctorate's Thesis. University of Sussex (1970).
11. L. Neel, *J. Phys. Rad.*, **5**, 241 (1944).
12. R. R. Birss and B. C. Hegarty, *Brit. J. Appl. Phys.*, **17**, 1241 (1966).
13. R. R. Birss, Symmetry and Magnetism, p. 140, North Holland Publishing Company (1966).
14. W. H. Kleiner, *Phys. Rev.*, **142**, 318 (1966).
15. E. Ascher, *Phil. Mag.*, **17**, 149 (1968).
16. J. Sivardiere. Private communication.
17. T. H. O'Dell, *Phil. Mag.*, **16**, 487 (1967).
18. G. R. Lee, M. Mercier and P. Bauer, *C.R. Coll. Int. Terres Rares, Paris, Grenoble,* p. 390 (1970).
19. M. Mercier, *J. Phys. Appl.*, **2**, 109 (1967).
20. M. J. Cardwell, *Phil. Mag.*, **167**, 1087 (1969).
21. D. J. Craik, *Brit. J. Appl. Phys.*, **18**, 1355 (1967).

22. R. L. White and T. G. Phillips, *J. Appl. Phys.*, **39**, 579 (1968).
23. R. Hornreich and S. Shtrikman, *Phys. Rev.*, **161**, 506 (1967).
24. G. T. Rado, *Phys. Rev.*, **128**, 2546 (1962).
25. A. Kiel, *Phys. Rev.*, **148**, 247 (1966).
26. M. Weger and E. Fehel, *Proc. Int. Conf. on Paramagnetic Resonance*, p. 628, Academic Press N.Y. (1963).
27. A. M. Clogston, *J. Phys. Rad.*, **20**, 151 (1959).
28. M. J. Cardwell, *Phys. Stat. Sol. (b)*, **45**, 597 (1971).
29. R. Pauthenet, *J. Appl. Phys.*, **29**, 252 (1958).
30. E. E. Anderson, *Proc. Int. Conf. on Magnetism*, Nottingham. I.P. and P.S. London (1965).
31. R. E. Pearson and D. Lacklison, Private communication to G. R. Lee.
32. F. W. Harrison, J. F. A. Thompson and G. K. Lang, *J. Appl. Phys.*, **36**, 1014 (1963).

Discussion

L. M. HOLMES Do you have data on the variation of the magnitude of β with composition in the garnets?

M. MERCIER For these three garnets the maximum value of β (corrected value) at room temperature is of the order of 10^{-6} in Gaussian units.

SPIN ORDERING IN MAGNETOELECTRICS †‡

D. E. COX

Physics Department, Brookhaven National Laboratory, Upton, New York 11973

A systematic survey of known magnetic structures determined by neutron diffraction techniques has been made in order to identify materials which should exhibit linear magnetoelectric effects. This can be accomplished from a knowledge of the magnetic point symmetry, and the limitations of the neutron diffraction technique in this respect are discussed. Consideration is given to the general features of magnetic ordering in the more important structural types, with particular emphasis on systems in which cation ordering can occur with a consequent lowering of symmetry. Rutile, perovskite, spinel, corundum and olivine lattices are discussed in some detail. In addition to the approximately twenty-five known magnetoelectric materials, another twenty or so are suggested as potential candidates.

The purpose of this paper is to make a systematic survey of known magnetic structures, and on the basis of their magnetic symmetry to predict specifically in which compounds magnetoelectric (ME) effects may be expected. From consideration of some of the general features of magnetic ordering in a number of important crystal classes, it is also possible to predict that ME phenomena are unlikely to be present in many systems. The survey is restricted to linear effects, which appear in the free energy expression as terms of the type $\alpha_{ij}E_iH_j$, where E and H are the electric and magnetic fields, and α the ME susceptibility tensor. Non-linear terms such as $\beta_{ijk}E_iH_jH_k$ have not been considered.

Although several hundred magnetic structures are now known, the survey is considerably simplified as a result of two well-known symmetry restrictions. No ME effect is allowed if the *magnetic* point symmetry includes an inversion center, or if the Bravais lattice contains an antitranslation, since terms of the type EH are not invariant under the action of either of these symmetry elements. The operation of an inversion center reverses the direction of an electric vector but not that of a magnetic vector, while in the case of an antitranslation the reverse applies. In addition, experimental observation of the effect is dependent upon the material having a reasonably high resistivity so that for practical purposes metals and alloys can be ignored.

In the course of the survey, two recent compilations of magnetic structures have been referred to extensively.[1,2] The table of magnetic space groups compiled by Kucab[3] has also been very helpful, although there are a number of instances in which this is at variance with Ref. 1. The articles by Schmid[4] and Bertaut and Mercier[5] contain numerous references to various kinds of ME materials, including discussion of those showing higher order effects, while surveys by Hornreich[6] contain predictions for a number of compounds listed in the present paper.

In order to make any predictions about ME effects in a particular material, it is necessary to know the magnetic symmetry, and it is worth mentioning some of the limitations, both inherent and practical, of neutron diffraction techniques—in particular powder techniques—by which virtually all existing data on magnetic structures have been obtained. For example, in a polycrystalline sample (and also a multi-domain crystal), it is not possible to determine the direction of the moments in a cubic material, nor the orientation within the basal plane of tetragonal or hexagonal materials. Moreover, in some high symmetry structures, there may be two or more models whose calculated intensities are identical. These ambiguities can sometimes be resolved by application of a magnetic field if this is available. A second factor is that since the magnetic intensity is proportional to the square of the components of the moments, a small canting may not be detected so that the full symmetry is not always revealed. Finally, a lower magnetic symmetry is frequently not reflected in a detectable distortion of the crystal lattice. The assignment of a particular magnetic space group must then be regarded with a certain degree of caution. Indeed, it is in just such cases that ME data can be very helpful in enabling a definite choice to be made.

A survey of the common types of magnetic structures

† Work performed under the auspices of the U.S. Atomic Energy Commission.

‡ This paper was presented at the Symposium on Magnetoelectric Interaction Phenomena in Crystals, Seattle, Washington, May 21–24, 1973.

TABLE I

Magnetic structural data for trirutile type compounds. Space group $P4_2/mnm$, magnetic ions in 4(e) positions. ME type AAIII.[a]

Compound	T_N(°K)	Spin direction	Magnetic space group	Magnetic point group	Form of ME tensor	Maximum α_{obs}	Reference
Cr_2TeO_6	105	⊥[001]	$Pn'nm$(?)	$m'mm$	α_{23}, α_{32}	–	7, 10
Fe_2TeO_6	219	[001]	$P4_2/m'n'm'$	$4/m'm'm'$	$\alpha_{11} = \alpha_{22}, \alpha_{33}$	3×10^{-5}	7–9, 70
Cr_2WO_6	69	⊥[001]	$Pn'nm$(?)	$m'mm$	α_{23}, α_{32}	–	7, 10
V_2WO_6	–	⊥[001]	$Pn'nm$(?)	$m'mm$	α_{23}, α_{32}	–	7
$(Fe_{0.5}Cr_{0.5})_2WO_6$	–	⊥[001]	$Pn'nm$(?)	$m'mm$	α_{23}, α_{32}	–	7

[a] The notation and classification of magnetoelectric materials suggested by Schmid[4] have been followed throughout the present paper.

makes it clear that a great many magnetic materials will not show ME effects. Nature very often places magnetic atoms on inversion centers which are retained in the magnetic state, as for example, in most substances with rock-salt, nickel arsenide, rutile or calcite structures. However, in some cases substitution of an appropriate cation results in chemical ordering with lowering of the symmetry and the appearance of ME effects. A good example is provided by rutile-type compounds, in which the magnetic structure is generally a simple body-centered antiferromagnetic arrangement having the magnetic point symmetry $4'/mmm'$, which is not magnetoelectric. By appropriate cation substitution the ordered trirutile structure is obtained, as for example in Fe_2TeO_6.[7,8] This has the symmetry $4/m'm'm'$ and should therefore exhibit magnetoelectricity, as first pointed out by Hornreich[6] and subsequently confirmed by Buksphan *et al.*[9] Other materials of this type studied by neutron diffraction which should show ME effects are listed in Table I. Only in Fe_2TeO_6 is the magnetic symmetry known with certainty; in the other compounds, the orientation of the moments within the basal plane is not known, but a reasonable guess is $m'mm$ symmetry.

A great many compounds have structures related to that of perovskite and have been extensively studied by neutrons. The ideal perovskite structure is cubic, with $Pm3m$ symmetry, but this is seldom realized, and a number of distorted variants are known. One of the most widely studied is the orthorhombic $GdFeO_3$ structure, which has $Pbnm$ symmetry with the Gd and Fe ions in 4(c) and 4(b) positions, site symmetries m and $\bar{1}$ respectively. When a magnetic ion occupies the inversion sites, the magnetic order usually observed is the so-called G type, in which a given moment is coupled antiparallel to its six nearest neighbors, and the inversion symmetry is retained with no possibility of a ME effect. However, when the inversion site contains a diamagnetic ion, the rare earth ions can form ordered arrays which do permit ME behavior. Known compounds of this sort are listed in Table II. An interesting exception is $TbFeO_3$ as discussed by Hornreich.[6] The low temperature symmetry appears to be $P2_1'2_1'2$, but an ME effect, although possible, was not observed.[12,17]

$BiFeO_3$ is an interesting material which has been the center of much controversy. The most recent studies[20,21,22] confirm that it undergoes a ferroelectric transition at 850°C from the cubic structure to a rhombohedral one with $R3c$ symmetry, and an

TABLE II

Magnetic structural data for perovskite type compounds. Space group $Pbnm$, rare earth ions in 4(c) positions, transition metal ions in 4(b) positions. ME type AA III.

Compound	T_N(°K)	Spin direction	Magnetic space group	Magnetic point group	Form of ME tensor	Maximum α_{obs}	Reference
$DyAlO_3$	3.5	⊥[001]	$Pb'n'm'$	$m'm'm'$	$\alpha_{11}, \alpha_{22}, \alpha_{33}$	2×10^{-3}	11–13
$GdAlO_3$	4.0	[010]	$Pb'n'm'$	$m'm'm'$	$\alpha_{11}, \alpha_{22}, \alpha_{33}$	1×10^{-4}	14
$TbAlO_3$	4.0	⊥[001]	$Pb'n'm'$	$m'm'm'$	$\alpha_{11}, \alpha_{22}, \alpha_{33}$	1×10^{-3}	12, 15–17
$DyCoO_3$[a]	8.8	⊥[001]	$Pb'n'm'$	$m'm'm'$	$\alpha_{11}, \alpha_{22}, \alpha_{33}$	–	18
$HoCoO_3$[a]	2.4	⊥[001]	$Pb'n'm'$	$m'm'm'$	$\alpha_{11}, \alpha_{22}, \alpha_{33}$	not obs.	12, 18
$TbCoO_3$[a]	3.3	⊥[001]	$Pbnm'$	mmm'	α_{12}, α_{21}	3×10^{-5}	12, 16, 18
$TbRhO_3$	1.9	[010]	$Pb'n'm'$	$m'm'm'$	$\alpha_{11}, \alpha_{22}, \alpha_{33}$	–	19

[a] Co is in the diamagnetic low-spin state.

antiferromagnetic transition at 375°C to a G type structure, with the same magnetic symmetry. ME behavior should therefore be displayed (see Table X).

Substituted perovskites of the type $A_2BB'O_6$ and $A_3BB''_2O_6$ have been extensively studied, particularly by Soviet groups.[22,23] In the former, cation ordering on (111) layers frequently occurs and the resulting cell has *Fm*3*m* symmetry. In compounds of the latter type, in which B'' is Nb or Ta, the divalent *B* ions tend to order on every third (111) layer and the structure is hexagonal ($P\bar{3}m1$). In neither case is an ME effect expected. However, it is known that the cubic structure distorts in a great many instances,[24,25] and many of these materials have been reported to have both ferroelectric and magnetic transitions. Magnetoelectricity has been reported in Pb_2FeNbO_6 and Pb_2MnNbO_6,[26] at low temperatures. The former has been found to have a G type magnetic structure, but no long-range chemical order.[27] Very few neutron measurements or detailed structural data have been obtained on these materials. When B' is diamagnetic, ordering of the second kind (next-nearest-neighbors antiparallel) is likely,[28,29] but when B' is paramagnetic, ordering of G type is to be expected. Because of differences in the moments of the *B* and B' ions, however, the compound will be ferrimagnetic. Ferrimagnetism of this sort may also occur if there is only partial ordering of diamagnetic B' ions, but in none of the above cases is the inversion center destroyed. Without a more detailed knowledge of the crystal structure of these compounds, it is difficult to make predictions about possible ME phenomena, and this certainly seems to be an area which warrants further attention.

Another large group of compounds crystallize with the cubic spinel structure. In this, the space group is *Fd*3*m* with cations in the tetrahedral 8(a) and octahedral 16(d) positions, usually called *A* and *B* sites with symmetries $\bar{4}3m$ and $\bar{3}m$ respectively. When both sets of positions are occupied by magnetic ions, there is strongly negative *A–B* exchange and the collinear ferrimagnetic Néel arrangement is usually found, the inversion center being retained. However, substitution of a diamagnetic ion into the octahedral sites renders the *A–B* coupling inoperative and leads to a nearest-neighbor antiparallel array on the tetrahedral sites related by anti-inversion symmetry.[30] Known structures of this kind are listed in Table III. The direction of the moments has been determined only for $MnGa_2O_4$,[31] in which they lie along (111), and the most likely space group is therefore $R\bar{3}'m'$. For an (001) direction, an appropriate choice would be $I4'_1/a'm'd$. Both groups allow an ME effect, but the form of the tensor differs, and ME measurements should allow a choice to be made.

With the appropriate ionic substitutions, it is possible to make *B–B* interactions competitive, in which case rather complicated non-collinear structures are found, usually with enlarged or even incommensurate magnetic cells. However, there is a group of tetragonally distorted spinels with $I4_1/amd$ symmetry, which at low temperatures have the simple Yafet–Kittel canted arrangement. A reasonable choice of space group appears to be the orthorhombic one $Fd'd2'$, in which case an ME effect is expected. $NiCr_2O_4$ has a more complex structure with monoclinic symmetry $P2'$. Data for these materials are listed in Table IV. ME data could prove to be very valuable in these cases.

Another possibility is the substituted spinel $Li_{0.5}Fe_{2.5}O_4$†, in which the Li ions order chemically on one-quarter of the octahedral sites, thereby lowering the symmetry to $P4_332$[40]. The magnetic order consists of the ferrimagnetic Néel arrangement[41] with the moments directed along the [111] axis[42], and the magnetic point symmetry can therefore be assumed to be no higher than $32'$. If the latter occurs this would be the first example of a FM II type of material (Table X).

Ferrimagnetism is also found in two other im-

† I am indebted to Dr. S. Shtrikman for pointing this out.

TABLE III

Magnetic structural data for spinel type compounds. Space group *Fd*3*m*, magnetic ions in 8(a) and 16(d) positions. ME type AA III.

Compound	T_N(°K)	Spin direction	Magnetic space group	Magnetic point group	Form of ME tensor	Reference
$CoAl_2O_4$	4	?	see text	–	–	32
$CoCo_2O_4$[a]	40	?	see text	–	–	30
$MnAl_2O_4$	6.4	?	see text	–	–	32
$MnGa_2O_4$	33	[111]	$R\bar{3}'m'$	$\bar{3}'m'$	$\alpha_{11} = \alpha_{22}, \alpha_{33}$	31

[a] Octahedral Co is in the diamagnetic low-spin state.

TABLE IV

Magnetic structural data for hausmannite type compounds. Space group $I4_1/amd$, magnetic ions in 4(a) and 8(d) positions. ME type FE I / FM I.

Compound	T_C(°K)	Spin direction	Magnetic space group	Magnetic point group	Form of ME tensor	Reference
$CuCr_2O_4$[a]	135	⊥[001]	*Fd'd2'*	*m'm2'*	α_{23}, α_{32}	33–35
$NiCr_2O_4$	65	⊥[100]	*P2'*	2'	α_{13}, α_{31} α_{23}, α_{32}	36, 37
$CoMn_2O_4$	100	⊥[1$\bar{1}$0]	*Fd'd2'*	*m'm2'*	α_{23}, α_{32}	38
$CrMn_2O_4$	65	⊥[1$\bar{1}$0]	*Fd'd2'*	*m'm2'*	α_{23}, α_{32}	39

[a] Crystal space group probably $I\bar{4}2d$

portant classes of materials related to spinels, the garnets and hexagonal ferrites. These have been studied extensively by neutron diffraction techniques, but although a wide variety of magnetic structures have been found, the inversion symmetry is invariably retained. Induced second-order ME effects have, however, been observed in garnets following application of a magnetic field.[115]

The symmetry found in corundum and related type structures offers more encouraging prospects for magnetoelectricity since the magnetic ions are not situated on the inversion centers which are present and ME effects have in fact been observed in some of these materials. Of the simple sesquioxides, only Cr_2O_3 is ME active, since in the magnetic structures of Fe_2O_3 and V_2O_3, the inversion symmetry is retained. Ti_2O_3 has not been found to order magnetically,[43] so that an explanation of the ME behavior previously reported[44] must be sought elsewhere. By appropriate cation substitution, the chemically ordered ilmenite type compounds can be obtained. Most of these have doubled magnetic cells, but $MnTiO_3$ qualifies as an ME candidate with $R\bar{3}'$ symmetry. A different kind of chemical order is found in compounds of the type $Mn_4Nb_2O_9$, which were first identified by Hornreich[6] as potential ME compounds, and an effect has since been discovered.[53] Data for the corundum group are listed in Table V.

Another group of well-characterized ME materials have the substituted olivine or triphylite type of structure. In the parent compound Fe_2SiO_4, the space group is *Pbnm* and the iron atoms occupy both 4(a) and 4(c) sites with $\bar{1}$ and *m* symmetry respectively. In the magnetic state the former is retained and an ME effect is ruled out. However, the diamagnetic Li ion can be substituted into the 4(a) sites, the coupling scheme is changed, and the resulting magnetic structure permits an ME effect. Data are summarized in Table IV.

The rare-earth manganites $RMnO_3$ are an interesting series of compounds for which neutron data but not ME data exist.[62–64] They are ferroelectric,[65] and below the ferroelectric transition temperature have the hexagonal space group $P6_3cm$, with the Mn ions in 6(c) positions forming a simple hexagonal net. In the magnetic state, the moments order in a simple triangular arrangement in the basal plane, but powder neutron data do not permit a choice to be made between two possible models which differ as to the relative orientation of moments in adjacent planes, corresponding to symmetry operators 6_3 and $6'_3$ respectively. However, the absolute orientation of the

TABLE V

Magnetic structural data for corundum type compounds. ME Type AA III.

Compound	Space group	T_N(°K)	Spin direction	Magnetic space group	Magnetic point group	Form of ME tensor	Maximum α_{obs}	Reference
Cr_2O_3	$R\bar{3}c$	318	[111]	$R\bar{3}'c'$	$\bar{3}'m'$	$\alpha_{11} = \alpha_{22}, \alpha_{33}$	8×10^{-5}	45–49, 70, 71
$MnTiO_3$	$R\bar{3}$	64	[111]	$R\bar{3}'$	$\bar{3}'$	$\alpha_{11} = \alpha_{22}, \alpha_{33}$ $\alpha_{12} = -\alpha_{21}$	—	50, 51
$Nb_2Co_4O_9$	$P\bar{3}c1$	27	[001]	$P\bar{3}'c'1$	$\bar{3}'m'$	$\alpha_{11} = \alpha_{22}, \alpha_{33}$	2×10^{-5}	52, 53
$Nb_2Mn_4O_9$	$P\bar{3}c1$	110	[001]	$P\bar{3}'c'1$	$\bar{3}'m'$	$\alpha_{11} = \alpha_{22}, \alpha_{33}$	2×10^{-6}	52, 53
$Ta_2Co_4O_9$	$P\bar{3}c1$	21	[001]	$P\bar{3}'c'1$	$\bar{3}'m'$	$\alpha_{11} = \alpha_{22}, \alpha_{33}$	1×10^{-4}	53
$Ta_2Mn_4O_9$	$P\bar{3}c1$	104	[001]	$P\bar{3}'c'1$	$\bar{3}'m'$	$\alpha_{11} = \alpha_{22}, \alpha_{33}$	1×10^{-5}	53

TABLE VI

Magnetic structural data for triphylite type compounds. Space group *Pbnm*, magnetic ions in 4(c) positions. ME type AA III.

Compound	T_N(°K)	Spin direction	Magnetic space group	Magnetic point group	Form of ME tensor	Maximum α_{obs}	Reference
$LiCoPO_4$	23	[001]	*Pb'nm*	*m'mm*	α_{23}, α_{32}	6×10^{-4}	54–56
$LiFePO_4$	50	[001]	*Pb'nm*	*m'mm*	α_{23}, α_{32}	1×10^{-4}	57, 58
$LiMnPO_4$	35	[010]	*Pb'n'm'*	*m'm'm'*	$\alpha_{11}, \alpha_{22}, \alpha_{33}$	2×10^{-5}	55, 56, 58, 60
$LiNiPO_4$	23	[100]	*Pbnm'*	*mmm'*	α_{12}, α_{21}	4×10^{-5}	54, 55, 61

moments with respect to the hexagonal axes within a given plane can be determined. These differ by 90° depending on which of the two models is chosen, and in the case of $HoMnO_3$ and $YMnO_3$ the possible symmetries are $P6_3cm$ or $P6_3'cm'$†. In the other compounds, some intermediate basal plane direction is observed, and the symmetry is accordingly lower, $P6_3$ or $P6_3'$. The first of the two possibilities in each case would allow a linear ME term, but not the latter. In addition, in $HoMnO_3$, there is a spin reorientation at about 50°K involving a change in symmetry. ME measurements could greatly clarify the situation in these materials.

The zircon structure, typified by $DyPO_4$, offers a wide range of compounds of potential interest. These include a great many of the combinations of a rare-earth ion with one of the group V ions P, As, or V. The rare-earth ions have site symmetries $\bar{4}2m$ and in the compounds studied to date (Table VII), there is simple antiparallel coupling of nearest-neighbor moments. Of particular interest is $TbPO_4$, which has the highest ME susceptibility yet observed.[66] However, some compounds of this type have an anticentered magnetic lattice in which no ME effect is to be expected.[73]

† In neither case is a weak ferromagnetic component permitted. The presence of a weak spontaneous moment reported in Ref. 62 was later attributed to impurities.[64]

TABLE VII

Magnetic Structural data for zircon type compounds. Space group $I4_1/amd$, magnetic ions in 4(a) positions. ME Type AA III.

Compound	T_N	Spin Direction	Magnetic Space Group	Magnetic Point Group	Form of ME Tensor	Maximum α_{obs}	Reference
$DyPO_4$	3.4	[001]	$I4_1'/a'm'd$	$4'/m'm'm$	$\alpha_{11} = -\alpha_{22}$	1×10^{-3}	68, 69
$GdVO_4$	2.4	[001]	$I4_1'/a'm'd$	$4'/m'm'm$	$\alpha_{11} = -\alpha_{22}$	3×10^{-4}	70
$HoPO_4$	1.4	[001]	$I4_1'/a'm'd$	$4'/m'm'm$	$\alpha_{11} = -\alpha_{22}$	not stated	72
$TbPO_4$	2.2	$\perp[1\bar{1}0]$	–	$2/m'$?	$\alpha_{11}, \alpha_{22}, \alpha_{33}$ α_{12}, α_{21}	1×10^{-2}	66, 67

TABLE VIII

Magnetic structural data for miscellaneous ferromagnetoelectric compounds. ME type FE I/FM I.

Compound	Space group	T_C(°K)	Spin direction	Magnetic space group	Magnetic point group	Form of ME tensor	Maximum α_{obs}	Reference
$Ni_3B_7O_{13}I$	$Pca2_1$[a]	60[b]	?	$Pc'a2_1'$	$m'm2'$	α_{23}, α_{32}	3×10^{-4}	74, 75, 77–79
$Co_3B_7O_{13}Cl$	$R3c$[c]	22[b]	?	$Bb(?)$	m	$\alpha_{12}, \alpha_{21}, \alpha_{23}, \alpha_{32}$	not stated	75–77
$Ni_3B_7O_{13}Cl$	$Pca2_1$[a]	20[b]	?	$Pc'a2'_1$	$m'm2'$	α_{23}, α_{32}	not stated	75, 77, 80
β-$NaFeO_2$	$Pna2_1$	723[b]	[001]	$Pn'a2_1'$	$m'm2'$	α_{23}, α_{32}	–	6, 81, 82
$MnGeN_2$	$Pna2_1$	448[b]	[001]	$Pn'a2_1'$	$m'm2'$	α_{23}, α_{32}	–	83
$FeGaO_3$[d]	$Pc2_1n$	305[e]	[010]	$Pc'2_1'n$	$m'2'm$	α_{23}, α_{32}	4×10^{-4}	84–86
Na_2NiFeF_7	$Imm2$	88[e]	[100]	$Imm'2'$	$mm'2$	α_{13}, α_{31}	–	87

[a] At high temperature, transforms to $F\bar{4}3c$.
[b] Weak ferromagnetism observed.
[c] At high temperature, transforms to $Pca2_1$, via a monoclinic structure, then to $F\bar{4}3c$.
[d] Fe:Ga ratio can vary.
[e] Ferrimagnetic.

TABLE IX

Magnetic structural data for miscellaneous antiferromagnetoelectric compounds. ME Type AA III.

Compound	Space group	T_N(°K)	Spin direction	Magnetic space group	Magnetic point group	Form of ME tensor	Maximum α_{obs}	Reference
$CoGeO_3$	*Pbca*	31	–	–	*m'mm*	α_{23}, α_{32}	1×10^{-4}	70
$CoCs_3Cl_5$	*Pbcn*	0.5	[001]	*I*4'/*m'cm'*	4'/*m'mm'*	$\alpha_{11} = -\alpha_{22}$	–	88
$CrTiNdO_5$	*Pbam*	13	–	*Pbam'*	*mmm'*	α_{12}, α_{21}	1×10^{-5}	70, 89
$CrUO_4$	*Pbcn*	–	[010]	*Pbc'm*	*mm'm*	α_{13}, α_{31}	–	90
$LiCuCl_3 . 2H_2O$	$P2_1/c$	4.5	⊥[010]	$P2_1'/c$	2'/*m*	$\alpha_{11}, \alpha_{22}, \alpha_{23}, \alpha_{32}$	–	91
DyOOH	$P2_1/m$	7.2	⊥[010]	$P2_1/m'$	2/*m'*	$\alpha_{11}, \alpha_{22}, \alpha_{33}, \alpha_{12}, \alpha_{21}$	1×10^{-4}	92, 114
ErOOH	$P2_1/m$	4.1	[010]	$P2_1'/m$	2'/*m*	$\alpha_{13}, \alpha_{31}\alpha_{23}, \alpha_{32}$	5×10^{-4}	93, 114
$FeGeO_3$	*Pbca*	14	–	–	–	–	1×10^{-5}	70
α-FeOOH	*Pnma*	403	[010]	*Pnma'*	*mmm'*	α_{12}, α_{21}	–	94, 95
γ-FeOOH	*Cmcm*	75	⊥[001]	$P2_1/m'$	2/*m'*	$\alpha_{11}, \alpha_{22}, \alpha_{33}, \alpha_{12}, \alpha_{21}$	–	96
$Ca(Fe_{1-x}Cr_x)_2O_4$	*Pnam*	–	[001]	*Pna'm*	*mm'm*	α_{13}, α_{31}	–	97
$MnGeO_3$	*Pbca*	16	[010]	*Pb'ca*	*m'mm*	α_{23}, α_{32}	2×10^{-6}	98–100
$MnNb_2O_6$	*Pbcn*	4.4	[100]	*Pb'cn*[a]	*m'mm*	α_{23}, α_{32}	3×10^{-6}	101, 102
TbOOH	$P2_1/m$	10.0	–	–	–	–	4×10^{-4}	114
UOTe	*P*4/*nmm*	157	[001]	*P*4/*n'm'm'*	4/*m'm'm'*	$\alpha_{11} = \alpha_{22}, \alpha_{33}$	–	103, 104

[a] Neutron data favor *P*2'/*c*

Other compounds in which ME effects have been found, or on the basis of available neutron diffraction evidence should exist, are listed in the remaining tables. Tables VIII and IX contain data for FE I/FM I and AA III type materials, and Table X lists the few which fall into other categories. Most of these compounds have rather low-symmetry structures not representative of any widely studied group. An exception is the family of ferromagnetoelectric boracites $M_3B_7O_{13}X$, where M is a divalent transition metal ion and X a halide ion, which have been studied extensively by Schmid, Ascher, and co-workers.[74–77] In these, the cations are in "mixed" octahedral coordination consisting of two halogen and four oxygen anions, the latter being nearly coplanar. The compounds undergo a ferroelectric transition from $F\bar{4}3c$ to $Pca2_1$ symmetry at various temperatures. At still lower temperatures, there is an antiferromagnetic transition which is in most cases accompanied by the appearance of weak ferromagnetism, and the magnetic point group $m'm2'$ has been assigned. In some of the compounds there is an intermediate transition from $Pca2_1$ via a monoclinic structure to $R3c$ symmetry,[4] the latter also being ferroelectric. An ME effect has been observed in the Co material, and based on this and electrical and magnetic measurements it has been concluded that the magnetic point symmetry is *m*. However, as far as is known, no neutron diffraction investigations have

TABLE X

Magnetic structural data for miscellaneous compounds, various ME Types.

Compound	Type	Space group	T_N(°K)	Spin direction	Magnetic space group	Magnetic point group	Form of ME tensor	Maximum α_{obs}	Reference
$BiFeO_3$	FE II	*R*3*c*[a]	375	[111]?	*R*3*c*	3*m*	$\alpha_{12} = -\alpha_{21}$	–	20–22
$FeSb_2O_4$	FE II	$P4_2/mbc$	46	–	$Pmc2_1$	*mm*2	α_{12}, α_{21}	not stated	105–107
FeS	AA I	$P\bar{6}2c$[b]	600	[001]	$P\bar{6}'2c'$	$\bar{6}'m'2$	$\alpha_{11} = \alpha_{22}, \alpha_{33}$	–	108–109
$CuFeS_2$	AA I	$I\bar{4}2d$	815	[001]	$I\bar{4}2d$	$\bar{4}2m$	$\alpha_{11} = -\alpha_{22}$	–	110
$Li_{0.5}Fe_{2.5}O_4$	FM II	$P4_332$	923[c]	[111]	–	32'(?)	$\alpha_{12} = -\alpha_{21}$	–	40–42
$MnNb_3S_6$	FM II	$P6_322$	33[d]	⊥[001]	$C22'2_1'$?	22'2'	α_{23}, α_{32}	–	111

[a] At 850°K transforms to *Pm*3*m*.

[b] At 458°K transforms to $P6_3/mmc$.

[c] Ferrimagnetic.

[d] Ferromagnetic.

been carried out on any of these substances, although Heinrich and Zitkova have suggested a model for the magnetic structure of Ni–I boracite based on negative exchange via the iodine ions which is consistent with the symmetry.[78]

A neutron diffraction investigation of the $Ca(Fe_{1-x}Cr_x)_2O_4$ system has revealed very unusual behavior.[97] These compounds have orthorhombic symmetry *Pnam*, and for $x \leqslant$ *ca.* 15%, two distinct magnetic phases appear to coexist over at least part of the temperature range. One of these should be magnetoelectric, the other not, and ME measurements should throw further light on the magnetic behavior of this system.

The sulfides listed in Table X have rather unusual magnetic symmetry, and the ME behavior would be of interest. However, it is doubtful whether their resistivities will be high enough for measurements to be made.

Finally, it should be mentioned that while an ME effect has been reported in $BaCoF_4$,[112] the magnetic symmetry determined in a subsequent neutron study appears to be incompatible with this.[113]

ACKNOWLEDGMENTS

I would like to thank Dr. R. M. Hornreich for making the results of some unpublished work available.

REFERENCES

1. A. Oles, A. Bombik, M. Kucab, W. Sikora and F. Kajzar, "Tables of Magnetic structures Determined by Neutron Diffraction," Reports 1/PS, 4/PS, 7/PS, 8/PS, 11/PS, 12/PS, 24/PS, Institute of Nuclear Techniques, Cracow (1970–2).
2. D. E. Cox, "Table of Antiferromagnetic Materials Studied by Neutron Diffraction," Report BNL 13822, Brookhaven National Laboratory, Upton, New York (1969). See also *IEE Transactions on Magnetics* **MAG 8**, 161 and 798 (1972).
3. M. Kucab, "Magnetic Space Groups of Magnetic Structures Determined by Neutron Diffraction," Report 25/PS, Institute of Nuclear Techniques, Cracow (1972).
4. H. Schmid, *Int. J. Magn.* **4**, 337 (1973).
5. E. F. Bertaut and M. Mercier, *Mat. Res. Bull.* **6**, 907 (1971).
6. R. M. Hornreich, *Solid State Commun.* 7, 1081 (1969). *IEEE Transactions on Magnetics* **MAG 8**, 584 (1972).
7. W. Kunnmann, S. LaPlaca, L. M. Corliss and J. M. Hastings, *J. Phys. Chem. Solids* **29**, 1359 (1968).
8. M. C. Montmory, M. Belakhovsky, R. Chevalier and R. Newnham, *Solid State Commun.* **6**, 317 (1968).
9. S. Buksphan, E. Fischer and R. M. Hornreich, *Solid State Commun.* **10**, 657 (1972).
10. M. C. Montmory and R. Newnham, *Solid State Commun.* **6**, 323 (1968).
11. R. Bidaux and P. Meriel, *J. de Phys.* **29**, 220 (1968).
12. M. Mercier and P. Bauer, "Les Elements des Terres Rares," Colloques Internationaux du CNRS, No. 180, Vol. 2, 377, Paris (1970).
13. L. M. Holmes, L. G. Van Uitert and G. W. Hull, *Solid State Commun.* **9**, 1373 (1971).
14. M. Mercier and G. Velleaud, *J. de Phys.* **32C**, 499 (1971).
15. J. Bielen, J. Mareschal and J. Sivardiere, *Z. Angew. Phys.* **23**, 243 (1967).
16. J. Mareschal, J. Sivardiere, G. F. DeVries and E. F. Bertaut, *J. Appl. Phys.* **39**, 1364 (1968).
17. M. Mercier and B. Cursoux, *Solid State Commun.* **6**, 207 (1968).
18. A. Kappatsch, S. Quezel-Ambrunaz and J. Sivardiere, *J. de Phys.* **31**, 369 (1970).
19. J. Sivardiere and S. Quezel-Ambrunaz, *Compt. Rend.* **B273**, 619 (1971).
20. C. Michel, J. Moreau, G. D. Achenbach, R. Gerson and W. J. James, *Solid State Commun.* **7**, 701 (1969).
21. J. R. Teague, R. Gerson and W. J. James, *Solid State Commun.* **8**, 1073 (1970).
22. Yu. N. Venevtsev, V. N. Lyubimov, V. V. Ivanova and G. S. Zhadanov, *J. de Phys.* **C2**, 255 (1972).
23. G. A. Smolensky, V. A. Bokov, V. A. Isupov, N. N. Krainik, R. E. Pas'inkov and M. S. Shur, *Segnetoelektriki i antisegnetoelektriki 'Izdatel'* (Stvo "Nauka," Leningrad, 1971).
24. F. S. Galasso, *Structure, Properties, and Preparation of Perovskite-Type Compounds* (Pergamon Press, New York, 1969).
25. G. Blasse, *J. Inorg. Nucl. Chem.* **27**, 993 (1965).
26. D. N. Astrov, B. I. Al'shin, R. V. Zorin and L. A. Drobyshev, *Zh. Eksp. Teor. Fiz.* **55**, 2122 (1968). English transl.: *Soviet Phys.–JETP* **28**, 1123 (1969).
27. G. M. Drabkin, E. I. Mal'tsev and V. P. Plakhtii, *Fiz. Tverd. Tela* **7**, 1241 (1965). English transl.: *Soviet Phys.–Solid State* **7**, 997 (1965).
28. G. Blasse, Proceedings of the International Conference on Magnetism, Nottingham, 350 (1964).
29. D. E. Cox, G. Shirane and B. C. Frazer, *J. Appl. Phys.* **38**, 1459 (1967).
30. W. L. Roth, *J. Phys. Chem. Solids* **25**, 1 (1964).
31. B. Boucher and A. Oles, *J. de Phys.* **27**, 632 (1966).
32. W. L. Roth, *J. de Phys.* **25**, 507 (1964).
33. E. Prince, *Acta Cryst.* **10**, 554 (1957).
34. R. Nathans, S. J. Pickart and A. Miller, *Bull. Amer. Phys. Soc.* **6**, 54 (1961).
35. P. Meriel, CEN Saclay DOC/714/DD (1968); cited in Ref. 3.
36. E. Prince, *J. Appl. Phys.* **32**, 68S (1961).
37. E. F. Bertaut and J. Dulac, *Acta Cryst.* **A28**, 580 (1972).
38. B. Boucher, R. Buhl and M. Perrin, *J. Appl. Phys.* **39**, 632 (1968).
39. B. Boucher, R. Ruhl and M. Perrin, *J. Phys. Chem. Solids* **32**, 1471 (1971).
40. P. B. Braun, *Nature* **170**, 1123 (1952).
41. E. Prince, *J. de Phys.* **25**, 503 (1965).
42. G. A. Petrakovskii, V. N. Seleznev, K. A. Sablina and L. M. Protopopova, *Fig. Tverd. Tela* **11**, 11 (1969). English transl.: *Soviet Phys.–Solid State* **11**, 7 (1969).

43. R. M. Moon, T. Riste, W. C. Koehler and S. C. Abrahams, *J. Appl. Phys.* **40**, 1445 (1969).
44. B. I. Al'shin and D. N. Astrov, *Zh. Eksp. Teor. Fiz.* **44**, 1195 (1963). English transl.: *Soviet Phys.–JETP* **17**, 809 (1963).
45. B. N. Brockhouse, *J. Chem. Phys.* **21**, 961 (1953).
46. L. M. Corliss, J. M. Hastings, R. Nathans and G. Shirane, *J. Appl. Phys.* **36**, 1099 (1965).
47. D. N. Astrov, *Zh. Eksp. Teor. Fiz.* **38**, 984 (1960). English transl.: *Soviet Phys.–JETP* **11**, 708 (1960).
48. G. T. Rado and V. J. Folen, *J. Appl. Phys.* **33**, 1126 (1962).
49. T. J. Martin and J. C. Anderson, *Phys. Letters* **2**, 109 (1964).
50. G. Shirane, S. J. Pickart and Y. Ishikawa, *J. Phys. Soc. Japan* **14**, 1352 (1959).
51. J. Akimitsu, Y. Ishikawa and Y. Endoh, *Solid State Commun.* **8**, 87 (1970).
52. E. F. Bertaut, L. M. Corliss, F. Forrat, R. Aleonard and R. Pauthenet, *J. Phys. Chem. Solids* **21**, 234 (1961).
53. E. Fischer, G. Gorodetsky and R. M. Hornreich, *Solid State Commun.* **10**, 1127 (1972).
54. R. P. Santoro, D. J. Segal and R. E. Newnham, *J. Phys. Chem. Solids* **27**, 1192 (1966).
55. M. Mercier, J. Gareyte and E. F. Bertaut, *Compt. Rend.* **B264**, 979 (1967).
56. M. Mercier, J. Gareyte and B. Fouilleux, *Solid State Commun.* **5**, 139 (1967).
57. R. P. Santoro and R. E. Newnham, *Acta Cryst.* **22**, 344 (1967).
58. M. Mercier, P. Bauer and B. Fouilleux, *Compt. Rend.* **B267**, 1345 (1968).
59. R. E. Newnham, R. P. Santoro and M. J. Redman, *J. Phys. Chem. Solids* **26**, 445 (1965).
60. M. Mercier, E. F. Bertaut, G. Quezel and P. Bauer, *Solid State Commun.* **7**, 149 (1969).
61. M. Mercier and P. Bauer, *Compt. Rend.* **B267**, 465 (1968).
62. E. F. Bertaut, M. Mercier and R. Pauthenet, *J. de Phys.* **25**, 550 (1964).
63. W. C. Koehler, H. L. Yakel, E. O. Wollan and J. W. Cable, *Phys. Letters* **9**, 93 (1964). Proceedings of the Rare Earth Conference, Phoenix, 63 (1964).
64. E. F. Bertaut, R. Pauthenet and M. Mercier, *Phys. Letters* **18**, 13 (1965).
65. P. Coeure, P. Guinet, J. C. Penzin, G. Buisson and E. F. Bertaut, Proceedings of the International Meeting on Ferroelectricity, Prague, 332 (1966).
66. G. T. Rado and J. M. Ferrari, *AIP Conference Proceedings* **10**, 1417 (1972).
67. S. Spooner, J. N. Lee and H. W. Moos, *Solid State Commun.* **9**, 1143 (1971).
68. W. Scharenberg and G. Will, *Int. J. Magnetism* **1**, 277 (1971).
69. G. T. Rado, *Phys. Rev. Lett.* **23**, 644 (1969); *Solid State Commun.* **8**, 1349 (1970).
70. R. M. Hornreich, *Int. J. Magn.* **4**, 321 (1973).
71. T. H. O'Dell, *IEEE Transactions on Magnetics* **MAG 2**, 449 (1966).
72. A. H. Cooke, S. J. Swithenby and M. R. Wells, *Int. J. Magn.* **4**, 309 (1973).
73. W. Schafer and G. Will, *J. Phys. C.* **4**, 3224 (1971).
74. E. Ascher, H. Rieder, H. Schmid and H. Stossel, *J. Appl. Phys.* **37**, 1404 (1966).
75. C. Quezel and H. Schmid, *Solid State Commun.* **6**, 447 (1968).
76. H. Schmid, *Phys. Stat. Solidi* **37**, 209 (1970).
77. E. Ascher, *J. Phys. Soc. Japan Suppl.* **28**, 7 (1970).
78. B. Heinrich and J. Zitkova, *Czech. J. Phys.* **B19**, 48 (1969).
79. B. I. Al'shin, D. N. Astrov and Yu. M. Gufan, *Fiz. Tverd. Tela* **12**, 2666 (1970). English transl.: *Soviet Phys.–Solid State* **12**, 2143 (1971).
80. J. P. Rivera, J. M. Moret and H. Schmid, unpublished work; cited in Ref. 4.
81. E. F. Bertaut, A. Delapalme and G. Bassi, *J. de Phys.* **25**, 545 (1964).
82. E. F. Bertaut, *Helv. Phys. Acta* **41**, 683 (1968).
83. M. Wintenberger, *Solid State Commun.* **11**, 1485 (1972).
84. E. F. Bertaut, G. Bassi, G. Buisson, J. Chappert, A. Delapalme, R. Pauthenet, H. P. Rebouillat and R. Aleonard, *J. de Phys.* **27**, 433 (1966).
85. G. T. Rado, *Phys. Rev. Lett.* **13**, 335 (1964); *J. Appl. Phys.* **37**, 1403 (1966).
86. A. Delapalme, *J. Phys. Chem. Solids* **28**, 1451 (1967).
87. G. Heger and R. Viebahn-Hänsler, *Solid State Commun.* **11**, 1119 (1972).
88. J. Hammann, *Physica* **43**, 277 (1969).
89. G. Buisson, *J. Phys. Chem. Solids* **2831**, 1171 (1970).
90. M. Bacmann, E. F. Bertaut and G. Bassi, *Bull. Soc. Franc. Mineral. Crist.* **88**, 214 (1965).
91. S. C. Abrahams, *J. Chem. Phys.* **39**, 2923 (1963).
92. A. N. Christensen, S. Quezel and M. Belakhovsky, *Solid State Commun.* **9**, 925 (1971).
93. A. N. Christensen and S. Quezel, *Solid State Commun.* **10**, 765 (1972).
94. J. B. Forsyth, I. G. Hedley and C. E. Johnson, *J. Phys. C* **1**, 179 (1968).
95. A. Szytula, A. Burewicz, Z. Dimitrijevic, S. Krasnicki, H. Rzany, J. Todorovic, A. Wanic and W. Wolski, *Phys. Stat. Solidi* **26**, 429 (1968).
96. A. Oles, A. Szytula and A. Wanic, *Phys. Stat. Solidi* **41**, 173 (1970). Space group given in Refs. 1 and 3.
97. L. M. Corliss, J. M. Hastings and W. Kunnmann, *Phys. Rev.* **160**, 408 (1967).
98. P. Herpin, A. Whuler, B. Boucher and M. Sougi, *Phys. State. Solidi* **b44**, 71 (1971).
99. L. M. Holmes and L. G. Van Uitert, *Solid State Commun.* **10**, 853 (1972).
100. G. Gorodetsky, R. M. Hornreich and B. Sharon, *Phys. Letters* **39A**, 155 (1972).
101. H. Weitzel, *Zeit. Anorg. Allgem. Chem.* **380**, 119 (1971).
102. L. M. Holmes, A. A. Ballman and R. R. Hecker, *Solid State Commun.* **11**, 409 (1972).
103. A. Murasik and J. Niemiec. *Bull. Acad. Pol. Sic., Ser. Sci. Chim.* **13**, 291 (1965).
104. J. Przystawa, *Phys. Stat. Solidi* **30**, K115 (1968).
105. J. A. Gonzalo, D. E. Cox and G. Shirane, *Phys. Rev.* **147**, 415 (1966).
106. F. Varret, P. Imbert, A. Gerard and F. Hartmann-Boutron, *Solid State Commun.* **6**, 889 (1968).
107. G. Gorodetsky, M. Sayer and S. Shtrikman, *Mat. Res. Bull.* **5**, 253 (1970).
108. A. F. Andresen, *Acta Chem. Scand.* **14**, 919 (1960).
109. A. F. Andresen and P. Torbo, *Acta Chem. Scand.* **21**, 2841 (1967).

110. G. Donnay, L. M. Corliss, J. D. H. Donnay, N. Elliott and J. M. Hastings *Phys. Rev.* **112**, 1917 (1958).
111. B. Van Laar, H. M. Rietveld and D. J. W. Ijdo, *J. Solid State Chem.* **3**, 154 (1971).
112. B. I. Al'shin, D. N. Astrov, A. V. Tishchenko and S. V. Petrov, *ZhETF Pis. Red.* **12**, 206 (1970). English transl.: *JETP Lett.* **12**, 142 (1970).
113. M. Eibschütz, L. Holmes, H. J. Guggenheim and D. E. Cox, *Phys. Rev.* **B6**, 2677 (1972).
114. A. N. Christensen, R. M. Hornreich and B. Sharon, private communication.
115. G. Lee, M. Mercier and P. Bauer, "Les Elements des Terres Rares," Colloques Internationaux du CNRS, no. 180, Vol. 2, 389, Paris (1970).

Discussion

Dr. T. P. SRINIVASAN Could you comment on some of the antiferromagnetic rare-earth chalcogenides?

Dr. D. E. COX Most of those which have been studied by neutrons have the rock-salt type of structure and would not be expected to show magnetoelectricity.

Dr. V. J. FOLEN In view of the reported observation of a linear ME effect in $BaCoF_4$, could there exist an ambiguity in the neutron diffraction determination of the magnetic structure of this material?

Dr. D. E. COX I think the neutron analysis is straightforward enough in this case, although there is the complication of two coexisting magnetic phases. However, both of these involve doubling of at least one axis, giving one of the colored magnetic lattices, which would appear to preclude an ME effect. Of course, there is the possibility of a crystallographic distortion and a lowering of the symmetry, but this was not reflected in any significant change in the nuclear diffraction peaks.

Dr. R. M. HORNREICH I should like to make the following comments:

1) Regarding the trirutiles in Table I ordering in the basal plane–if powder specimens of these compounds are used for ME measurements, it will be necessary to take into account the four or more possible antiferromagnetic domain configurations.

2) Regarding the ferromagnetic materials of Table IV and also those of Table VIII that have only off-diagonal ME tensor elements in the plane perpendicular to the magnetic moment, it will not be possible to observe an ME effect in powder specimens of these materials as the net effect will always to zero.

3) Other potentially ME materials are

a) $Cu_{0.5}In_{0.5}Cr_2S_4$ (R. Plumier, F. K. Lotgering, and R. P. van Stapele, *J. de Phys.* **C32**, 324 (1971).

b) $MnTa_2O_6$ (H. Weitzel and S. Klein, *Solid State Commun.* **12**, 113 (1973)) which has the same magnetic structure as $MnNb_2O_6$.

4) Our group has searched for ME effects in $GdCrO_3$, $GdFeO_3$, and $DyFeO_3$ below the rare-earth ordering temperature, but with negative results.

W. OPECHOWSKI *Comment:* In his survey of magnetic structures, Dr. Cox has assumed that only those can exhibit magnetoelectric effect whose symmetry is described by magnetic space groups with no primed translations. This assumption can in fact be justified by a very plausible physical argument concerning the relation between the symmetry at atomic level (magnetic space groups) and the symmetry at macroscopic level (magnetic point groups). However, one could imagine that the argument is incorrect and for that reason it would be of interest to check for magnetoelectric effect a few magnetically ordered substances whose magnetic space groups do contain primed translations.

G. T. RADO *Comment:* Your table of magnetoelectric susceptibilities $\overleftrightarrow{\alpha}$ does not contain a value for $TbPO_4$. As reported recently [G. T. Rado and J. M. Ferrari, in *Magnetism and Magnetic Materials–1972*, AIP Conference Proceedings No. 10, edited by C. D. Graham, Jr., and J. J. Rhyne (American Institute of Physics, New York, 1973), p. 1417] we measured the value $|\alpha_{aa}|_{max} = |P_a/H_a|_{max} = 1.2 \times 10^{-2}$ Gaussian units for a component of $\overleftrightarrow{\alpha}$ in a magnetoelectrically annealed $TbPO_4$ crystal whose symmetry is still under investigation. This value is considerably larger than any component of $\overleftrightarrow{\alpha}$ previously reported for any material.

ON A MAGNETOELECTRIC CLASSIFICATION OF MATERIALS†‡

HANS SCHMID

Battelle, Geneva Research Centre, 1227 Carouge-Geneva, Switzerland

Among the 122 Shubnikov point groups, 13 "magnetoelectric types" have been set off. Each "type" except one is described by one or several characteristic terms permitted by symmetry in the Taylor expansion of the stored free enthalpy. Linear and second-order magnetoelectric effects and related crystal properties are considered. For all groups, the permitted magnetic and electric ordering is indicated. The groups permitting "weak ferroelectricity" are obtained by analogy with those permitting "weak ferromagnetism" due to a symmetric arrangement of groups and types. Various kinds of domain switching are discussed; piezoelectric and second-order magnetoelectric switching is proposed. Modulation of various coercive fields by magnetoelectric and other means is discussed. The importance of ferroelasticity for magnetoelectric research is pointed out. Several selected topics of examples of compounds, representing particular magnetoelectric types, are discussed. Accent is put on ferromagnetoelectric compositions. Extended tables on materials of various magnetoelectric types are presented.

1 INTRODUCTION

Following the theoretical prediction by Landau and Lifshitz (1956)[1] of the linear magnetoelectric effect in crystals, Dzyaloshinsky[2] predicted it to exist in antiferromagnetic Cr_2O_3, and shortly afterwards Astrov[3] and Folen, Rado and Stalder[4,5] confirmed the phenomenon experimentally. In the subsequent years, various other magnetoelectric compounds were found,[6,7] and it was realized that the magnetoelectric effect is a quite common phenomenon. Parallel to this, the search for materials with co-existing magnetic and electric ordering was pursued at various places. The original impetus to this line of research is doubtlessly due to Smolensky and Joffe,[8] who announced at the 1958 Grenoble Conference on Magnetism the synthesis of antiferromagnetic ferroelectric perovskites. In the beginning, there was probably the intuitive feeling that magnetoelectric interactions in such compounds with co-existing magnetic and electric ordering would be strong and that interesting coupling between the magnetic and electric polarizations and antipolarizations would occur. We know today that these premises were indeed correct to a large extent.

With a view to facilitating and encouraging future research on linear and second-order magnetoelectric phenomena, to finding and developing new appropriate materials, and to conceiving possible technological applications, this paper tries to make an inventory of what Nature offers from the point of view of symmetry and structure.

First, we shall discuss a way of classifying magnetoelectrics on the basis of symmetry, and second, we shall discuss some examples of compounds pertaining to the various magnetoelectric types that we want to define.

2 SYMMETRY TYPES OF MAGNETO-ELECTRICS

2.1 *Scope of Classification*

We want to consider primarily linear and second-order magnetoelectric effects of crystals with the combinations of properties as shown in Figure 1.

In addition to these combinations of properties, every crystal is either ferroelastic, antiferroelastic, orthoelastic or paraelastic (see Section 2.5). Moreover, it may be stationary or in motion. The magnetoelectric classification that we propose pertains essentially to stationary crystals. For effects of motion, see Ref. 23.

2.2 *Principles of Classification*

The Shubnikov groups permitting the linear magnetoelectric effect (see e.g. Ref. 10) and the second-order magnetoelectric effects[11] as well as the corresponding tensors of the coefficients are well known. Certain tabulations of Shubnikov groups for crystals with particular property combinations

† Supported by Battelle Institute.

‡ Presented at the Symposium on Magnetoelectric Interaction Phenomena in Crystals, Seattle, Washington, May 21–24, 1973.

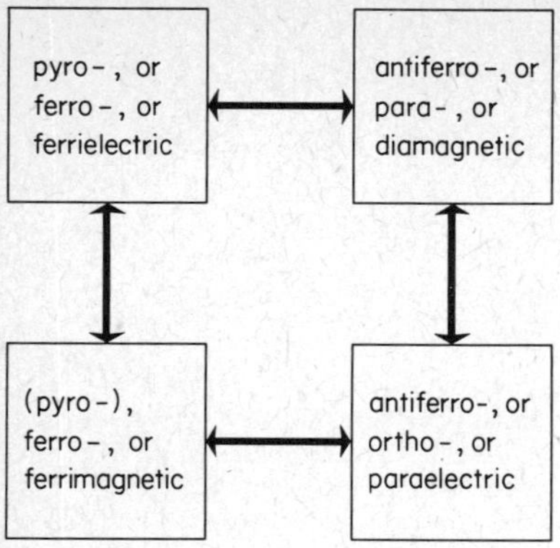

FIGURE 1 Frequently encountered magnetic-electric property combinations occurring in a material (one of the properties in a box combines with one of the properties in either of the adjacent boxes. For the definition of "paraelectric" and "orthoelectric" see Section 2.3.2.

have appeared in the literature.[12–14]† A useful one is that by Koptsik;[13] however, it does not contain the magnetoelectric properties. Because it may be useful for the experimentalist to obtain a rapid survey of the various possible combinations of magnetic and electric ordering and of the simultaneously permitted physical effects or properties among the 122 magnetic Shubnikov point groups, we have tried to set up a table from which the magnetoelectric (linear and second-order) and several other crystal properties can be perceived at one glance.

For this purpose, we have assembled certain Shubnikov point groups into "magnetoelectric types," each type having in common one or several terms permitted by symmetry in the Taylor expansion of the density of stored free enthalpy.

The density of stored "free enthalpy" g of a given (single domain) crystal phase is a convenient means of deriving the macroscopic symmetry-controlled properties of various crystallo-physical phenomena. It is a function dependent on the constraints of the system: temperature T, electric field E_i, magnetic field H_i and the tensor of mechanical stress T_{ij}:

$$g = g(T, E_i, H_i, T_{ij}) \tag{1}$$

The derivatives of g with respect to T, E_i, H_i and T_{ij} define the entropy, the polarization, the magnetization, and the tensor of the mechanical deformations, respectively:

$$-\frac{\partial g}{\partial T} = s \quad -\frac{\partial g}{\partial E_i} = P_i \quad -\frac{\partial g}{\partial H_i} = M_i \quad \frac{\partial g}{\partial T_{ij}} = s_{ij} \tag{2}$$

The function g is the free enthalpy g_0, from which the contributions of the electric and magnetic fields in the vacuum have been subtracted:

$$g = g_0 - \frac{\varepsilon_0}{2}E^2 - \frac{\mu_0}{2}H^2 \tag{3}$$

Here, ε_0 and μ_0 are the electric and magnetic permittivity of free space, respectively.

The Taylor expansion of g defines the vectorial and tensorial properties. In Table I we have written down (in vertical sequence) all of the terms of g by developing up to third order. A term of the stored free enthalpy is permitted by a point group, if it remains invariant under all the symmetry operations of the group.

The meaning of the terms and the names of the coefficients—also given in Table I—become immediately clear by calculating the various types of "deformation" (electrical, magnetic and mechanical):

$$P_i = -\frac{\partial g}{\partial E_i} = \ldots \kappa_i{}^0 + \alpha_{ij}H_j + \alpha_{ijk}H_jE_k + \tfrac{1}{2}\beta_{ijk}H_jH_k + d_{ijk}T_{jk} + \ldots \tag{4}$$

$$M_i = -\frac{\partial g}{\partial H_i} = \ldots \chi_i{}^0 + \alpha_{ik}E_k + \tfrac{1}{2}\alpha_{ijk}E_jE_k + \beta_{ijk}E_jH_k + g_{ijk}T_{jk} + \ldots \tag{5}$$

$$s_{ij} = \frac{\partial g}{\partial T_{ij}} = \ldots s_{ij}^0 + s_{ijkl}T_{kl} + d_{ijk}E_k + g_{ijk}H_k + \pi_{ijkl}E_kH_l + \tfrac{1}{2}\alpha_{ijkl}E_kE_l + \tfrac{1}{2}\beta_{ijkl}H_kH_l + \ldots \tag{6}$$

For example, Eq. (4) shows that the total polarization P_i may contain, in a more complicated case, a spontaneous polarization $\kappa_i{}^0$, a magnetic field induced polarization $\alpha_{ij}H_j$, second-order magnetoelectric polarizations $\alpha_{ijk}H_jE_k$ and $\frac{1}{2}\beta_{ijk}H_jH_k$, and a stress-induced piezoelectric polarization.

The corresponding expressions for the total magnetization M_i and the total deformation s_{ij} are given in (5) and (6), respectively.

For the purposes of classifying the 122 Shubnikov groups according to "magnetoelectric types,"

† For errata in these references, see Appendix.

TABLE I

Terms of the density of stored "free enthalpy" g and the meaning of the corresponding physical phenomena and properties. (*Legend:* E_i = electric field, H_i = magnetic field, T_{ij} (T_{kl}, T_{lm}) = stress tensor)

Terms of the density of stored "free enthalpy" $-g=$	Corresponding phenomena[a]	Name of coefficient
$\ldots + \kappa_i^0 E_i$	Pyro-, or ferro-, or ferrielectricity	Spontaneous polarization
$+\chi_i^0 H_i$	Pyro-, or ferro-, or ferrimagnetism	Spontaneous magnetization
$+s_{ij}^0 T_{ij}$	Pyro-, or ferroelasticity	Spontaneous deformation
$+\frac{1}{2}\kappa_{ik} E_i E_k$	Induced polarization; Brillouin, Raman, Raleigh scattering	Electric susceptibility
$+\frac{1}{2}\chi_{ik} H_i H_k$	Induced magnetization	Magnetic susceptibility
$+\frac{1}{2}s_{ijkl} T_{ij} T_{kl}$	Elasticity (Hooke's law, harmonic approximation)	Elastic compliance
$+\alpha_{ik} E_i H_k$	Magnetoelectric effect	Magnetoelectric susceptibility
$+d_{ijk} T_{ij} E_k$	Piezoelectricity	Piezoelectric coefficient
$+g_{ijk} T_{ij} H_k$	Piezomagnetism	Piezomagnetic coefficient
$+\frac{1}{6}\kappa_{ijk} E_i E_j E_k$	Electro-optic (Pockels) effect, optical electric rectification, frequency multiplication, optical mixing, hyper Raman effect, hyper Raleigh scattering	Non-linear electric susceptibility
$+\frac{1}{6}\chi_{ijk} H_i H_j H_k$	Magneto-optic ("Mockels")[15] effect, optic magnetic rectification, frequency multiplication, etc.	Non-linear magnetic susceptibility
$+\frac{1}{6}s_{ijklmn} T_{ij} T_{kl} T_{mn}$		"Third-order" elastic constants
$+\frac{1}{2}\alpha_{ijk} H_i E_j E_k$	Second-order magnetoelectric effect (I), optical rectification, frequency multiplication, etc.	First non-linear magnetoelectric susceptibility
$+\frac{1}{2}\beta_{ijk} E_i H_j H_k$	Second-order magnetoelectric effect (II), optical rectification, frequency multiplication, etc.	Second non-linear magnetoelectric susceptibility
$+\frac{1}{2}\alpha_{ijkl} T_{ij} E_k E_l$	Quadratic piezoelectric effect	Quadratic piezoelectric coefficient
$+\frac{1}{2}\beta_{ijkl} T_{ij} H_k H_l$	Quadratic piezomagnetic effect	Quadratic piezoelectric coefficient
$+\frac{1}{2}s'_{ijklm} T_{ij} T_{kl} E_m$		
$+\frac{1}{2}s''_{ijklm} T_{ij} T_{kl} H_m$		
$+\pi_{ijkl} T_{ij} E_k H_l$	Piezomagnetoelectricity[122, 123]	Piezomagnetoelectric coefficient

[a] For the sake of clarity, frequency is not specified for a subdivision of the effects. The interested reader may consult e.g. Ref. 16.

satisfying the various property combinations of practical interest given in Figure 1, it was found sufficient to use only the invariant terms *H*, *E*, *EH*, *HEE* and *EHH*.

These terms permit respectively pyro-, ferro- or ferrimagnetism ($\chi_i^0 H_i$)/pyro, ferro- or ferrielectricity ($\kappa_i^0 E_i$)/magnetoelectric effect ($\alpha_{ik} E_i H_k$)/second-order magnetoelectric effects I ($\frac{1}{2}\alpha_{ijk} H_j E_j E_k$)/ and II ($\frac{1}{2}\beta_{ijk} E_i H_j H_k$). By assembling those groups having in common the same term(s) or the absence of all five terms, we can distinguish 13 types of groups (see Table II).

Owing to the fact that the 66 Shubnikov groups allowing for the *HEE* term are identical to the groups permitting piezomagnetism, and those 66 groups allowing for the *EHH* term permit piezoelectricity,[11] the notation *EHH* and *HEE* used in Table II to describe the occurrence of the second-order magnetoelectric effects is synonymous with the permission of piezoelectricity and piezomagnetism, respectively. Moreover, because the piezoelectric and piezomagnetic point groups are the same as those permitting *EEE* and *HHH* terms, respectively, the permission of *EHH* and *HEE* also indicates the possibility of the linear electro-optic (Pockels) effect and the linear magneto-optic ("Mockels"[15]) effect, respectively. Other related effects which can be described by these terms are given in Table I.

2.3 *Comments on the Classification Table*

2.3.1 Arrangement of groups and magnetoelectric types (Table II) We have set off in particular the ferroelectric (*FE* I–IV), the ferromagnetic (*FM* I–IV), and the antiferromagnetic–antiferroelectric (*AA* I–VI) types.

TABLE II

Classification of the 122 Shubnikov groups according to "magnetoelectric types"

Magneto-electric type	Type of ordering: Magnetic	Type of ordering: Electric	Permitted terms of stored free enthalpy	Shubnikov point groups: V_s not permitted	Shubnikov point groups: V_s permitted	Number of Shubnikov groups
FE IV	*D*	*P*	*E* *EHH*	¦1′¦, ¦21′¦, ¦*m*1′¦, ¦*mm*21′¦, 41′, 4*mm*1′, 31′, 3*m*1′, 61′, 6*mm*1′		10
FE III	$\bar{M}$	*P*	*E* *HEE* *EHH*	6′, 6′*mm*′		2
FE II	$\bar{M}$	*P*	*E* *EH* *HEE* *EHH*	4′, 4′*mm*′	¦*mm*2¦, 4*mm*, 3*m*, 6*mm*	6
FE I/*FM* I	*M*	*P*	*E* *H* *EH* *HEE* *EHH*	¦[*m*′*m*′2]¦, 3*m*′, 4*m*′*m*′, 6*m*′*m*′	¦[1]¦, ¦[2]¦, 3, 4, 6, ¦[*m*]¦, ¦[2′]¦, ¦[*m*′]¦, ¦[*mm*′2′]¦	13 (*FE* IV to *FE* I: 31)
FM II	*M*	$\bar{P}$	*H* *EH* *HEE* *EHH*	$\bar{4}$, $\bar{4}2'm'$	[2′2′2], 42′2′, 32′, 62′2′	6
FM III	*M*	$\bar{P}$	*H* *HEE* *EHH*	$\bar{6}$, $\bar{6}m'2'$		2
FM IV	*M*	*O*	*H* *HEE*	[$\bar{1}$], [2/*m*], [2′/*m*′], [*m*′*m*′*m*], 4/*m*, 4/*mm*′*m*′, $\bar{3}$, $\bar{3}m'$, 6/*m*, 6/*mm*′*m*′		10 (*FM* I to *FM* IV: 31; *FE* IV to *FM* IV: 49)
AA I	$\bar{M}$	$\bar{P}$	*EH* *HEE* *EHH*	222, 422, $\bar{4}2m$, 4′22′, $\bar{4}'2m'$, 32, 622, $\bar{6}'m'2$, 23, $\bar{4}'3m'$	$\bar{4}'$, $\bar{4}'2'm$, $\bar{6}'$, $\bar{6}'m2'$	14
AA II	$\bar{M}$	$\bar{P}$	*HEE* *EHH*	$\bar{6}m2$, 6′2′2		2
AA III	$\bar{M}$	$\bar{P}$	*EH*	*m*′*m*′*m*′, 4′/*m*′, 4′/*m*′*m*′*m*, 4/*m*′*m*′*m*′, $\bar{3}'m'$, 6/*m*′*m*′*m*′, 432, *m*′3, *m*′3*m*′	$\bar{1}'$, 2/*m*′, 2′/*m*, *mmm*′, 4/*m*′, 4/*m*′*mm*, $\bar{3}'$, $\bar{3}'m$, 6/*m*′, 6/*m*′*mm*	19
AA IV	$\bar{M}$	*O*	*HEE*	*mmm*, 4′/*m*, 4*mmm*, 4′/*mmm*′, 6/*mmm*, $\bar{3}m$, 6′/*m*′, 6′/*m*′*m*′*m*, *m*3, *m*3*m*′		10
	$\bar{M}$	$\bar{P}$		4′32′		1
AA V	$\bar{M}$	$\bar{P}$	*EHH*	$\bar{4}3m$		1
	D	$\bar{P}$		2221′, $\bar{4}1'$, 4221′, $\bar{4}2m1'$, 6221′, 321′, $\bar{6}1'$, $\bar{6}m21'$, 231′, $\bar{4}3m1'$		10
AA VI	$\bar{M}$	*O*		*m*3*m*		1
	D	$\bar{P}$		4321′		1
	$\bar{M}$	$\bar{P}$		6′/*m*, 6′/*mm*′*m*, *m*′3*m*		3
	D	*O*		$\bar{1}1'$, 2/*m*1′, *mmm*1′, 4/*m*1′, 4/*mmm*1′, $\bar{3}1'$, $\bar{3}m1'$, 6/*m*1′, 6/*mmm*1′, *m*31′, *m*3*m*1′		11 (*AA* I to *AA* VI: 73; total: 122)

Legend: [ ] "Weak ferromagnetism" (Dzialoshinsky 1957)[17] permitted, corresponding Shubnikov groups determined by Tavger[18] for nearly uniaxial antiferromagnets.

¦ ¦ "Weak ferroelectricity" permitted.

Type of order: *M* = pyro-, ferro-, or ferrimagnetic; *P* = pyro-, ferro-, or ferrielectric; $\bar{M}$ = antiferromagnetic; $\bar{P}$ = antiferroelectric or orthoelectric;
D = diamagnetic, or paramagnetic, or antiferromagnetic; *O* = orthoelectric, or paraelectric, or antiferroelectric.
V_s = invariant velocity vector.[21,22]
H: spontaneous magnetization permitted; *E*: spontaneous polarization permitted; *EH*: linear magnetoelectric effect permitted; *EHH*: second-order magnetoelectric effect (I) ("paramagnetoelectric effect"),[19] piezoelectricity, Pockels effect, etc., permitted (see Table II); *HEE*: second-order magnetoelectric effect (I), piezomagnetism, "Mockels" effect,[15] etc., permitted (see Table I).

The symmetric arrangement of the ferromagnetic (*FM*) and ferroelectric (*FE*) types relative to the ferromagnetoelectric (*FM* I/*FE* I) groups is intentional. It allows various analogies in the number of groups of ferromagnetic and ferroelectric types to be seen. Furthermore, the form of the tensor of the coefficients of the *HEE* and *EHH* effects are the same for corresponding groups. Corresponding means here that, e.g., the fourth group of row *FE* IV and the fourth group of row *FM* IV have the same tensor form of the piezoelectric and piezomagnetic (or second-order magnetoelectric I and II) coefficients, respectively. The same is true for the types *AA* IV and *AA* V. A more general table of corresponding groups was established by Ascher.[23]

Another useful result of the symmetric arrangement is the following: inscription in Table II of the groups permitting "weak ferromagnetism"[17,18] yielded by analogy the groups permitting "weak ferroelectricity," so far not reported. An example of a weak ferroelectric is certainly $Gd_2(MoO_4)_3$, with point group $mm21'$ and the structure of which has been described as that of a canted antiferroelectric[20] with a very small polarization. Magnetoelectric weakly ferroelectric, and weakly ferromagnetic materials may become of interest for magnetoelectrically modulating the respective electric and magnetic coercive field by magnetic and electric fields, respectively (see also Section 2.4).

In Table II, the groups permitting an invariant velocity vector[21,22] V_s are also indicated. A crystal moving along V_s will not change symmetry, whereas motion in a direction different from that of V_s, or motion of a crystal with a symmetry not permitting V_s, will lead to a decrease of its symmetry. Related kineto-electric, kineto-magnetic and kineto-magnetoelectric effects are discussed by Ascher.[23]

2.3.2 Magnetic and electric ordering Because for certain groups antiferromagnetism *or* para- *or* diamagnetism (e.g. type *FE* IV), and for others antiferroelectricity *or* para- *or* orthoelectricity is permitted, some more subtle information about the possible kinds of combinations has been indicated in the column "Type of Ordering."

The symbol P stands for spontaneous polarization; this includes pyro-, ferro-, and ferrielectrics. M stands for spontaneous magnetization and includes pyro-, ferro- and ferrimagnetics. $\bar{P}$ and $\bar{M}$ stand for "necessarily antiferroelectric" and "necessarily antiferromagnetic" ordering, respectively. Here "necessarily" pertains to the ordering, but not to domain switching. It is common use to call "pyroelectric" a polar material, the spontaneous polarization of which cannot be reversed by electric fields[24] (e.g. tourmaline), whereas the term ferroelectric pertains to a pattern of polar domains and (or) to their switching. There exists, however, no analogy, to our knowledge, between a pyroelectric in the described sense and a pyromagnetic, because all known pyromagnetics are ferromagnetics, the magnetization of which can always be switched by accessible fields.

The same absence of analogy holds between "antiferroelectrics" ($\bar{P}$) and "antiferromagnetic" ($\bar{M}$): a material having an antiferroelectric ordering does not necessarily allow the reversal of its dipoles, whereas this would be expected to be always possible for energetic reasons for the spins of an antiferromagnetic, namely by magnetoelectric, or piezomagnetic switching (see Section 2.4), or by thermally activated fluctuations near the Néel point.

The symbol D stands for diamagnetic, or paramagnetic or antiferromagnetic. This choice is possible in the 32 gray point groups. There exists a correspondence with 32 groups permitting paraelectricity or orthoelectricity or antiferroelectric ordering. They are designated by the symbol O. "Paraelectricity" pertains to a material with orientable dipoles and a corresponding dielectric Curie–Weiss law. This definition concurs with that of Gränicher and Müller.[9] However, we shall use the term "orthoelectricity"—proposed for the "normal" dielectric behaviour[9]—in a more restrictive sense, i.e. only for materials with antiferroelectric ordering but without a antiferroelectric phase transition in the temperature range of existence of the crystal.

2.3.3 "Induced" versus "second-order" magnetoelectric effects Table II allows us easily to perceive the point groups permitting an "induced magnetoelectric" effect in the sense of O'Dell,[25] i.e. "non-centrosymmetric paramagnetic crystals which will show the magnetoelectric effect when placed in a constant magnetic field." These are the ten polar groups of type *FE* IV plus the ten non-polar antiferroelectric ones of type *AA* V containing the 1′ operation.† By strict analogy, "the centrosymmetric paraelectric (orthoelectric) crystals which will show a magnetoelectric effect when placed in a constant electric field" would be the ten ferromagnetic groups of type *FM* IV plus the ten antiferromagnetic, not necessarily antiferroelectric ones

† For errata in Ref. 25, see Appendix.

of type AA IV. Inspection of Table II shows clearly, however, that these cases of "induced effects" are only very special ones among the overall possibilities offered by the 66 groups permitting HEE terms and the 66 ones permitting EHH terms.[11]

2.4 *Domain Switching and the Magnetoelectric Modulation of Coercive Fields*

Tables I and II contain intrinsically also general information about the possibilities of domain switching for a crystal of a given point group by means of a magnetic field, an electric field, a mechanical constraint,† or combinations thereof. The stored free enthalpy is a convenient means of analysing such phenomena. Particular situations will not, however, be analysed because they would necessitate knowledge of the prototypic symmetry.[21, 26]‡ In order to switch an orientation state (domain) within a given coordinate system from one position into another, equivalent one, it is necessary that the symmetry allows an energy difference to be created between these states in the field(s) of the externally applied constraint(s): magnetic field, electric field, stress.

We have to write down the stored free enthalpy for the two or more domains between which we want to switch. If a switching is possible, it is necessary that the coefficient of the term which we want to use for switching changes sign for that operation in a fixed coordinate system.

As a simple example, let us mention pure *ferromagnetic switching* for the external fields $E_i = 0$ and $T_{ij} = 0$, or for magnetoelectric and (or) piezomagnetic terms not permitted in the stored enthalpy expansion:

$$-g^{(\text{domain }+)} = +\chi_i{}^0(+H_i) + \chi_{ik} H_i H_k + \ldots$$
$$-g^{(\text{domain }-)} = -\chi_i{}^0(+H_i) + \chi_{ik} H_i H_k + \ldots$$
$$\Delta g^{(+,-)} = 2\chi_i{}^0 H_i$$

All terms other than $\chi_i{}^0 H_i$ cancel out. In this case, H_i is the coercive field for switching the spontaneous magnetization $\chi_i{}^0$. In Table III, some terms which allow domain switching are summarized.

Second-order magnetoelectric switching of magnetic domains via EHH terms or via HEE terms has not been reported so far, but it can be expected to be possible for all spin-ordered piezoelectrics, and for piezomagnets, all of which are spin ordered. For an experimental demonstration of EHH or HEE switching, it would be most advantageous to take a crystal with a symmetry permitting one or both of these terms only (types AA II, IV and V). In case of strong spin-orbit-lattice coupling of an antiferroelectric piezoelectric, it might be possible to switch also antiferroelectric domains via the EHH term. This should also be true for antiferroelectrics permitting the EH term.

In an antiferromagnetic material with a symmetry permitting EHH or HEE switching, the switching energies are probably comparable to those of other antiferromagnetics (see e.g. Ref. 30). This being so, and the second-order magnetoelectric coefficients being known to be very small,[6, 7] the coercive fields to be expected in these cases will, however, be very high.

Another feature of interest is *the modulation of electric, magnetic and ferroelectric coercive fields* by means of adequate terms. In order to expect that such a modulation is possible, we have to postulate that the switching energy Δg at a given temperature does not depend in first approximation on the applied fields. Then the Δg for modulation of the ferroelectric, ferromagnetic and ferroelastic coercive fields $E_{i(c)}$, $H_{i(c)}$ and $T_{ij(c)}$ read as follows:

$$\Delta g_E = 2E_{i(c)}(\kappa_i{}^0 + \alpha_{ij} H_j + \alpha_{ijk} E_j H_k + \tfrac{1}{2}\beta_{ijk} H_j H_k + d_{ijk} T_{jk} + \ldots)$$

$$\Delta g_H = 2H_{i(c)}(\chi_i{}^0 + \alpha_{ij} E_j + \tfrac{1}{2}\alpha_{ijk} E_i E_k + \beta_{ijk} E_j H_k + q_{ijk} T_{jk} + \ldots)$$

$$\Delta g_T = 2T_{ij(c)}(s_{ij}^0 + q_{ijk} H_k + d_{ijk} E_j + \pi_{ijkl} E_k H_l + \ldots)$$

As an example, let us consider a ferromagnet permitting the EH term (types FM I and FM II) in the absence of stress T_{jk}; if, besides the EH term, the second-order magnetoelectric coefficients are small and the corresponding terms can be neglected, then the switching energy is:

$$\Delta g_H = 2H_{i(c)}(\chi_i{}^0 \pm \alpha_{ij} E_j)$$

If $\alpha_{ij} E_j$ is parallel to, and of the same or opposite sign as, $\chi_i{}^0$, it can be seen that the coercive field $H_{i(c)}$ will decrease or increase, respectively. In order that this effect—so far not demonstrated—may be technically of interest, i.e. that $H_{i(c)}$ could be modulated by means of an electric field E_j by 50% or more, we would need a material with a spontaneous

† More details about mechanical switching are given in Section 2.4.

‡ Observed or inferred high temperature symmetry.

TABLE III
Different kinds of switching

Switching energy Δg	Type of switching	Remarks
$2\kappa_i{}^0 E_i$	Ferroelectric	Well known
$2\chi_i{}^0 H_i$	Ferromagnetic	
$2s_{ij}^0 T_{ij}$	Ferroelastic	E.g. demonstrated on ferroelectric/ferroelastic $Gd_2(MoO_4)_3$[27] and boracite[28]
$2\alpha_{ij} E_i H_j$	Magnetoelectric	Demonstrated on Cr_2O_3[29, 30]
$2g_{ijk} H_i T_{jk}$	Piezomagnetic	Demonstrated on CoF_2[31]
$2d_{ijk} E_i T_{jk}$	Piezoelectric	Demonstrated on $Gd_2(MoO_4)_3$;[27] apparently not yet demonstrated experimentally on a piezoelectric antiferroelectric
$2\alpha_{ijk} H_i E_j E_k$	Second order magnetoelectric	Not yet demonstrated experimentally
$2\beta_{ijk} E_i H_j H_k$		

magnetization $\chi_i{}^0$ comparable with $\alpha_{ij} E_j$ for reasonable fields E_j. Weak ferromagnets with high α_{ij} would be good candidates. Analogously, we should have magnetoelectric "weak ferroelectrics" (see Table II) in order to obtain a noticeable modulation of the ferroelectric coercive field by means of a magnetic field.

In addition to the kinds of modulation described, another kind—due to coupling between the directions of spontaneous polarization and magnetization—is possible for certain symmetries. An example is the ferromagneto electric phase $m'm2'$ of $Ni_3B_7O_{13}I$.[59, 60]

It is not the purpose of this paper to analyse particular switching situations. If this were intended, we would need to know the prototypic symmetry[21, 26] of the low temperature symmetry of interest (for possible "species" see Aizu[32] and Cracknell[33]) and write down explicitly all vectorial and tensorial components. Ferromagnetoelectric twinning and switching is discussed by Janovec and Shuvalov[34] (see also Ref. 21).

2.5 *Magnetoelectric Ferroelastics, Antiferroelastics and Para- (or Ortho-) Elastics*

In analogy to ferromagnetism and ferroelectricity, Aizu[32] has termed as "ferroelasticity" the change of orientation state (domain orientation) of the spontaneous strain tensor s_{ij}^0 under the influence of an external stress T_{ij}, whereas, in case of "antiferroelasticity," different orientation states of the crystal cannot be interchanged by applying stress T_{ij}. By phenomenological analogy with "paraelectric" and "orthoelectric,"[9] we might designate those phases as "paraelastic" and "orthoelastic," the elastic susceptibility s_{ijkl} (see Table I) of which would follow a Curie law or be temperature-independent, respectively.

One obvious impact that the concept of ferroelasticity has in the field of magnetoelectricity (as well as in other disciplines) resides in the possibility of *producing single domain* specimens by applying an adequate stress T_{ij}. We have to distinguish between cases (i) where the application of *stress is a necessary* condition from the point of view of symmetry in order to obtain a single domain, and (ii) where the application of stress is not necessary for reasons of symmetry but only because of special situations.

We encounter the first situation—condition (i)—in the cases of "partial ferroelectricity" and "partial ferromagnetism" (Aizu's nomenclature), i.e. where the high temperature/low temperature symmetry relationship is such that an electric field and a magnetic field alone is not able to produce a ferroelectric and ferromagnetic single domain, respectively. Only in the case of "full ferrogmagnetism," "full ferroelectricity" and "full ferroelasticity" does a respective magnetic field H_i, electric field E_i and stress field T_{ij} *alone* suffice to obtain a ferromagnetic, ferroelectric and ferroelastic single domain, respectively. By analogy, we can single out those magnetoelectrics with EH terms (and other term combinations) for which a polydomain sample can be oriented by the simultaneous application of E_i *and* H_j fields alone, and those for which this is *not* possible. We might call them "full and partial magnetoelectrics," respectively. For the former case, Cr_2O_3 is an example with the Aizu species $\bar{3}m1'$ F $\bar{3}'m'$ and for which the "poling" by simultaneous application of E_i and H_j fields alone has been demonstrated.[29, 30] The domains cannot be influenced by pressure alone, i.e. the crystal is antiferroelastic.

Examples of the latter case would be the Aizu species $4/mmm1'$ F mmm' and $4/mmm1'$ F $2'/m$, where magnetoelectric single domains cannot be obtained by the simultaneous application of E_i and H_j alone, but where additional ferroelastic switching is necessary for those domains which are related by a $\pi/4$ rotation around the lost 4-fold axis.

The information on full and partial magnetoelectrics is implicitly contained in the Tables of Aizu.[32] Apart from the linguistic point of view—the terminology of "fullness" and "partiality" is unfortunately highly confusing—the distinction made by Aizu is extremely helpful for the experimentalist.

Special situations relating to item (ii) would be: the electric coercive field of a ferroelastic ferroelectric is too high to obtain a poling by means of an electric field; the electric resistivity of a ferroelastic ferroelectric is too low for switching P_s; the coupling spontaneous magnetization/spontaneous polarization of a ferromagnetoelectric is too small to obtain ferroelectric poling by means of a magnetic field alone. If the crystal allows full ferroelasticity in such cases, the problem can be solved in either case by the application of stress alone.

Because of the above-mentioned importance of ferroelasticity for study of magnetoelectric (and other) structures, one would be tempted to incorporate into Table II also partial and full ferroelasticity into the set of properties. Unfortunately, this is not possible in a simple compact way because *ferroelasticity is not described by a single Shubnikov group but pertains to domain switching*. Domain switching always depends on both the high (prototypic) and low temperature symmetry group (see Ascher[21, 26]). These pairs of groups, "species," have been tabulated by Aizu.[32] Recently, Cracknell[33] has represented the ferroelastic species in a very compact manner by using the sub-group tables of Janner and Ascher;[25] the information about "fullness" and "partiality" is, however, lacking.

3 STRUCTURAL EXAMPLES OF SOME MAGNETOELECTRIC TYPES

A number of structural examples of the magnetoelectric types as defined in Table II has been assembled in Tables IV, IVa, V and VI. In this chapter we shall discuss certain aspects of a small choice of examples.

3.1 *Type* FE *I*/FM *I*, *Pyro-, Ferro- or Ferrielectric/ Pyro-, Ferro- or Ferrimagnetic* (E, H, EH, EHH, HEE)

3.1.1 Introductory remarks Among the different possibilities of combining electric and magnetic long-range order in one and the same material, that of simultaneous ferroelectricity and ferromagnetism has always created the greatest fascination; new kinds of network element were expected to be realizable, and interesting magnetoelectric interactions were hoped to be discovered in such materials. We know today that ferromagnetic† ferroelectrics *do* exist. Even two of them—Congolite and Ericaite (Fe-Cl-boracites)[36] and Chambersite (Mn-Cl-boracite)[37]—are found in Nature. Since the beginning of research in this field, the accent had to be put, however, on "molecular engineering." In principle, two lines of approach are open to the chemist:

1) modifying ferro-, ferri- or antiferromagnets, or

2) modifying diamagnetic ferro-, ferri- or antiferroelectrics.

All successful results achieved so far were obtained with approach 2.‡ This is not surprising, because the aimed at introduction of a lattice instability—necessary for ferroelectricity and antiferroelectricity—into an ordinary ferro-, ferri- or antiferromagnetic structure is a delicate enterprise.

During the early investigations in this field, an apparent obstacle appeared in the form of conjectures, formulated by Matthias[38, 39] and Smolensky,[40, 41] saying approximately the following: in the oxyoctahedral compounds (e.g. $BaTiO_3$), ions with small ionic radius, a high charge and a rare gas configuration are responsible for the ferroelectricity. This was essentially correct§ but excluded, however, the use of paramagnetic ions. Consequently, Smolensky and Joffe[8] chose the compromise: replacing the ferroelectrically active ions (e.g. Ti^{4+}) only partially by paramagnetic ones, for example by forming Pb $(Fe^{3+}_{1/2}Nb^{5+}_{1/2})O_3$. Here, Nb^{5+} has the desired rare gas configuration, and Fe^{3+} bears the magnetic moment. Many different perovskite

† So far, only ferrimagnetic and weakly ferromagnetic ferroelectrics are known.

‡ Attempts at modifying garnets were unsuccessful[44] because adjacent perovskite phases were thermodynamically more stable.

§ Details on cations having empty d orbitals and causing ferro- and antiferroelectric spontaneous distortions in perovskites, are given by Goodenough and Longo (Ref. 83, pp. 142–144).

compounds have been prepared along this line of approach[42] by sinter techniques. The first compound in which simultaneous ferroelectric and ferromagnetic properties seem to have been discovered was prepared by Smolensky, Isupov, Krainik and Agranovskaya.[43] It has the composition:

$$(1-x)\,Pb(Fe_{2/3}W_{1/3})\,O_3 - xPb(Mg_{1/2}W_{1/2})O_3$$

With $x=0.3$ the ferroelectric and ferromagnetic Curie points are $-100°C$ and $-80°C$, respectively. Unfortunately, no single crystals could be obtained.

Favourable structure elements for simultaneous electric and magnetic ordering. Because of the low magnetic spin ordering temperatures which could be foreseen for such compounds with "mutually diluting" properties, Janner and Ascher[45] suggested the use of anisotropic structural elements, e.g. mixed octahedra with four co-planar anions and two different ions at opposite apexes, permitting an electrostatic double potential well for the paramagnetic metal ion centred in the octahedron (see Figure 2) to be obtained. Such a double potential well was considered favourable to permit ferroelectric and antiferroelectric displacements of the central ion, even were it to have low polarizability as is the case for the paramagnetic $3d$-transition metal ions. In the first experiments along this line, it was tried to replace the central ion in perovskites completely by ions with an open $3d$-shell, and the mixed octahedra were hoped to be formed by introducing fluorine: $BaCr(O_2F)$, $BaFe(O_2F)$, $Pb(CrO_2)F$ and $PbFe(O_2F)$. These experiments were unsuccessful,[44] but urged for the search for a ferroelectric material possessing already mixed octahedra. This was found in the form of boracite $Mg_3B_7O_{13}Cl$, in which the diamagnetic Mg^{2+}—centred in the octahedron—can be easily replaced by paramagnetic ions[46] which lead to weakly ferromagnetic properties at low temperatures.[47] Most of these ferromagnetic compositions are simultaneously ferroelectric[48] (see also Table IV).

A few more examples of ferroelectrics are known, in which a local chemical anisotropy seems to lead to double potential wells creating or enhancing ferroelectricity:

i) In the antiferromagnetic ferroelectrics of the type $YMnO_3$[49] (see Table IV), the Mn^{3+} occupies a five-fold coordinated site within a bipyramid (Figure 3) which probably creates a double potential well responsible for the observed ferroelectricity.

ii) The pyrochlore $Cd_2Nb_2O_7$ is ferroelectric at below 185°K only.[50] By replacing the oxygen partly by sulphur ($\rightarrow Cd_2Nb_2O_6S$), the ferroelectric properties are extended to high temperatures.[50, 51] This is

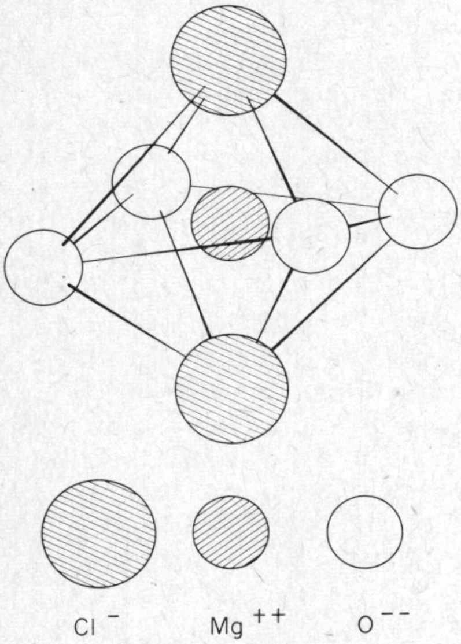

FIGURE 2 Anisotropic Cl–O-octahedron of Mg–Cl-boracite, responsible for an electrostatic double potential well with its two minima along the Cl–Mg–Cl axis.

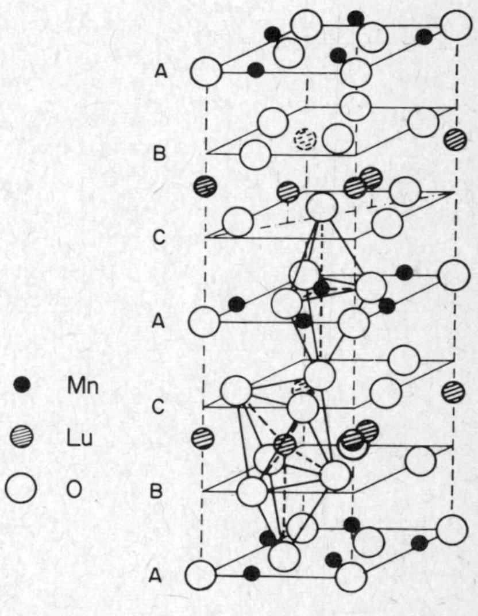

FIGURE 3 Structure of $MnREO_3$ (RE = Y, Ho, Er, Tu, Yb, Lu, Sc), showing five-fold coordinated Mn-sites (according to Ref. 49) probably creating electrostatic double potential wells for the manganese.

probably due to the enhancement by the sulphur of the depth of a double potential well for the eight-fold coordinated Cd^{2+} along the S–Cd–S axis.[52]

iii) The partial replacement of O^{2-} and Al^{3+} in $BaAl_2O_4$ by F^- and Li^+, respectively, leads to ferroelectricity.[53] This may possibly be due to mixed O–F octahedra.

iv) A non-linear magnetoelectric effect was reported for a pseudo-ilmenite type compound with formula $Li(Fe_{1/2}Ta_{1/2})O_2F$, a ferroelectric Curie point of 560°C and a Néel temperature of 610°C.[54] Possibly, the mixed O–F octahedra may here be responsible for the ferroelectricity; however, the magnetic transition temperature of 610°C, astonishingly high for a 50% occupation of the octahedral sites by Fe^{3+}, sheds some doubt on this compound being single phase.

v) In antiferromagnetic/ferroelectric $BiFeO_3$ (see Refs. 55–57 and Table IV), double potential wells are

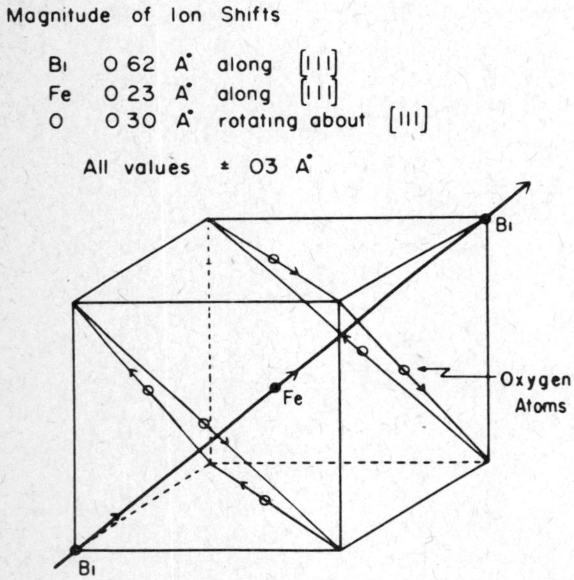

FIGURE 4 Local anisotropic environment of iron in $BiFeO_3$ and ion shifts superposed on the ideal perovskite cell (according to Refs. 55, 57).

apparently created along the Bi–Fe–Bi axis parallel ⟨111⟩ (see Figure 4).

We can conclude that *one* effective way of creating or enhancing ferroelectricity by means of molecular engineering is the introduction of structural elements liable to give double or multiple electrostatic potential wells, between the minima of which the ferroelectrically active ions can change place. These mobile ions should be paramagnetic and responsible for the magnetic ordering, if an effective coupling between the spontaneous electric and magnetic polarization or antipolarizations is to be obtained.

The more detailed structural conditions for obtaining magnetic ordering and high magnetic ordering temperatures in non-metals are quite well understood,[58]† so that we can omit a discussion in this paper.

3.1.2 Ferroelectric–ferromagnetic boracites Boracites have the general formula $M_3B_7O_{13}X$, where M stands for bivalent ions of metals such as Mg, Ca, Mn, Fe, Cr, Ni, Cu, Zn, Cd and X for OH, Cl, Br, I, S. In the composition $Ni_3B_7O_{13}I$, the occurrence of both ferroelectricity and weak ferromagnetism, and the coupling between spontaneous polarization and magnetization (<64°K) was demonstrated[59,60] for the first time, but most of the other compositions containing paramagnetic ions are also both ferroelectric and ferromagnetic at low temperatures (see Table IV).

The ideal‡ high temperature (prototypic) phase of all boracites has point group $\bar{4}3m1'$, and there occur ferroelectric phases in most compositions with decreasing temperature: $mm21' \rightarrow m1' \rightarrow 3m1'$ (see Table IV). In some compositions, the $3m1'$ and (or) $m1'$ phase does not occur. Most of the paramagnetic compositions become weak ferromagnets at low temperatures.[47]

For the time being, we know about the occurrence of two ferromagnetic–ferroelectric Shubnikov groups among boracites: $m'm2'$ and m. Although experimental data are still scarce, we shall discuss some aspects of the respective compounds $Ni_3B_7O_{13}I$ and $Co_3B_7O_{13}Cl$.

$Ni_3B_7O_{13}I$ is ferroelectric *and* ferromagnetic below about 64°K. It is not yet clear whether the two properties set in exactly at the same temperature. At about 40°K, the point group $m'm2'$ was deduced from ferroelectric and ferromagnetic switching experiments and simultaneous polarization optical inspection.[59,60] By using the co-ordinates x, y, z and defining their orientation by $m_x m'_y 2'_z$,§ the linear magnetoelectric coefficients are α_{xz} and α_{zx} and the second-order magnetoelectric coefficients

† For magnetic ordering in perovskites see Ref. 83, pp. 207–254.

‡ Sometimes non-cubic growth sectors are formed.[60]

§ At present, it is not yet known whether x or y is parallel to the crystallographic a-parameter ($a < b < c$).

are α_{zzx}, α_{yzx}, α_{xxx}, $\alpha_{xzz} = \alpha_{zxz}$, $\alpha_{yxy} = \alpha_{xyy}$; β_{zxx}, β_{zyy}, β_{zzz}, $\beta_{xxz} = \beta_{xzx}$, $\beta_{yzy} = \beta_{yyz}$.

Not considering the piezoelectric, the piezomagnetic and the second-order magnetoelectric contributions, the equations for the total polarizations and magnetizations along the three coordinate directions read:

$$P_x = \kappa_{xx} E_x + \alpha_{xz} H_z$$
$$P_y = \kappa_{yy} E_y$$
$$P_z = \kappa_z^0 + \kappa_{zz} E_z + \alpha_{zx} H_x; \quad \kappa_z^0 = P_s$$

$$M_x = \chi_x^0 + \chi_{xx} H_x + \alpha_{zx} E_z; \quad \chi_x^0 = M_s$$
$$M_y = \chi_{yy} H_y$$
$$M_z = \chi_{zz} H_z + \alpha_{xz} E_x$$

So far, only the coefficient α_{zx} has been measured by means of the magnetic field induced effect.[59–63]

During magnetoelectric measurements and other magnetic switching experiments, various authors[59,62,64] have reported *asymmetries in the magnetic coercive field* for reversing the spontaneous magnetization χ_x^0. Furthermore, asymmetries in the coercive field for reversing the spontaneous polarization κ_z^0 have been observed.[59,64] These results were explained by a kind of spontaneous magnetoelectric interaction between an internal electric anisotropy field E_0, acting via the magnetoelectric coefficient α to give a magnetic transfer field asymmetry $\Delta H \sim \alpha\varepsilon E_0/\chi$, where ε is the dielectric constant and χ the magnetic susceptibility.[65] We shall show that such interpretation is, however, erroneous. By forming the energy difference Δg, e.g. for opposite ferromagnetic domains with the magnetization along x, the magnetization χ_x^0 and the coefficient α_{zx} change sign if we keep the spontaneous polarization κ_z^0 and the coordinates fixed in space:[59,60]

$$-g^+ = \ldots + \chi_x^0 H_x + \alpha_{zx} E_z + \chi_{xx} H_x H_x + \ldots$$
$$-g^- = \ldots - \chi_x^0 H_x - \alpha_{zx} E_z + \chi_{xx} H_x H_x + \ldots$$
$$\Delta g = 2H_x(\chi_x^0 + \alpha_{zx} E_z) \qquad \text{here } H_x = H_{x(c)}$$

In the expression for Δg, H_x is the magnetic coercive field. Then it can be seen that Δg represents the switching energy. Admitting that it remains in first approximation independent of applied magnetic and electric fields, it can be seen that the magnetic coercive field $H_{x(c)}$ can be higher or lower if the applied electric field E_z is positive or negative, respectively. By replacing phenomenologically E_z by an internal fictitious field E_0 as was done by Alshin *et al.*,[65] it must, however, be noted that the sign of E_0 *does not* and *cannot* change during the process of reversing the spontaneous magnetization χ_x^0. Therefore, it can easily be seen that $H_{x(c)}$ is symmetric for opposite sign of H_x, because χ_x^0 *and* α_{zx} both change sign upon reversal of χ_x^0.

Analogous considerations hold for the spontaneous polarization κ_z^0 and the connected coercive field which is not affected by a fictitious "spontaneous magnetoelectric" interaction either.

The conclusion is that the observed asymmetries of H_c and E_c must be of other origin. The most probable causes are:

ad asymmetric H_c: pinning of domains by defects to certain spin directions, as observed for Cr_2O_3;[30]

ad asymmetric E_c: stress-induced anisotropies due to ferroelastic behaviour; stress induced by growth dislocations,[84] etc. Such "internal bias" fields are also often observed in paramagnetic or diamagnetic ferroelectrics, showing that no magnetoelectric explication is necessary.

$Co_3B_7O_{13}Cl$ is a boracite which runs through the phases $\bar{4}3m1' \rightarrow mm21' \rightarrow m1' \rightarrow 3m1' \rightarrow m$ with decreasing temperature. The phase with symmetry m—below 22°K—is both ferroelectric and weakly ferromagnetic. The ferroelectric structure of that phase is certainly that of the trigonal $3m1'$ phase.[66] A ferroelectric–ferromagnetic single domain of the m-phase can be obtained as follows: poling a $(111)_{cubic}$-cut platelet (e.g. at room temperature) in such a way that the spontaneous polarization of the $3m1'$ phase and its optic axis lie perpendicular to the platelet.[66] After cooling below 22°K without an external magnetic field, antiparallel ferromagnetic domains are formed with their magnetization lying within the plane of the platelet ($\perp P_s$) and oriented perpendicular to one or more of the three pseudo-cubic (110) planes being perpendicular to the plate (Figure 5).

By choosing a coordinate system with z pointing along the spontaneous polarization (parallel to a $\langle 111 \rangle_{cub}$ direction), x parallel to the spontaneous magnetization (perpendicular to a $(110)_{cub}$ plane) and y perpendicular to both (Figure 5) the linear magnetoelectric coefficients are α_{yx}, α_{zx}, α_{xy} and α_{xz} and the second-order magnetoelectric coefficients are α_{xxx}, α_{yyx}, α_{zzx}, $\alpha_{xyy} = \alpha_{yxy}$, $\alpha_{xyz} = \alpha_{yxz}$, $\alpha_{xzz} = \alpha_{zxz}$, $\alpha_{xzy} = \alpha_{zxy}$, $\alpha_{yzx} = \alpha_{zyx}$; β_{xxy}, β_{xxz}, β_{yyy}, β_{yyz}, β_{zzy}, β_{zzz}, $\beta_{xyz} = \beta_{yxx}$, $\beta_{xzx} = \beta_{zxx}$.

Then the equations for the total polarization and magnetization along the three coordinate directions read as follows:

$$P_x = \kappa_{xx} E_x + \alpha_{xy} H_y + \alpha_{xz} H_z$$

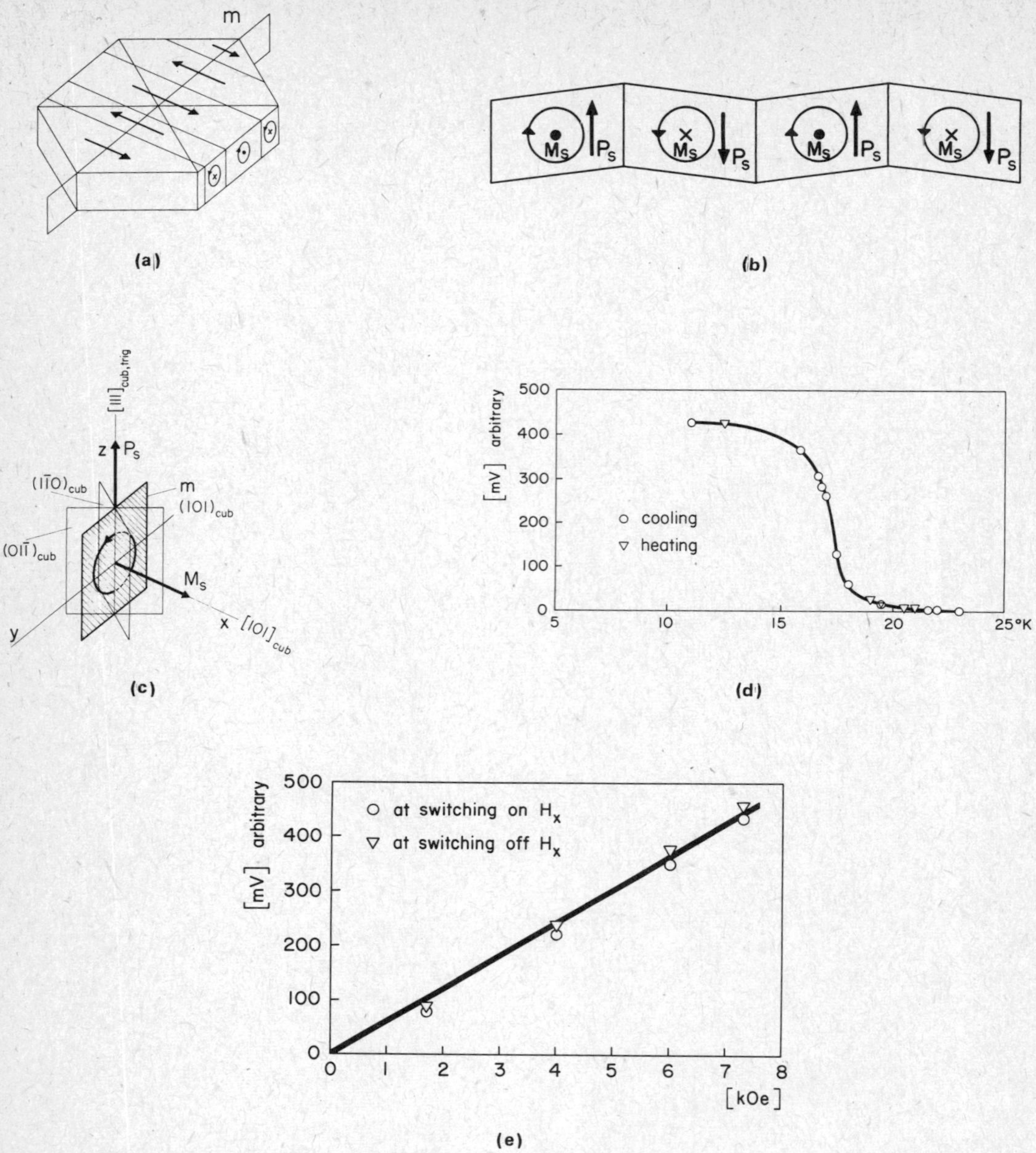

FIGURE 5 (a) Schematic representation of one of the observed three equivalent patterns of antiparallel ferromagnetic domains occurring at zero magnetic field in one (of the four equivalent) ferroelectric single domains of the pseudotrigonal phase of Shubnikov group *m* of Co–Cl-boracite. (b) Schematic cross-section through the antiparallel magnetic domains; the shear—permitted by symmetry and exaggerated on the drawing—is probably extremely small. The domains were observed with red/green contrast on a 25 μm thick ferroelectric single domain platelet, with nearly crossed polarizers and the *z*-axis tilted a few degrees from the microscope axis around the *y*-axis (see (c)). (c) Representation of the co-ordinates chosen for a ferroelectric–ferromagnetic single domain (correspondence with the equations in the text). (d) Temperature dependence of the $(ME)_H$ signal (in arbitrary mV units) of a 95% ferroelectric–ferromagnetic single domain of a Co–Cl boracite platelet (thickness 39 μm, electrode diameter 1.6 mm, $H_x = 7.3$ kOe). (e) $(ME)_H$ signal of α_{zx} at 11° K versus magnetic field (ferromagnetic coercive field $\ll$ 250 Oe).

$$P_y = \kappa_{yy} E_z + \alpha_{yx} H_x$$
$$P_z = \kappa_z{}^0 + \kappa_{zz} E_z + \alpha_{zx} H_x; \quad \kappa_z{}^0 = P_s$$
$$M_x = \chi_x{}^0 + \chi_{xx} H_x + \alpha_{yx} E_y + \alpha_{zx} E_z; \quad \chi_x{}^0 = M_s$$
$$M_y = \chi_{yy} H_z + \alpha_{xy} E_x$$
$$M_z = \chi_{zz} H_z + \alpha_{xz} E_x$$

The magnetization can be easily rotated in the plane perpendicular to P_s, whereas there is a strong anisotropy tending to keep M_s within that plane. In case of a strong coupling of the easy plane to the lattice, it would be possible—in principle—to switch the spontaneous polarization into another one of the three remaining possible orientations by means of a magnetic field applied at an angle different from zero relative to the easy plane. Owing to the high electric coercive fields, spin flop is, however, expected to occur before P_s will move. Contrary to that, the reorientation of the easy plane of M_s due to an electric field induced switching of P_s will not be hindered.

The magnetic field induced $(ME)_H$ effect due to the coefficient α_{zx} has been measured qualitatively between 10 and 22°K[68] (see Figure 5a–e).

3.1.3 Perovskite/(Bi_2O_2)-layer compounds $Bi_9(Ti_3Fe_5)O_{27}$ has been reported to be a ferroelectric weak ferromagnet with the highest respective Curie temperatures, 830°C and 130°C, so far known[69] for a ferromagnetoelectric material.

A magnetic field induced magnetoelectric polarization has been measured on a polycrystalline sample. There seems to be a superposition of a linear and higher order magnetoelectric effect. The compostion is a representative of a remarkable series of ferroelectric compounds (see Table IVa) which can be represented by the general formula $Bi_2M_{n-1}R_nO_{3n+1}$, where M = Bi^{3+}, Pb^{2+}, Na^+, K^+, Sr^{2+}, Ba^{2+}, RE^{3+}, R = Ti, Nb, Ta, Fe, Ga, W, Cr, and $n = 1, 2, 3, 4, 5, 6, 7, 8 \ldots \infty$. The M-site is dodecahedrally surrounded by oxygen, and the R-site is of the oxy-octahedral type. The structure can be considered to consist of perovskite blocks with the formula $M_{n-1}R_nO_{3n+1}$, separated by $(Bi_2O_2)^{2+}$ layers (see Figure 6). Thus, $Bi_9(Ti_3Fe_5)O_{27}$ is a compound with 8-layered perovskite blocks separated by $(Bi_2O_2)^{2+}$ layers. Several representatives of the series are given in Table IVa. By replacing more and more Ti^{4+} on the R-site, the number n of perovskite block layers rises. With the ratio 3Ti:5Fe, the magnetic ordering temperature reaches 130°C. By further increase of n to quasi infinite, we end up

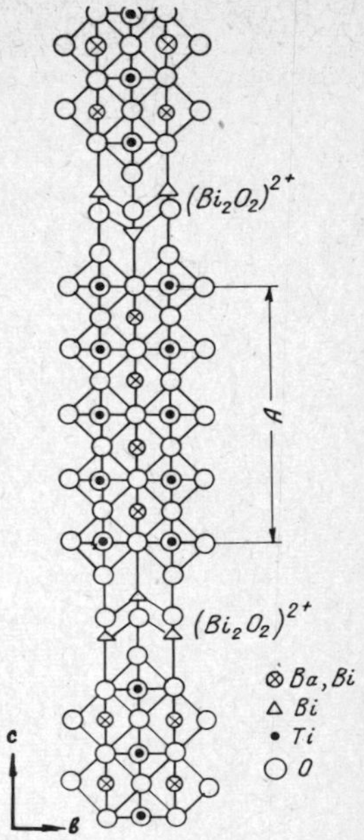

FIGURE 6 Projection of the unit cell of $Ba_2Bi_4Ti_3O_{18}$ on (100) (according to Ismailzade[77]).

with the ferroelectric compensated antiferromagnet $BiFeO_3$ (see type *FE* II). Most of the compounds of the series are orthorhombic with $mm21'$ symmetry, whereas $Bi_9(Ti_3Fe_5)O_{27}$ must have Shubnikov group $m'm'2$ or $m'm2'$ if the reported weak ferromagnetism can be confirmed. Owing to the fact that the orthorhombic c-parameter (polar axis) rises rapidly with increasing n (about 75 Å for $n = 8$), the tendency towards disorder of Ti and Fe will increase. Consequently, synthesis of unperturbed single crystals of these compounds can be considered as a great challenge for crystal growers.

For the interesting compounds containing iron and with $n = 4$, 5 and 8, unfortunately only powder data are known.

3.1.4 Perovskite compounds $PbFe_{1/2}Nb_{1/2}O_3$ *and* $PbMn_{1/2}Nb_{1/2}O_3$ have recently been obtained in single crystalline form.[75,121] Owing to this fact, it was possible to demonstrate antiferromagnetic ordering with weak ferromagnetism below 9° and 11°K, respectively.[75,81] An enhancement of the magnetic moment by the application of an electric field in the case of $PbMn_{1/2}Nb_{1/2}O_3$ at 1.5°K was observed. After cooling through the transition temperatures in simultaneous electric and magnetic fields, the ferromagnetic character of the two compositions was well brought out. Therefore, there is strong evidence for simultaneous ferroelectricity.[75]

It is noteworthy that $PbFe_{1/2}Nb_{1/2}O_3$ was among the first perovskite compositions that Smolensky and co-workers had imagined and synthesized[8] in 1957 with a view to obtaining ferroelectric ferromagnets. Further measurements on these interesting compounds should be made under a microscope in order to clarify the exact symmetry, and the electric and magnetic domain configurations. It is supposed that only by taking such precautions will valid magnetoelectric measurements be possible.

3.2 *Type* FE *II*, *Ferroelectric/antiferromagnetic* (E, EH, HEE, EHH)

$BiFeO_3$ has for a long time been the subject of a controverse as to the magnetic and electric state of ordering because only more or less pure highly conducting sintered specimens were available. The ferroelectric behaviour is, however, now clearly established by hysteresis cycles on single crystals at liquid nitrogen temperature.[56] The structure at room temperature is a slightly deformed perovskite type with space group C_{3v}^6–$R3c$.[55] By neutron diffraction on single crystals,[57] powder[70] and by Mössbauer[71] measurements, a compensated antiferromagnetism with the spins lying in the (111) plane was established (Néel temperature: 375°C[72]). The Shubnikov point group is therefore most probably $3m$, which permits the linear magnetoelectric effect (Table II). The extrapolated ferroelectric Curie temperature is 850°C and coincides with the decomposition temperature. The phase sequence is therefore:

$$\underset{\text{(prototype?)}}{m3m1'} \xleftrightarrow{850°C} 3m1' \xleftrightarrow{375°C} 3m\,(?)$$

From Figure 4, it becomes obvious that the Fe^{3+} ion, situated in the centre of a slightly deformed cube, probably moves in a double potential well along the polar Bi–Fe–Bi axis. The strong off-centre displacement of the Fe^{3+} (0.23 Å![55]) relative to the ideal cubic cell explains the high ferroelectric Curie point and high coercive field (~55 kV/cm[56]).

3.3 *Various Magnetoelectric Types*

As regards examples of compounds of the other magnetoelectric types set off in Table II, we refer to Tables IV, IVa, V and VI, as well as to the tables of the concomitant paper by Cox.[90]

CONCLUSIONS

By dividing the 122 Shubnikov groups into "magnetoelectric types" and by classifying a large number of materials correspondingly, it was possible to obtain a clearer image of the possibilities that are offered to us by symmetry and real materials for exploiting magnetoelectric and related crystal properties.

As it is known that high electric and magnetic susceptibilities are necessary for obtaining the desirable high magnetoelectric coefficients,[119,120] magnetically ordered ferroelectrics and antiferroelectrics are certainly good candidates. It is therefore noteworthy that the survey has shown that such materials are by far not rare, but their magnetoelectric properties are not or hardly investigated. Here lies a great challenge for generations of crystal growers, and neutron and X-ray crystallographers, because serious progress will be possible only by measurements on well-defined single crystals.

ACKNOWLEDGEMENTS

The author wishes to thank Dr. Edgar Ascher for many helpful discussions and for critical advice about the manuscript. His gratitude is also due to Mrs. L. Amiguet for careful typing of the manuscript and tables.

Appendix

Errata in Refs. 11, 12, 13, 14, 25

ad 11: page 154, (α): replace $4/mmm$ by $4'/mmm'$, replace $m3m$ by $m3m'$
page 154, (β): replace $41'$ (at 2nd place) by $\bar{4}1'$, replace $6224'$ by $6221'$, replace $\bar{6}'m'2'$ by $\bar{6}m'2'$
page 155, 7. (β): replace $41'$ by $\bar{4}1'$
page 156, 10. (α): replace $4/mmm$ by $4'/mmm'$
page 156, 15. (β): replace $\bar{6}m2'$ by $\bar{6}'m2'$

ad 12: The cubic groups 432, $\bar{4}3m$ and $m3m$ are erroneously described as piezomagnetic. Correct tables are given in Ref. 10.

ad 13: In the last line of Table 20, p. 85, replace $m3m'$ by $m'3m$.

ad 14: The magnetic and electric order is given erroneously for many groups.

ad 25: page 147, 10th line: replace 18 by 20
page 147, 14th line: delete group 432 and add groups 32, $\bar{6}$, $\bar{6}m2$, $\bar{4}3m$ (note here the use of non-Shubnikov groups as in Ref. 25).
page 268, 4th line from bottom: eleven groups are indicated as compatible with dielectric crystals in motion. There are however thirty-one, namely those permitting V_s (see Table II).

TABLE IV

Examples of pyro- and ferroelectric magnetoelectric types. (*Legend:* *PY* = pyroelectric/non-ferroelectric, *PT* = prototype, *FE* = ferro-, or ferrielectric, *FM* = ferro- or ferrimagnetic, *WFM* = weak ferromagnet)

Type *FE* I-*FM* I (*E, H, EH, EHH, HEE*) Pyro-, or ferro-, or ferrielectric / ferro-, or ferrimagnetic

Compound	Phase transitions	α_{max} obs. (Gaussian units)	Ref.
$Ga_{2-x}Fe_xO_3$	$mm21'$ (*PY*) $\overset{305°K}{\text{———}}$ $m'2'm$ (*PY*)	4×10^{-4}	73
Boracites ($M_3B_7O_{13}X$) (spin-ordered phases: *FE*/*WFM*)			
M–X	*PT*		
Cu–Cl	$\bar{4}3m1' \xleftrightarrow{365°K} mm21' \xleftrightarrow{20°K} m'm2'$ (?)		46–48
Cu–Br	$\bar{4}3m1' \xleftrightarrow{226°K} mm21' \xleftrightarrow{24°K} m'm2'$ (?)		46, 47
Ni–Cl	$\bar{4}3m1' \xleftrightarrow{610°K} mm21' \xleftrightarrow{20°K} m'm2'$		46, 47
Ni–Br	$\bar{4}3m1' \xleftrightarrow{398°K} mm21' \xleftrightarrow{40°K} m'm2'$ (?)		46, 47
Ni–I	$\bar{4}3m1' \xleftrightarrow{64°K} m'm2' \xleftrightarrow{?} m$ (?)	3×10^{-4}	46, 59
Co–Cl	$\bar{4}3m1' \xleftrightarrow{623°K} mm21' \xleftrightarrow{538°K} m1' \xleftrightarrow{468°K} 3m1' \xleftrightarrow{22°K} m$		46, 66, 48
Co–Br	$\bar{4}3m1' \xleftrightarrow{458°K} mm21' \longleftrightarrow$?		46, 47
Co–I	$\bar{4}3m1' \xleftrightarrow{197°K} mm21' \longleftrightarrow$?		46, 47
Fe–Cl	$\bar{4}3m1' \xleftrightarrow{609°K} mm21' \xleftrightarrow{543°K} m1' \xleftrightarrow{528°K} 3m1' \xleftrightarrow{11.5°K} m$ (?)		46–48
Fe–Br	$\bar{4}3m1' \xleftrightarrow{495°K} mm21' \xleftrightarrow{405°K} 3m1' \xleftrightarrow{15°K} m$ (?)		46–48
Fe–I	$\bar{4}3m1' \xleftrightarrow{349°K} mm21' \xrightleftharpoons[218°K]{203°K} m1' \xrightleftharpoons[205°K]{191°K} 3m1' \xleftrightarrow{30°K} m$ (?)		46–48
Mn–Cl	$\bar{4}3m1' \xleftrightarrow{680°K} mm21' \longleftrightarrow$?		46
Mn–Br	$\bar{4}3m1' \xleftrightarrow{564/69°K} mm21' \longleftrightarrow$?		46
Mn–I	$43m1' \xleftrightarrow{412°K} mm21' \longleftrightarrow$?		46
β-$NaFeO_2$ (<760°K)	*PT* $mm21' \xleftrightarrow{723°K} m'm2'$ (?), $T_c \equiv T_N$		83, pp. 22–23

TABLE IV (continued)

Compound	Phase transitions		Ref.
Perovskites			
$(1-x)Pb(Fe_{2/3}W_{1/3})O_3$–$x\ Pb(Mg_{1/2}W_{1/2})O_3$	$T_c(FE)$ (°K)	$T_c(FE)$ (°K)	
For ex.: $x = 0.3$	~173	~193	43
	$T_c(FE)$ (°K)	$T_c(FM)$ (°K)	
$Pb(Fe_{1/2}Ta_{1/2})O_3$	233	133	72
$Pb(Fe_{1/2}Mn_{1/4}W_{1/4})O_3$	263	143	72
$Bi_xPr_{1-x}FeO_3$			74
$0.75 < x < 0.80$			
	$T_c(FE)$ (°K)	$T_c(FM)$ (°K)	
$Pb(Mn_{1/2}Nb_{1/2})O_3$	293	11	75, 76
$Pb(Fe_{1/2}Nb_{1/2})O_3$	385	9	75, 76
Perovskite-related (see Table IVa)			
	$T_c(FE)$ (°K)	$T_c(FM)$ (°K)	
$Bi_9(Ti_3Fe_5)O_{27}$	1103	403	69, 77

Type *FE* II (E, *EH*, *EHH*, *HEE*) Pyro- or ferroelectric/antiferromagnetic

Compound	Phase transitions	Remarks	Ref.
$BiFeO_3$	*PT* $m3m1'$ (?) $\xleftrightarrow{1123°K}$ $3m1'$ $\xleftrightarrow{648°K}$ $3m$ (probable)		55, 56, 72
$Gd_2(MoO_4)_3$	*PT* $\bar{4}2m1'$ $\xleftrightarrow{432–36°K}$ $mm21'$ $\xleftrightarrow{\sim 0.5°K}$ $mm2$ (probable)		20, 85
$FeSb_2O_4$	*PT* $mm21'$ $\xleftrightarrow{46°K}$ $mm2$ (PY) (PY)	*ME*-effect observed	86

TABLE IV (continued)

Type *FE* III (*E, HEE, EHH*) Pyro- or ferroelectric/antiferromagnetic

Compound	Phase transition	Ref.
$YMnO_3$	*PT*[a] $6/mmm1' \xleftrightarrow{913°K} 6mm1' \xleftrightarrow{\sim 77°K} 6'mm'$?	49, 87–91
$HoMnO_3$	$6/mmm1' \xleftrightarrow{873°K} 6mm1' \xleftrightarrow{76°K} 6'mm'$?	87, 90–92
$ErMnO_3$	$6/mmm1' \xleftrightarrow{833°K} 6mm1' \xleftrightarrow{79°K} 6'$?	87, 91, 92
$TuMnO_3$	$6/mmm1' \xleftrightarrow{?} 6mm1' \xleftrightarrow{86°K} 6'$?	87, 91, 92
$YbMnO_3$	$6/mmm1' \xleftrightarrow{993°K} 6mm1' \xleftrightarrow{?} 6'$?	87, 91, 92
$LuMnO_3$	$6/mmm1' \xleftrightarrow{?} 6mm1' \xleftrightarrow{91°K} 6'$?	87, 91, 92
$ScMnO_3$	$6/mmm1' \xleftrightarrow{?} 6mm1' \xleftrightarrow{120°K} 6'$?	87, 91, 92

[a] The *PT* is not stated in the literature, however it is $6/mmm1'$ with great probability.

Type *FE* IV (*E, EHH*) Pyro- or ferroelectric/antiferromagnetic or paramagnetic or diamagnetic

Ferroelectric/antiferromagnetic

Compound	Phase transition	Remark	Ref.
$CaMn_2O_4$	*PT* $mmm1' \xleftrightarrow{225°K} mm21'$		93, 94
$TbCrO_3$	*PT* $mmm1' \xleftrightarrow{3.05°K} mm21'$		93, 95
$BiMn_2O_5$	*PT* $mmm1' \xleftrightarrow{52°K;\ 42°K} mm21'$		93, 96
$BaNiF_4$	*PT* $mm21' \xleftrightarrow{\sim 150°K} 21'$?		78, 79
$BaCoF_4$	$mm21' \xleftrightarrow{69.6°K} 21', m1'$?	*ME*-effect observed, 81[b]	78, 80–82, 117
$BaFeF_4$	$mm21' \xleftrightarrow{?}$?		78
$BaMnF_4$	$mm21' \xleftrightarrow{?}$?		78

[b] The *ME*-effect reported is probably due to the *EHH*-term

TABLE IV (continued)

Ferroelectric/paramagnetic or diamagnetic (most known ferroelectrics belong to this sub-type of *FE* IV)

Perovskites with general formula ABO_3($BaTiO_3$, etc.)

Perovskites with general formula $AB'_{1/3}B''_{2/3}O_3$

Compound	$T_c(FE)$ (°K)	Magnetism	Ref.
$PbCd_{1/3}Nb_{2/3}O_3$	523–73	Dia	42, 83, pp. 126–175
$PbZn_{1/3}Nb_{2/3}O_3$	413	Dia	
$PbMg_{1/3}Nb_{2/3}O_3$	261	Dia	
$PbCo_{1/3}Nb_{2/3}O_3$	203	Para	
$PbNi_{1/3}Nb_{2/3}O_3$	153	Para	
etc.			
$PbMg_{1/3}Ta_{2/3}O_3$	175	Dia	
$PbCo_{1/3}Ta_{2/3}O_3$	133	Para	
$PbNi_{1/3}Ta_{2/3}O_3$	93	Para	

Perovskites with general formula $AB'_{1/2}B''_{1/2}O_3$

Compound	$T_c(FE)$ (°K)	$T_c(FM)$ (°K)	Ref.
$PbFe_{1/2}Nb_{1/2}O_3$	233		42
$PbFe_{1/2}Ta_{1/2}O_3$	233	133	42, 81[c]

[c] In disaccord with Ref. 42, see Table *FE* I–*FM* I.

Boracite family, $M_3B_7O_{13}X$; X = OH, Cl, Br, I, S

Paramagnetic: M = Cr, Mn, Fe, Co, Ni, Cu, Pt, etc., see type *FE* I–*FM* I

Diamagnetic: M = Mg, Zn, Cd, Li ($Li_4B_7O_{12}X$)

Bronzes of the potassium wolframate type

Compound	$T_c(FE)$ (°K)	Magnetism	Ref.
$Ba_4Nd_2Fe_2Nb_8O_{30}$	~393	Para	42, 97, p. 97
$Ba_2Sm_4Fe_3Nb_7O_{30}$	403	Para	42, 97, p. 97
$PbTa_2O_6$	523	Dia	42, 97, p. 97
etc.			

TABLE IV (continued)

Ferroelectric/antiferromagnetic without attribution of type

Compound	Phase transitions		Ref.
Perovskites	$T_c(FE)$ (°K)	$T_N(AFM)$ (°K)	
$Pb(Fe_{1/2}Nb_{1/2})O_3$	385	143	42, p. 399
$Pb(Fe_{2/3}W_{1/3})O_3$	178	263	42, p. 399
$Pb(Co_{1/2}W_{1/2})O_3$		68	42, p. 399
Sr_2CuWO_6	1193	43	72
Pseudo-ilmenite			
$Li(Fe_{1/2}Ta_{1/2})O_2F$	833	883[d]	54

[d] The Néel temperature is unusually high for the small iron content, possibly an artefact due to an impurity.

TABLE IVA

Various *FE*-types with general formula $M_{n+1}R_nO_{3n+3}$ (stacking of perovskite and $(Bi_2O_2)^{2+}$-layers).
($M = Bi^{3+}, Pb^{2+}, Na^+, K^+, Sr^{2+}, Ca^{2+}, Ba^{2+}, RE^{3+}$ (rare earth);
R = Ti, Nb, Ta, Fe, Ga, W, Cr, etc.; n = number of perovskite layers in one block of perovskite)

Composition (examples)	$T_c(FE)$ (°K)	T_N (°K)	Symmetry of proto-typic phase	Symmetry of low temperature phase	Lattice parameters a (Å)	b (Å)	c (Å)	ε at 25°C	Ref.
WO_3									87, p. 88
Bi_2WO_6	1223			$mm21'$ (C_{2v}^{17}–$B2cb$)	5.458	5.438	15.434	60	98
(Russelite)				Type *FE* IV					
$Bi_2Nb(O_5F)$	303			Type *FE* IV					98
$Bi_2Ta(O_5F)$	283			Type *FE* IV					98
$Bi_3(TiNb)O_9$	(I) <1223								
	(II) <1023			$mm21'$ (C_{2v}^{12}?–$A2_1am$) Type *FE* IV	5.431	5.389	25.050	100	98

TABLE IVA (continued)

n	Composition (examples)	$T_c(FE)$ (°K)	T_N (°K)	Symmetry of prototypic phase	Symmetry of low temperature phase	Lattice parameters a (Å)	b (Å)	c (Å)	ε at 25°C	Ref.
3	$Bi_4Ti_3O_{12}$	948		$4/mmm1'$	Type *FE* IV	5.448(2)	5.411(2)	32.83(1)	180	98
4	$Bi_5(Ti_3Fe)O_{15}$	1023[c]			Type *FE* IV				200	98, 77
		853			$mm21'$ (C_{2v}^{12}–$A2_1am$)	5.455	5.445	41.31[c]		
	$Pr_{(Bi)}Bi_4(Ti_3Fe)O_{15}$	≈1023			Type *FE* IV				175	77
	$La_{(Pr)}Bi_4(Ti_3Fe)O_{15}$	803–963			Type *FE* IV				225	77
5	$Bi_6(Ti_3Fe_2)O_{18}$[b]	1083			Type *FE* IV	5.490	5.500	50.185[c]		77
6	No example known									
7	No example known									
8	$Bi_9(Ti_3Fe_5)O_{27}$	1103	403 *WFM*		Type *FM* I–*FE* I	5.491	5.502	76.20		54
Quasi ∞	$BiFeO_3$	1123	648	$m3m1'$	$3m1' \xleftrightarrow{648°K} 3m$ Type *FE* II	a_{hex} 5.59	—	c_{hex} 13.7		56, 99

[a] Some more representatives of the series $M_{n+1}R_nO_{3n+1}$ are given on p. 192 of Ref. 83 and on pp. 107/113 of Ref. 87.
[b] A non-linear magnetoelectric $(ME)_H$-effect has been measured on sintered material.[69, 77]
[c] Data from Ref. 100.

TABLE V

Ferromagnetic magnetoelectric types

Type *FM* 1-*FE* I (*E, H, EH, EHH, HEE*) Ferromagnetic/ferroelectric or polar only

See Table V

Type *FM* II (*H, EH, HEE, EHH*) Ferromagnetic/antiferroelectric or orthoelectric

No example known

Type *FM* III (*H, HEE, EHH*) Ferromagnetic/antiferroelectric or orthoelectric

No example known

Type *FM* IV (H, HEE) Ferromagnetic/antiferroelectric or paraelectric or orthoelectric

Ferromagnetic/orthoelectric

Compound	Phase transitions	β_{ijk}(max) obs.	Ref.
Ferrimagnetic garnets	*PT*		
$Gd_3Fe_5O_{12}$	$m3m1' \xleftrightarrow{\sim 563°K} \bar{3}m'$	4×10^{-6}	101
$Y_3Fe_5O_{12}$	$m3m1' \xleftrightarrow{\sim 563°K} \bar{3}m'$	5×10^{-6}	101
$Dy_3Fe_5O_{12}$	$m3m1' \xleftrightarrow{\sim 563°K} \bar{3}m'$	10^{-5}	101
Weakly ferromagnetic orthoferrites	*PT*		
$TbFeO_3$	$mmm1' \longleftrightarrow mm'm'$	10^{-5}	102
$YbFeO_3$	$mmm1' \longleftrightarrow mm'm'$	10^{-6}	102

Ferromagnetic/antiferroelectric Without attribution of magnetoelectric type

Compound	$T_c(AFE)$ (°K)	$T_c(FM)$ (°K)	Ref.
$BiMnO_3$	773	103	72
$Pb_2Mn^{2+}Re^{6+}O_6$	368	123	72
$Pb_2Mn^{3+}Re^{5+}O_6$	393	103	72
$BiCrO_3$	(*AFE* probable)	123	103
$Cu(HCOO)_2 . 4H_2O$	$m1' \xleftrightarrow{235°K} AFE/\text{Para} \xleftrightarrow{17°K} WFM/AFE$ possibly *FM* IV		97, p. 180

TABLE VI

Examples of magnetoelectric structure types with the property combinations

Antiferroelectric or paraelectric or orthoelectric/antiferromagnetic or para magnetic or diamagnetic

(No literature search concerning types *AA* I, II and IV has been undertaken)

Type *AA* III (EH)

Orthoelectric/antiferromagnetic

Compound	*PT*	Phase transitions		α_{max}	Ref.
Cr_2O_3	$3m1'$	$\xleftrightarrow{318°K}$	$3'm'$	8×10^{-5}	104, 105
Ti_2O_3	$3m1'$	$\longleftrightarrow$	?		106
$LiMnPO_4$	$mmm1'$	$\xleftrightarrow{35°K}$	$m'm'm'$	2×10^{-5}	107
$LiCoPO_4$	$mmm1'$	$\xleftrightarrow{23°K}$	$m'mm$	6×10^{-4}	107
$LiNiPO_4$	$mmm1'$	$\xleftrightarrow{23°K}$	mmm'	4×10^{-5}	107
$LiFePO_4$	$mmm1'$	$\xleftrightarrow{50°K}$	$m'mm$	1×10^{-4}	108
$TbAlO_3$	$mmm1'$	$\xleftrightarrow{4.0°K}$	$m'm'm'$	1×10^{-3}	109
$TbCoO_3$	$mmm1'$	$\xleftrightarrow{3.3°K}$	mmm'	3×10^{-5}	102
$DyAlO_3$	$mmm1'$	$\xleftrightarrow{3.5°K}$	$m'm'm'$	2×10^{-3}	102, 110
$GdAlO_3$	$mmm1'$	$\xleftrightarrow{4.0°K}$	$m'm'm'$	1×10^{-4}	111
$DyPO_4$	$4/mmm1'$	$\xleftrightarrow{314°K}$	$4'/m'mm'$	1×10^{-3}	112
$GdVO_4$	$4/mmm1'$	$\longleftrightarrow$	$4'/m'mm'$	3×10^{-4}	113
Fe_2TeO_6	$4/mmm1'$	$\xleftrightarrow{219°K}$	$4/m'm'm'$	0.3×10^{-4}	114
$GeMnO_3$	$mmm1'$	$\xleftrightarrow{16°K}$	$m'mm$	1.5×10^{-6}	113
$Nb_2Mn_4O_9$	$\bar{3}m1'$	$\xleftrightarrow{110°K}$	$\bar{3}'m'$	2×10^{-6}	115
$Ta_2Mn_4O_9$	$\bar{3}m1'$	$\xleftrightarrow{104°K}$	$\bar{3}'m'$	1×10^{-5}	115
$Nb_2Co_4O_9$	$\bar{3}m1'$	$\xleftrightarrow{21°K}$	$\bar{3}'m'$	2×10^{-5}	115
$Ta_2Co_4O_9$	$\bar{3}m1'$	$\xleftrightarrow{21°K}$	$\bar{3}'m'$	1×10^{-4}	115

TABLE VI (continued)

Type *AA* V (EHH)

i) Antiferroelectric/antiferromagnetic (only $\bar{4}3m$; see Table II), no example or orthoelectric

ii) antiferroelectric/antiferromagnetic (see Table II), or orthoelectric or paramagnetic or diamagnetic

Orthoelectric/paramagnetic

Compound	Shubnikov point group	Ref.
$NiSO_4.6H_2O$	$4221'$	19
$NiSeO_4.6H_2O$		19
$Ni(BrO_3)_3.9H_2O$		19
Antiferroelectric (?)/paramagnetic ("paramagnetoelectric")		
$Ni_3B_7O_{13}Cl$ and other paramagnetic cubic boracites	$\bar{4}3m1'$ (above ~560°K)	116
Antiferroelectric (?)/diamagnetic		
$Mg_3B_7O_{13}Cl$	$\bar{4}3m1'$ (above ~538°K)	117

Antiferroelectric/antiferromagnetic without attribution of type

Compound	$T_N(AFE)$ (°C)	$T_N(AFM)$ (°C)	Ref.
$Pb(Mn_{1/2}Nb_{1/2})O_3$	293	11	72
$Pb(Mn_{2/3}W_{1/3})O_3$	473	93	72
$Cd(Fe_{1/2}Nb_{1/2})O_3$	723	48	72
$Pb(Co_{1/2}W_{1/2})O_3$	293	9	42, p. 383
$Pb(Yb_{1/2}Nb_{1/2})O_3$	573		42, p. 384
$Pb(Yb_{1/2}Ta_{1/2})O_3$	558		42, p. 384
$Pb(Lu_{1/2}Nb_{1/2})O_3$	543		42, p. 384
$Pb(Lu_{1/2}Ta_{1/2})O_3$	551		42, p. 384

REFERENCES

1. L. D. Landau and E. M. Lifshitz, *Electrodynamics of Continuous Media* (Addison-Wesley, Reading, March 1960), translation of a Russian edition of 1958.
2. I. E. Dzyaloshinsky, *J. Exptl. Theor. Phys. (USSR)* **37**, 881 (1959) [*Sov. Phys. JETP* **37**, 628 (1960)].
3. D. N. Astrov, *J. Exptl. Theor. Phys. (USSR)* **38**, 984 (1960) [*Sov. Phys. JETP* **11**, 708 (1960)].
4. V. J. Folen, G. T. Rado and E. W. Stalder, *Phys. Rev. Letters* **6**, 607 (1961).
5. G. T. Rado and V. J. Folen, *Phys. Rev. Letters* **7**, 310 (1961).
6. E. F. Bertaut and M. Mercier, *Mat. Res. Bull.* **6**, 907 (1971).
7. R. M. Hornreich, *IEEE Trans. Mag.* **8** (*Proceedings of the INTERMAG* 1972), 584 (1972).
8. G. Smolensky and V. A. Joffe, Communication Nr. 71 du Colloque International de Magnétisme, Grenoble 1958.
9. H. Gränicher and K. A. Müller, *Mat. Res. Bull.* **6**, 977 (1971).
10. R. R. Birss, *Symmetry and Magnetism* (North Holland Publishing Company, Amsterdam, 1964).
11. E. Ascher, *Phil. Mag.* **17**, 149 (1968).
12. Y. Le Corre, *J. Phys. Radium* **19**, 750 (1958).
13. V. A. Koptsik, *Šubnikovskie gruppy* (Izdatel'stvo Moskovskovo Universiteta, 1966), Table 20, p. 85.
14. R. Schelkens, *Phys. Stat. Sol.* **37**, 739 (1970).
15. E. Ascher, *Helv. Phys. Acta* **39**, 466 (1966).
16. P. A. Franken and J. F. Ward, *Reviews of Mod. Physics* **35**, 23 (1963).
17. I. F. Dzialoshinsky, *Sov. Phys. JETP* **5**, 1259 (1957).
18. B. A. Tavger, *Kristallografiya* **3**, 339 (1958).
19. S. L. Hou and N. Bloembergen, *Phys. Rev.* **138A**, 1218 (1965).
20. W. Jeitschko, *Acta Cryst.* **328**, 60 (1972).
21. E. Ascher, *J. Phys. Soc. Jap.* **28** (suppl.) 7 (1970).
22. E. Ascher, *Helv. Phys. Acta* **39**, 40 (1966).
23. E. Ascher, this Symposium.
24. F. Jona and S. Shirane, *Ferroelectric Crystals* (Pergamon Press, Oxford, 1962).
25. T. H. O'Dell, *The Electrodynamics of Magneto-Electric Media* (North Holland, Amsterdam/London, 1970).
26. E. Ascher, *Group Theoretical Considerations on Ferroelectric Phase Transitions and Polarization Reversal*, Lecture Note, 1967.
27. A. Kumada, *Ferroelectrics* **3**, 115/121 (1972).
28. H. Schmid, unpublished.
29. T. J. Martin, *Phys. Letters* **17**, 83 (1965).
30. T. J. Martin and J. C. Anderson, *IEEE Trans. on Magnetics*, MAG-2, 466 (1966).
31. A. S. Borovik-Romanov, *J. Exptl. Theor. Phys. (USSR)* **38**, 1088 (1960) [*Sov. Phys. JETP* **11**, 786 (1960)].
32. K. Aizu, *Phys. Rev.* **B2**, 754 (1970); *J. Phys. Soc. Jap.* **27**, 1171 (1969).
33. A. P. Cracknell, *Acta Cryst.* **A28**, 597 (1972).
34. V. Janovec and L. A. Shuvalov, this Symposium.
35. E. Ascher and A. Janner, *Acta Cryst.* **18**, 325 (1965).
36. E. Wendling, R. V. Hodenberg and R. Kühn, *Kali und Steinsalz* **6**, 1 (1972).
37. R. M. Honea and F. R. Beck, *Amer. Mineralogist* **47**, 665 (1962).
38. B. T. Matthias, *Phys. Rev.* **75**, 1771 (1949).
39. B. T. Matthias, *Helv. Phys. Acta* **23**, 167 (1950).
40. G. A. Smolensky and N. V. Koževnikova, *Doklady AN SSSR* **76**, 519 (1951).
41. G. A. Smolensky, *Izvjestija AN SSSR* **20**, 163 (1956).
42. G. A. Smolensky, V. A. Bokov, V. A. Isupov, N. N. Krainik, R. E. Pas'inkov, M. S. Shur, *Segnetoelektriki i antisegnetoelektriki* (Izdatel'stvo "Nauka", Leningrad, 1971); Engl. transl.: G. A. Smolensky *et al.*, *Ferroelectrics and Antiferroelectrics* (National Technical Information Service, U.S. Dept. of Commerce, 5285 Port Royal Road, Springfield VA. 22151, 27 Jan. 1972, Nr. Ad-741037).
43. G. A. Smolensky, V. A. Isupov, N. N. Krainik and A. I. Agranovskaya, *Izvest. Akad. Nauk SSSR*, Ser. Fiz. **25**, 1333 (1961).
44. Report No. 1 for C.N.E.T. "Recherche sur des composés possédant des propriétés ferroélectriques et ferrimagnétiques simultanées". Battelle-Geneva, 1961.
45. See footnote on page 13 of Ref. 21.
46. H. Schmid, *J. Phys. Chem. Sol.* **26**, 973 (1965).
47. G. Quézel and H. Schmid, *Solid State Comm.* **6**, 447 (1968).
48. H. Schmid, H. Tippmann and J. Kobayashi, unpublished (1971).
49. E. F. Bertaut, M. Mercier and R. Pauthenet, *J. de Physique* **25**, 550 (1964).
50. D. Bernard, S. Le Montagne, J. Pannetier and J. Lucas, *Mat. Res. Bull.* **6**, 75 (1971).
51. J. Pannetier, Y. Calage and J. Lucas, *Mat. Res. Bull.* **7**, 57 (1972).
52. J. Pannetier and J. Lucas, *Mat. Res. Bull.* **5**, 797 (1970).
53. T. G. Dunne and N. R. Stemple, *Phys. Rev.* **120**, 1949 (1960).
54. I. H. Ismailzade, R. G. Yakupov and T. A. Melik-Shanazarova, *Phys. Stat. Sol.* (a) K63 (1971).
55. C. Michel, J. M. Moreau, G. D. Achenbach, R. Gerson and W. J. James, *Solid State Comm.* **7**, 701 (1969).
56. J. R. Teague, R. Gerson and W. J. James, *Solid State Comm.* **8**, 2073 (1790).
57. J. M. Moreau, C. Michel, R. Gerson and W. J. James, *J. Phys. Chem. Sol.* **32**, 1315 (1971).
58. J. B. Goodenough, *Magnetism and the Chemical Bond* (Interscience, New York, 1963).
59. E. Ascher, H. Rieder, H. Schmid and H. Stössel, *J. Appl. Phys.* **37**, 1404 (1966).
60. H. Schmid, *Growth of Crystals* **7**, 25 (1969), Consultants Bureau; translated from *Rost Kristallov* **7**, 32 (1967).
61. G. Winter, *Magnetoelektrischer und pyroelektrischer Effekt im Nickel Iod-Boracit* (Diplomarbeit, Techn. Hochschule Darmstadt, 1968).
62. W. von Wartburg, Thesis (Federal Polytechnic Institute, Zürich, 1973).
63. M. Mercier, *Qualitative Measurement of the* ME_H *and* ME_E *Effect*, private communication.
64. T. Miyashita and T. Murakami, *J. Phys. Soc. Jap.* **29**, 1092 (1970).
65. B. I. Al'shin, D. N. Astrov and Yu. M. Gufan, *Fiz. Tverd. Tela* **12**, 2666 (1970) [*Sov. Phys.—Solid State* **12**, 2143 (1971)].
66. H. Schmid, *Phys. Stat. Sol.* **37**, 209 (1970); *J. Phys. Soc. Jap.* **28**, (Suppl.), 354 (1970).
67. J. M. Trooster, *Phys. Stat. Sol.* **32**, 179 (1969).
68. H. Schmid, A. Zimmermann and W. v. Wartburg, from unpublished data, 1971.

69. I. H. Ismailzade, R. G. Yakupov and T. A. Melik-Shanazarova, *Phys. Stat. Sol.* (a) K85 (1971).
70. S. V. Kiselev, R. P. Ozerov and G. S. Zhdanov, *Dokl. Akad. Nauk SSSR* **145**, 1255 (1962) [*Soviet Physics-Doklady* **7**, 742 (1963)].
71. K. P. Mitrofanov, A. S. Viskow, M. V. Plotnikova, Yu. N. Venevtsev and V. S. Shpinel, *Izvest. Akad. Nauk SSSR* Ser. Fiz. **29**, 2029 (1965).
72. Yu. N. Venevtsev, V. N. Lyubimov, V. V. Ivanova and G. S. Zhdanov, *J. Phys.* **33** (1972) Nr. 4 (suppl.) C2-255.
73. G. T. Rado, *Phys. Rev. Lett.* **13**, 335 (1964).
74. A. S. Viskov *et al.*, *Inorg. Mat.* **4**, 71 (1968); *Iz. Akad. Nauk SSSR, Neorganicheskie Materialy* **4**, 88 (1968).
75. D. N. Astrov, B. I. Al'shin, R. V. Zorin and L. A. Drobyshev, *Zh. Eksp. Teor. Fiz.* **55**, 2122 (1968) [*Sov. Phys. JETP* **28**, 1123 (1969)].
76. L. A. Drobyshev, B. I. Al'shin, Yu. Ya. Tomashpolskii and Yu. Venevtsev, *Kristallografiya* **14**, 736 (1969) [*Sov. Phys. Cryst.* **14**, 634 (1970)].
77. I. H. Ismailzade, *J. Phys.* (Suppl.) **30**, Fasc. 4, C2-236 (1972).
78. M. DiDomenico Jr., M. Eibschütz, H. J. Guggenheim and I. Camlibel, *Solid State Comm.* **7**, 1119 (1969).
79. D. E. Cox, M. Eibschütz, H. J. Guggenheim and L. Holmes, *J. Appl. Phys.* **41**, 943 (1970).
80. E. T. Keve, S. C. Abrahams and J. L. Bernstein, *J. Chem. Phys.* **53**, 3279 (1970).
81. R. V. Zorin, B. I. Al'shin, D. N. Astrov and N. A. Nyunina, *Fiz. Tverd. Tela* **13**, 2152 (1971) [*Sov. Phys.—Solid State* **13**, 1807 (1792)].
82. B. I. Al'shin, D. N. Astrov, A. V. Tishchenko and S. V. Petrov, *ZhETF, Pis. Red.* **12**, 206 (1970); *JETP Letters* **12**, 142 (1970).
83. Landolt-Börnstein, *Zahlenwerte und Funktionen*, Neue Serie, Gr. III, Vol. 4 (1970), pp. 22–23.
84. G. Faivre and G. Saada, *Phys. Stat. Sol.* (b) **52**, 127 (1972).
85. R. A. Fisher, E. W. Hornung, G. E. Brodale and W. F. Giauque, *J. Chem. Phys.* **56**, 193 (1972); **56**, 5007 (1972); **56**, 6118 (1972).
86. G. Gorodetsky, M. Sayar and S. Shtrikman, *Mat. Res. Bull.* **5**, 253 (1970).
87. Landolt-Börnstein, *Zahlenwerte und Funktionen*, Neue Serie, Gr. III, Vol. 3 (1969), p. 94.
88. E. F. Bertaut, R. Pauthenet and M. Mercier, *Phys. Letters* **18**, 13 (1965).
89. I. Chappert, *Phys. Letters* **28**, 229 (1965).
90. D. E. Cox, this Symposium.
91. Ph. Coeuré, P. Guinet, I. C. Peuzin, G. Buisson and E. F. Bertaut, *Proc. Int. Meeting on Ferroelectricity, Prague, June 28/July 1* (1966).
92. W. C. Koehler, H. L. Yakel, E. O. Wollan and J. W. Cable, *Phys. Letters* **9**, 93 (1963).
93. S. Goshen, D. Mukamel, M. Shaked and S. Shtrikman, *J. Appl. Phys.* **40**, 1590 (1969).
94. Y. Allain and B. Boucher, *J. de Phys.* **26**, 789 (1965).
95. E. F. Bertaut, J. Mareschal and G. F. de Vries, *J. Phys. Chem. Sol.* **28**, 2143 (1967).
96. E. F. Bertaut, G. Buisson, S. Quézel-Ambrunaz and G. Quézel, *Solid State Comm.* **5**, 25 (1967).
97. Landolt-Börnstein, *Zahlenwerte und Funktionen*, Neue Serie, Gr. III, Vol. 3, p. 97.
98. R. E. Newnham, R. W. Wolfe and J. F. Dorian, *Mat. Res. Bull.* **6**, 2029 (1971).
99. Yu. Ye. Roginskaya, Yu. Ya. Tomashpol'skii, Yu. N. Venevtsev, V. M. Petrov and G. S. Zhdanov, *Dokl. Akad. Nauk SSSR* **153**, 1313 (1963); *Kristallografiya* **9**, 746 (1964); **12**, 252 (1967).
100. I. H. Ismailzade, V. I. Nesterenko, F. A. Mirishli and P. G. Rustamov, *Kristallografiya* **12**, 468 (1967) [*Sov. Phys. Cryst.* **12**, 400 (1967)].
101. G. Lee, M. Mercier and P. Bauer, Colloque International du CNRS, No. 180. *Les éléments des Terres Rares*, Tome II, p. 389, Paris-Grenoble (1969).
102. M. Mercier and P. Bauer, Colloque International du CNRS, No. 180. *Les éléments des Terres Rares*, Tome II, p. 377, Paris-Grenoble (1969).
103. F. Sugawara, S. Iida, Y. Syono and S. Akimoto, *J. Phys. Soc. Jap.* **25**, 1553 (1968).
104. D. N. Astrov, *Zh. Eksp. Teor. Fiz.* **38**, 984 (1960) [*Sov. Phys. JETP* **11**, 708 (1960)].
105. D. N. Astrov, *Zh. Eksp. Teor. Fiz.* **40**, 2035 (1961) [*Sov. Phys. JETP* **13**, 729 (1961)].
106. B. I. Al'shin and D. N. Astrov, *Zh. Eksp. Teor. Fiz.* (*USSR*) **44**, 1195 (1963) [*Sov. Phys. JETP* **17**, 809 (1963)].
107. M. Mercier, J. Gareyte and E. F. Bertaut, *Compt. Rend.* **264B**, 979 (1967).
108. M. Mercier, P. Bauer and B. Fouilleux, *Compt. Rend.* **267B**, 1345 (1968).
109. M. Mercier and B. Cursoux, *Solid State Comm.* **6**, 207 (1968).
110. L. M. Holmes, L. G. Van Uitert and G. W. Hull, *Solid State Comm.* **9**, 1373 (1971).
111. M. Mercier and G. Velleau, *J. de Phys.* **32**, C1, 499 (1971).
112. G. T. Rado, *Phys. Rev. Lett.* **23**, 644 (1969).
113. G. Gorodetsky, R. M. Hornreich and B. Sharon, *Phys. Letters* **A39**, 155 (1972).
114. S. Buksphan, E. Fischer and R. M. Hornreich, *Solid State Comm.* **10**, 657 (1972).
115. E. Fischer, G. Gorodetsky and R. M. Hornreich, *Solid State Comm.* **10**, 1127 (1972).
116. F. Jona, *J. Phys. Chem.* **63**, 1750 (1959).
117. A. S. Sonin and I. S. Zheludev, *Kristallografiya* **8**, 183 (1963).
118. M. Eibschütz, L. Holmes, M. J. Guggenheim and D. E. Cox, *Phys. Rev.* **B6**, 2677 (1972).
119. W. F. Brown Jr., R. M. Hornreich and S. Shtrikman, *Phys. Rev.* **168**, 574 (1968).
120. E. Ascher and A. G. M. Janner, *Phys. Letters* **A29**, 295 (1969).
121. L. A. Drobyshev, B. I. Al'shin, Yu. Ya. Tomashpol'skii and Yu. N. Venevtsev, *Kristallografiya* **14**, 736 (1969) [*Sov. Phys. Cryst.* **14**, 634 (1970)].
122. G. T. Rado and V. J. Folen, *J. Appl. Phys.* suppl. **33**, 1126 (1962).
123. G. T. Rado, *Phys. Rev.* **128**, 2546 (1962).

Discussion

G. T. RADO *Comment*: A type of switching which may well be added to Table III of Dr. Schmid's paper is "metamagnetic switching", a method for "time-reversing" an antiferromagnetic crystal. As demonstrated in $DyPO_4$ [G. T. Rado, *Phys. Rev. Letters* **23**, 644 (1969); **23**, 946E (1969)], the sign of the magnetoelectric susceptibility α (and thus the phase of the majority of the antiferromagnetic domains) can be changed by means of a metamagnetic transition. The sign of α depends uniquely on the sign of that static magnetic field which was used in the "most recent" metamagnetic transition prior to the measurement of α. The memory aspect of this switching phenomenon is probably due to impurities or imperfections which act as "signposts of time".

S. IIDA *Comment to Dr. Rado's comment*: In a study of α-Fe_2O_3, we found that the direction of the residual magnetization reverses when cooled down through its ferro-antiferromagnetic transition temperature ($\sim -20°C$) and heated up again. This process can be repeated indefinitely and we have explained this phenomenon in terms of nucleation and growth procedure. The detailed mechanism is reported in the Proceedings of the International Conference on Magnetism, Nottingham, England 1964 (the Institute of Physics and the Physical Society, London, 1965), and I believe that the phase of the antiferromagnetic state changes alternatively just as in the case of Dr. Rado's comment.

L. M. HOLMES *Question on Dr. Rado's comment*: Could you obtain metamagnetic "switching" of the magnetoelectric susceptibility in $DyPO_4$ on single-domain samples as well?

G. T. RADO We observed metamagnetic switching of the magnetoelectric susceptibility α in $DyPO_4$ on single-domain crystals as well as on multi-domain crystals. However, this switching caused the single-domain crystals to become multi-domain crystals, i.e., the switching caused the maximum of $|\alpha|$ to decrease.

MAGNETO ELECTRIC MEASUREMENTS ON HOLMIUM PHOSPHATE, $HoPO_4$†

A. H. COOKE, S. J. SWITHENBY, and M. R. WELLS

The Clarendon Laboratory, Parks Road, Oxford, England

Measurements of the electric field induced magneto-electric susceptibility in holmium phosphate, $HoPO_4$, have been made in the temperature range 0.5 to 1.7 K. The result confirms that the magnetic structure in the ordered state corresponds to a simple two sub-lattice Ising antiferromagnet. The form of the reduced magneto-electric susceptibility versus temperature curve is compared with the theory of Essam and Sykes for the Ising diamond ($z = 4$) lattice. Good correlation is obtained in the critical region just below the Néel point 1.39 K.

1 INTRODUCTION

Holmium phosphate, $HoPO_4$, is one of the isomorphous family of compounds with the general formula RXO_4, where R is a tri-valent rare earth ion and X is vanadium, arsenic or phosphorus. The properties of these compounds have been extensively studied in the past few years. At room temperature they possess the tetragonal zircon structure[1] and space group $I4_1/amd$, (D_{4h}^{19}), but at low temperatures the vanadates and arsenates of dysprosium, terbium and thulium undergo a co-operative Jahn-Teller distortion. A short review of these compounds has been given by Gehring.[2] $GdVO_4$[3] and $GdAsO_4$[4] order antiferromagnetically at temperatures in the liquid helium range. Amongst the phosphates $TbPO_4$ exhibits a magnetically controllable crystallographic phase transition at 3.5 K and an antiferromagnetic transition at 2.2 K[5] whereas $DyPO_4$ behaves like an almost ideal Ising antiferromagnet.[6–8]

In our previous measurements of the magnetic properties of $HoPO_4$ we have shown that at 1.4 K there is a transition to an antiferromagnetic state;[9] the axis of sub-lattice alignment corresponding to the tetragonal *c*-axis. Our analysis demonstrated that the magnetic behavior of $HoPO_4$ could be adequately described in terms of the Ising model for the ($z = 4$) diamond lattice as presented by Essam and Sykes.[10]

In order to extend the scope for comparison with this theoretical treatment, as well as to confirm the ordered spin arrangement proposed in our previous work[9] a study of the (electric field induced) magneto-electric susceptibility has been undertaken. The investigations presented here closely

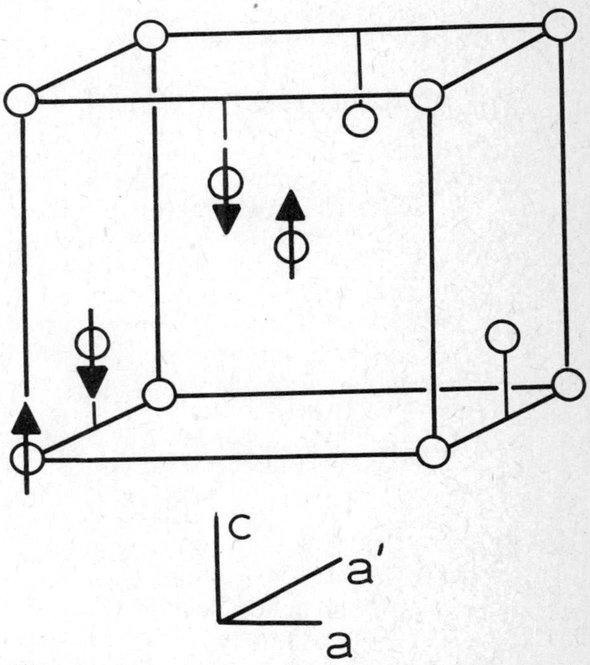

FIGURE 1 The Ho^{3+} atoms of a unit cell of $HoPO_4$ indicating the ordered antiferromagnetic structure.

† Part of this paper was presented at the "Symposium on Magnetoelectric Interaction Phenomena in Crystals" May 21, 1973 at Battelle Seattle Research Center, Seattle, Washington.

follow Rado's analysis[8] of the isomorphous compound $DyPO_4$.

The possible ordering modes for crystals with the tetragonal zircon structure have been discussed by Gorodetsky *et al.*[11] They conclude that in the ordered state, when the axis of sub-lattice alignment corresponds to the *c*-axis, a non-zero magneto-electric susceptibility is only possible if an origin ion has its nearest neighbors antiparallel to itself, an arrangement which is shown in Figure 1. The magneto-electric susceptibility tensor then has the form:

$$\begin{pmatrix} \alpha & 0 & 0 \\ 0 & -\alpha & 0 \\ 0 & 0 & 0 \end{pmatrix}$$

2 EXPERIMENTAL DETAILS

The single crystals used in the measurements were pure and optically clear, with typical dimensions $1.5 \times 1.6 \times 3.4$ mm, and had been flux grown in this laboratory.[12] Measurements have been carried out in the temperature range 0.5 K to 1.7 using a liquid helium three cryostat, which has been more fully described elsewhere.[13] A schematic diagram of the cryostat and probe assembly is shown in Figure 2. In our experiments the polarizing electric and magnetic fields were applied perpendicular to the crystalline *c*-axis, i.e. parallel to the *a*- or *a'*-axis. The electric field was applied across the sample through the axially symmetric probe assembly as shown in Figure 2. This arrangement eliminates any stray pick-up from the displacement currents in the crystal. An astatic pair of pick-up coils, with 5000 turns of 46 s.w.g. copper wire in each half, could be positioned to minimize the pick-up from the current fluctuations in a water-cooled solenoid capable of providing a magnetic field of 14 kOe. The signal from the pick-up coils was detected using a commercially available Brookdeal lock-in amplifier. The probe assembly containing the sample was positioned in the central tube of the cryostat, into which a small quantity of ^{3}He exchange gas could be admitted. This tube was surrounded by an annular reservoir of ^{3}He liquid; this arrangement provides good thermal contact between the liquid and the sample. This is of particular importance at temperatures below ~0.8 K because of the large specific heat contribution from the hyperfine interaction which might give rise to temperature inaccuracies. In the environment of low-pressure gas (less than 5 mmHg of ^{3}He gas at 0.5 K) the electric potential difference which could be maintained between the probe electrodes was limited to ~500 V because of the possibility of electrical discharge.

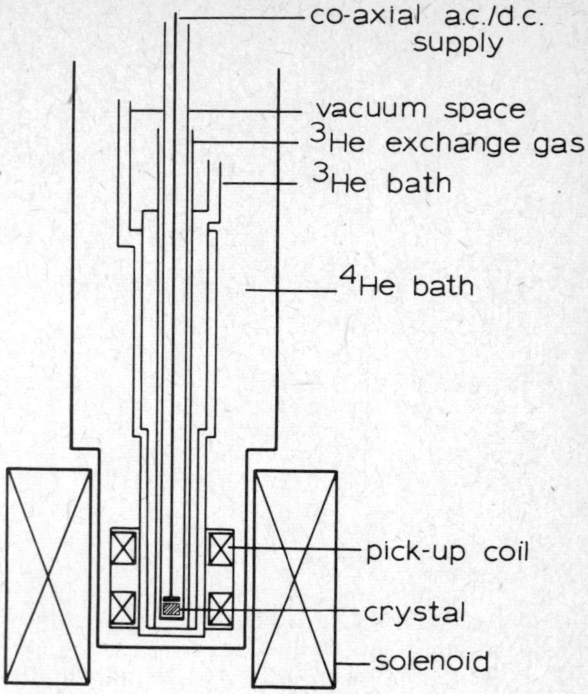

FIGURE 2 Schematic arrangement of the apparatus.

The technique employed in obtaining measurements was to cool the sample through its Néel temperature in the presence of static annealing fields of 1 kOe and 1500 V/cm and a modulating a.c. electric field of 2300 V/cm peak-to-peak at a frequency of 1470 Hz. At a temperature of 1.39 K, which corresponds to the Néel temperature for $HoPO_4$, a signal was observed. The static fields were reduced to zero at $T = 1.33$ K and measurements were recorded in this configuration until the sample had cooled to the lowest temperature, 0.5 K. The observed variation of α, in reduced form α/α_0, versus reduced temperature, T/T_N, is shown as the dashed curve in Figure 3. The curve shows that the magneto-electric susceptibility rises abruptly just below T_N, reaches a maximum value and then decreases, tending to zero at $T = 0$ K. The value of the magneto-electric susceptibility obtained was shown to be independent of the amplitude (0–2300 V/cm) and the frequency (300–3 kHz) of the

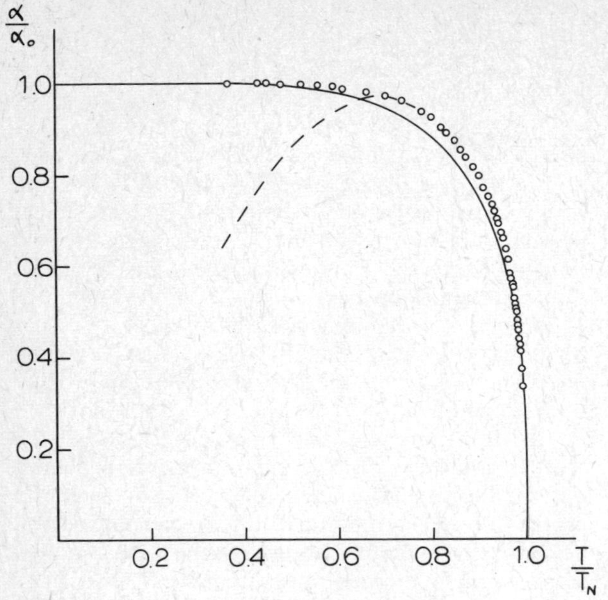

FIGURE 3 The reduced magneto-electric susceptibility, α/α_0, versus reduced temperature, T/T_N, for $HoPO_4$ (*a*-axis); ○ experimental points obtained in static fields of 1 kOe and 1500 V/cm; —— experimental curve obtained in zero static fields; —— theoretical curve of Essam and Sykes for the Ising model.

modulating electric field. The very presence of a signal for a polarizing field E parallel to the *a*- or *a'*-axis confirms our belief that the ordered state is as predicted from our previous measurements.[9]

This variation in the magneto-electric susceptibility was surprising. It was expected that the form of the curve would follow the reduced sub-lattice magnetization curve, as predicted and observed by Rado[8] in $DyPO_4$. Consequently an investigation of the value of α as a function of applied electric and magnetic fields was undertaken. At the lowest temperature, 0.5 K, the static d.c. electric field of 1500 V/cm was applied and the electric-field-induced magneto-electric susceptibility was measured as a function of applied magnetic field. It was found that a saturation value of α could be obtained for magnetic fields in excess of 1 kOe. If these static fields were reduced to zero, a lower value of α, identical to the original value, was observed without any suggestion of hysteresis.

It was assumed that this enhancement in the value of α, which required the presence of both electric and magnetic fields was caused by the reorientation of antiferromagnetic domains. Therefore, further experiments were conducted in which the static magnetic and electric fields of 1 kOe and 1500 V/cm respectively were maintained whilst the temperature was reduced from 1.7 to 0.5 K. The experimental points obtained are shown in Figure 3 and the form of the curve has the overall appearance of the sub-lattice magnetization curve for an antiferromagnetically ordered system. The saturation value of α (where $\alpha = M/E$) is $\approx 2 \times 10^{-4}$ in dimensionless Gaussian units.

3 DISCUSSION

Our previous results[9] have shown that $HoPO_4$ follows closely the Ising model predictions for a diamond ($z = 4$) lattice. As Essam and Sykes point out[10] this structure is particularly amenable to a theoretical treatment, as the power series expressions for the measurable thermodynamic quantities are convergent at all temperatures. The measured dependence of the magneto-electric susceptibility furnishes another field of comparison with this model.

Rado[8] has shown that, under certain conditions, the transverse magneto-electric susceptibility of an Ising antiferromagnet with $S = \frac{1}{2}$ is proportional to the sub-lattice magnetization:

$$\frac{\alpha}{\alpha_0} = \frac{M}{M_0}$$

where α_0 and M_0 are the values of the magneto-electric susceptibility and the sub-lattice magnetization at $T = 0$ K. Strictly, the validity of this relationship depends on the assumption that the magneto-electric effect arises from a single-ion mechanism, and further that the ground state splitting in the antiferromagnetic state at $T = 0$ K is small compared with the energy of the first excited state. Rado has expressed this by imposing an upper limit on the parameter Δ_0, which is proportional to the ratio of the ground state splitting to the energy separation of the first excited state at $T = 0$ K; the value of Δ_0 should be small compared with unity. Previous measurements[9] have shown the ground state splitting at $T = 0$ K to be 3 cm^{-1} and the results of optical absorption spectroscopy on dilute samples of Ho^{3+} in YPO_4 by Becker *et al.*[14] have shown the first excited state to be at 66 cm^{-1}. Using these figures we obtain a value of $\Delta_0 \simeq 0.03$, which is a somewhat smaller value than that found in $DyPO_4$, $\Delta_0 \simeq 0.1$[8] and $DyAlO_3$, $\Delta_0 \simeq 0.15$[15]. However, the value of Δ_0 clearly falls within the constraint set by Rado.

Essam and Sykes[10] have given the series expansion for the Ising model in terms of the parameter $\exp[(2J_e/kT)]$ for the temperature dependence of the sub-lattice magnetization. This series expansion is computed as a function of reduced temperature T/T_N and the form of the curve obtained is shown as the solid line in Figure 3. This theoretical curve does not fit the experimental data as closely as might have been expected in view of the high degree of success achieved in explaining the specific heat data[9] using a similar approach. A further point of comparison is provided by the theoretical predictions for the critical behaviour of the sub-lattice magnetization in the temperature range just below T_N. In this region the predicted dependence[10] is

$$\frac{M}{M_0} = \frac{\alpha}{\alpha_0} = B(1 - T/T_N)^{\beta}$$

Adopting the value of $T_N = 1.391$ K as was obtained from our specific heat data, a plot of log (α/α_0) versus log $(1 - T/T_N)$ produced the straight line shown in Figure 4. From the graph the critical parameters were evaluated as $\beta = 0.315 \pm 0.01$ and

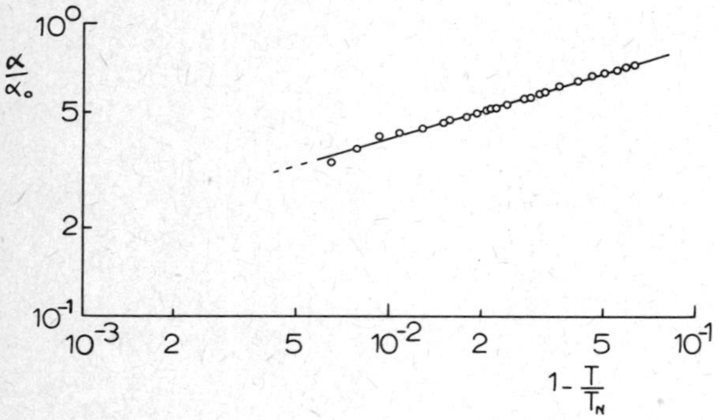

FIGURE 4 α/α_0 versus $(1 - T/T_N)$ for $HoPO_4$ in the temperature region just below T_N.

$B = 1.72 \pm 0.08$. The Ising model predictions of Baker and Daunt[16] give these quantities as $0.314 > \beta > 0.307$ and $B = 1.661 \pm 0.001$. By way of comparison Rado adopted the values of $\beta = 0.314$ and $B = 1.661$ in his analysis of $DyPO_4$. Using these figures a satisfactory fit to his data was obtained, upholding his supposition that the origin of the magneto-electric susceptibility in $DyPO_4$ is a single ion effect.

The non-hysteretic dependence of the magneto-electric susceptibility in the presence of static biassing fields prompted us to investigate $DyPO_4$. In this compound we found that the electric-field-induced magneto-electric susceptibility also diminished when the electric and magnetic annealing fields were reduced to zero. A saturation value of the magneto-electric susceptibility at low temperatures could only be achieved when static fields of 3 kOe and 2500 V/cm were applied. This contrasts with the results of Rado for the magnetically induced effect in which he reported no change in the signal on the removal of the static annealing fields. This apparent discrepancy may be the result of the different methods of inducing the magneto-electric effect. Further investigations into the effects of the static biassing fields on the magneto-electric susceptibility are being undertaken.

Measurements are also planned to investigate the correlation of the spin-flip phenomenon observed through the magneto-electric susceptibility at the lowest temperature, 0.5 K, which is well below the Néel temperature, with the data obtained from magnetic moment measurements.[9] The results of all these investigations will be reported in a later paper.

ACKNOWLEDGEMENTS

We wish to thank Mrs. B. M. Wanklyn of this laboratory who grew the crystals of $HoPO_4$. One of us (SJS) wishes to acknowledge the support of a SRC grant.

REFERENCES

1. R. W. G. Wyckoff, *Crystal Structures* (Interscience, London, 1965), Vol. 3.
2. K. A. Gehring, *A.I.P. Conf. Proc.* **10**, 1648 (1973).
3. J. D. Cashion, A. H. Cooke, L. A. Hoel, D. M. Martin, and M. R. Wells, *Colloques Internationaux du C.N.R.S.* No. 180, Tome II, 417 (1970).
4. J. H. Colwell, B. W. Mangum, and D. D. Thornton, *Phys. Rev. B* **3**, 3855 (1971).
5. H. C. Schopper, *Int. J. Magnetism*, **3**, 23, (1972).
6. J. C. Wright, H. W. Moos, J. H. Colwell, B. W. Mangum, and D. D. Thornton, *Phys. Rev. B* **3**, 843 (1971).
7. C. J. Ellis, M. J. M. Leask, D. M. Martin, and M. R. Wells, *J. Phys. C* **4**, 2937 (1971).
8. G. T. Rado, *Sol. State Comm.* **8**, 1349 (1970).
9. A. H. Cooke, S. J. Swithenby, and M. R. Wells, *J. Phys. C* **6**, 2209 (1973).
10. J. W. Essam and M. F. Sykes, *Physica* **29**, 378 (1963).
11. G. Gorodetsky, R. M. Hornreich, and B. M. Wanklyn, to be published in *Phys. Rev. B* (1973).
12. B. M. Wanklyn, *J. Mat. Sci.* **7**, 813 (1972).
13. J. D. Cashion, A. H. Cooke, T. L. Thorp, and M. R. Wells, *Proc. Roy. Soc.* **A318**, 473 (1970).
14. P.-J. Becker, H. G. Kahle, and D. Kuse, *Phys. Stat. Sol.* **36**, 695 (1969).
15. L. M. Holmes, L. G. Van Uitert, and G. W. Hull, *Sol. State Comm.* **9**, 1373 (1971).
16. G. A. Baker and D. S. Daunt, *Phys. Rev.* **155**, 545 (1967).

A Simple Method for the Study of Antiferromagnetic Domains Switching by Magnetoelectric Effect

B. TERRET and M. MERCIER

Institut Universitaire de Technologie, Av. A. Briand, 03107 Montluçon, France

and

J. C. PEUZIN

Centre d'Etudes Nucléaires, Av. des Martyrs, 38 Grenoble, France

Antiferromagnetic "180° domains" may be switched alternatively from one state to the other under pulsed electric field and constant magnetic field. We describe a very simple circuit that allows to separate the purely magnetoelectric from the purely dielectric charge that flows through the sample during this process. Application is made for Cr_2O_3.

I PRINCIPLE AND DESCRIPTION OF THE METHOD

Antiferromagnetic domain reversal in Cr_2O_3 under simultaneously applied electric and magnetic fields was first studied by Martin and Anderson,[1] then by Brown and O'Dell.[2] These authors both made use of the magnetoelectric effect to probe the domain state of a Cr_2O_3 sample.

The two methods consist basically in measuring the mean magnetoelectric susceptibility $\bar{\alpha}$ which is known to vary from $+\alpha$ to $-\alpha$ as the domain state is changed from positive to negative. Martin and Anderson's method is a static one: it does not allow us to measure the instantaneous value of a rapidly varying $\bar{\alpha}$.

In contrast, Brown and O'Dell's method is a dynamic one and allows detailed study of the switching process: in this method, the instantaneous mean magnetoelectric susceptibility is monitored by a sampling technique in which short magnetic field pulses are applied to the specimen and the resulting voltage pulses are detected.

Periodic switching is obtained by using a constant magnetic field and an alternating electric field, the period of which is large compared to switching time.

It was thought desirable to design a more simple method: firstly, we wanted to avoid the use of a pulse magnetic field and secondly, we wished to achieve excellent decoupling between voltage source and measured signal.

Consider a Cr_2O_3 sample, in the usual plane capacitor configuration, which is being switched from one domain state to the other. The instantaneous charge $Q(t)$ is given by:

$$Q(t)=\left(\int_0^t \frac{V(t')}{R}\,dt'\right)+CV(t)+G\bar{\alpha}(t)H \qquad (1)$$

where C and R are respectively the capacitance and the resistance of the sample, $\bar{\alpha}(t)$ its instantaneous mean magnetoelectric susceptibility, and G a geometrical factor. H is the applied magnetic field which is kept constant and $V(t)$ is the time dependent applied voltage.

If $V(t)$ is an alternating function of time, the product $H \cdot V(t)$ will reverse periodically and switching may occur.

It is seen that if one succeeds in eliminating the capacitive and the resistive terms in Eq. (1), the remaining charge will be directly proportional to $\bar{\alpha}(t)$ and thus will allow us to test domain state.

This may be done using a differential technique which is schematically illustrated by Figure 1: the driving voltage is simultaneously applied on the Cr_2O_3 sample and on an "attenuator inverter" which transforms the voltage $V(t)$ to $-AV(t)$ where A may be adjusted in the range $\{0-\frac{1}{50}\}$. This

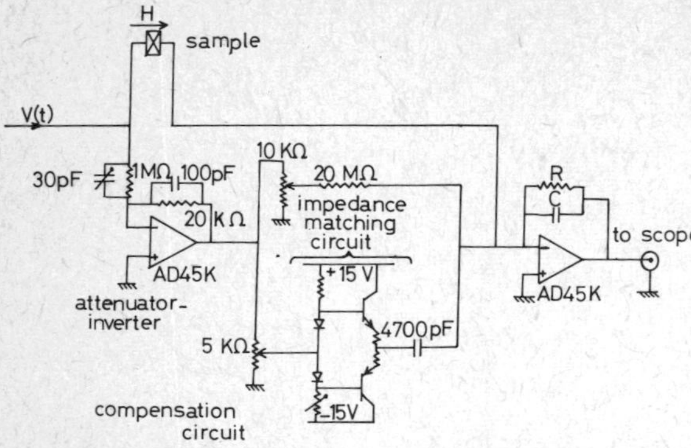

FIGURE 1 Schematic diagram of experimental set up.

attenuated-inverted voltage is then used to generate variable compensating currents by means of standard resistors and capacitors.

Compensating currents and sample current simultaneously feed a closed loop operational amplifier. This one may work as a current or a charge amplifier depending upon whether a resistor or a capacitor is used to close the loop.

Before any measurement can be made, it is necessary to adjust the compensating currents and this is done in the following way: the magnetic field is set equal to zero and voltage is applied to the crystal. In this way no antiferromagnetic domain reversal occurs and the only charge flowing through the sample is that due to its capacitance and resistance. Compensation currents may then be adjusted until the output of the amplifier is balanced at zero.

Then a magnetic field is applied and, if switching occurs, the output of the amplifier is just proportional to the instantaneous mean magnetoelectric susceptibility as required.

Several advantages are to be found, in our opinion, in the present method: it has inherently wide bandwidth and allows fast switching studies (in the present configuration switching times of 100 μs can be accurately measured. Improvement by a factor 10 is feasible). It provides excellent decoupling between driving source and signal to be measured. Finally, it is simple and economical.

II APPLICATION TO Cr_2O_3

It was thought that a detailed study of the switching transient shape would yield useful information on the domain behaviour. Analog studies have been made, for example, in ferroelectrics[3] and have not yet been performed in antiferromagnets.

For this purpose, crystals of Cr_2O_3 were used which were disk-shaped with a diameter of 6 mm and thicknesses ranging from 0.2 to 0.4 mm. Silver paste electrodes were deposited on the main faces of the samples and these were put inside a shielded cell between the poles of an electromagnet (0–12,000 Oe). Connections to the voltage source and the amplifier were made through coaxial cables.

The following experiment was then set up. The sequence of magnetic and electric fields shown in Figure 2a was applied on the sample. In this sequence, the electric field E remains always positive except during magnetic field reversal which takes about 1 s (t_2).

In this way the $E \times H$ product changes its sign at B and C and switching is expected to occur at these instants. Since the rise time of the $E \times H$ product is very short at time B, accurate study of the corresponding switching transient is possible, using the compensation method.

Antiferromagnetic domain reversal was effectively observed in our Cr_2O_3 samples using this technique (Figure 2b). The switching at $(C + t_2)$ is due to the reversal of H (when EH becomes sufficient) and the one at B is due to the application

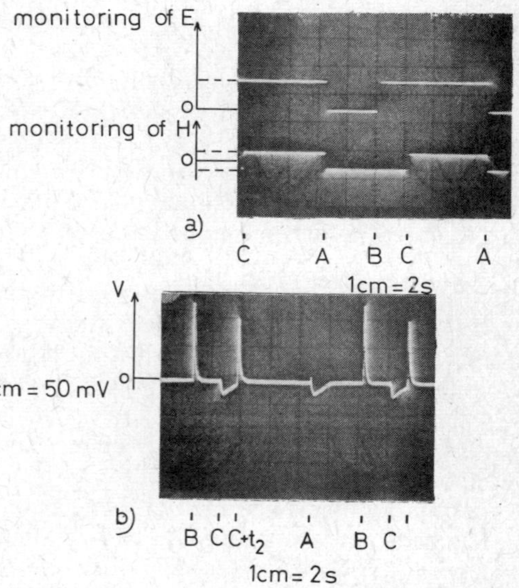

FIGURE 2 (a) Sequence of applied electric (E) and magnetic (H) fields.

$E \cdot H = 144$ MOe V/cm

(b) Magnetoelectric signal at $(C + t_2)$ and B.

of $+E$ that makes change the sign of EH. The increase of the peak is due to the switching of the domains state and the decrease to the time constant of the RC system. The switching time was found to be of the order of 10 ms just above switching threshold product EH.

III CONCLUSION

A very simple method which allows the study of fast domain switching in antiferromagnetic Cr_2O_3 samples has been described.

Detailed study of the switching transient shape has not been made but is now currently in progress in order to obtain quantitative results.

ACKNOWLEDGEMENTS

This work was made with the help of the Associated Center of the C.N.A.M. of Montluçon and the A.M.E.S.T.E.

REFERENCES

1. T. J. Martin and J. C. Anderson, "IEE Trans. on magnetics" *Mag.* **2** (3), 446 (1966).
2. Christopher A. Brown and T. H. O'Dell, "IEE Trans. on magnetics" *Mag.* **5** (4), 964 (1969).
3. See for example: J. C. Peuzin, Thèse d'Etat. Université de Grenoble (1969).

MÖSSBAUER HYPERFINE SPECTRA OF Fe–Br–BORACITE†

R. JAGANNATHAN, J. M. TROOSTER‡ and M. P. A. VIEGERS

Department of Physical Chemistry, University of Nijmegen, Nijmegen, The Netherlands

Mössbauer spectra of magneto-electric iron bromine boracite $Fe_3B_7O_{13}Br$ were measured and analysed. The results show that there are three sites with identical EFG-parameters, but different magnitude and orientation of the internal magnetic field. The Néel temperature was determined to be $T_N = 18.4 \pm 0.1$ K.

1 INTRODUCTION

To understand the microscopic mechanisms underlying the coupling between magnetic and electric polarization in iron boracites, knowledge of the spin- and orbital states of the iron ions in the magneto-electric phase is invaluable. The Mössbauer effect of ^{57}Fe gives information on these states through the measurement of the electric quadrupole- and magnetic hyperfine interaction of the iron nucleus. The Hamiltonian governing the energy-splitting of a nucleus with spin I for combined electric and magnetic interaction is given by:[1]

$$\mathscr{H} = \mathscr{H}_Q + \mathscr{H}_M \tag{1}$$

with

$$\mathscr{H}_Q = \frac{eQV_{zz}}{4I(2I-1)}\{3I_z^2 - I(I+1) + \tfrac{1}{2}\eta(I_+^2 + I_-^2)\}$$

$$\mathscr{H}_M = -g\mu_N H\{I_z \cos\vartheta + (I_x \cos\varphi + I_y \sin\varphi)\sin\vartheta\}$$

where

Q = electric quadrupole moment
$-e$ = charge of an electron
g = nuclear g-factor
μ_N = nuclear magneton
I = nuclear spin, I_x, I_y and I_z are the components of the nuclear spin operator
V_{zz} = $\partial^2 V/\partial_z^2$ with V = the electric potential at the nucleus
η = $(V_{xx} - V_{yy})/V_{zz}$ = asymmetry parameter of the electric field gradient
ϑ, φ = polar angles specifying the direction of the magnetic field H with respect to the principal axes of the electric field gradient tensor

The resulting energy level splitting of the ground state and first excited state of ^{57}Fe is illustrated in Figure 1. If $\mathscr{H}_M = 0$, the M_I degeneracy of the excited state is partly removed and the Mössbauer spectrum consists of two lines with a splitting 2ε, with

$$\varepsilon = \tfrac{1}{4}eQV_{zz}(1 + \tfrac{1}{3}\eta^2)^{1/2}$$

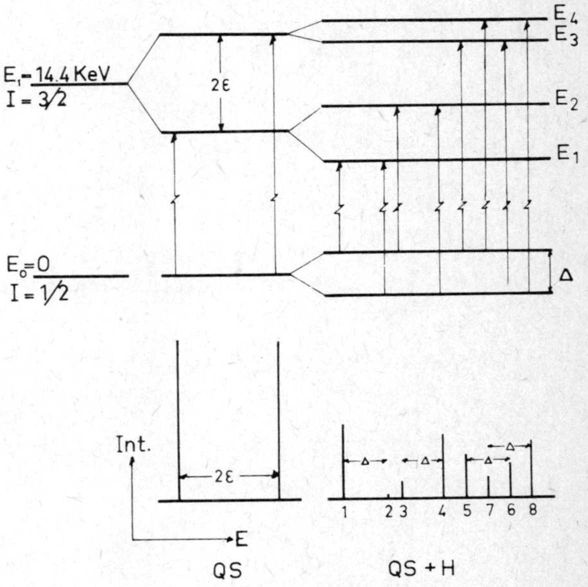

FIGURE 1 Energy levels of ^{57}Fe for combined quadrupole and magnetic hyperfine interaction. The energies and intensities are those of site B at 4.2 K (compare with Figure 2).

‡ Please address inquiries to this author.

† This paper has been presented in the Symposium on Magnetoelectric Interaction Phenomena in Crystals on May 21, 1973, at Battelle Seattle Research Center, Seattle, Washington.

For a powder absorber with random orientation of the crystal axis the two lines have equal intensities. If $\mathscr{H}_M \neq 0$ the degeneracy of both the excited and ground state is completely removed and the Mössbauer spectrum consists generally of eight lines. From the energies of these lines the values of the Isomer Shift (I.S.), ε, H, η, θ and ϕ can be derived. This will be discussed in Section 3.

2 EXPERIMENTAL

Mössbauer spectra were measured at temperatures between 4.2 K and 77 K. Measurements of 4.2 K were carried out on absorbers immersed in liquid He. For temperatures above 4.2 K source and absorber were placed in an evacuated tube inserted in liquid helium. The temperature of the absorber was measured with a calibrated germanium sensor and regulated with a small heater coil to better than 0.05°K. The source was placed on a stainless steel tube connected to a velocity transducer which was mounted on top of the He-cryostat. The velocity was changed periodically in a triangular way and the spectra recorded in a Multi Channel Analyser run in the time mode. One measurement gave thus two spectra containing 256 points each, which were analysed separately. The velocity was measured to high accuracy with a laser interferometer of the type described by Fritz and Schulze.[2] The absorbers were made from powdered crystals. The powder was tightly packed in lucite pill-boxes, wrapped in an aluminium foil to ensure uniform temperature distribution and clamped in a copper holder containing the germanium sensor.

3 ANALYSIS

Examples of spectra obtained on Fe–Br–boracite are given in Figure 2. From a cursory examination one concludes that these spectra contain at least three 8-line spectra. For a complete analysis a least square fitting program is indispensable. We have fitted the spectra with line positions and intensities as parameters. To get meaningful results it is important to take into account the relations between the positions and intensities of the lines belonging to one 8-line spectrum. From Figure 1 it is clear that the eight lines occur in pairs separated by Δ, the splitting of the ground state level. Moreover it can easily be shown that for powder absorbers the sum of the intensities of lines belonging to one pair is one-fourth of the total intensity. The line width was kept equal for all lines. Thus the least square fitting program gives as result the positions of lines 1, 3, 5 and 7 (see Figure 1) and their relative intensities, the value of the ground state splitting and the total intensity of each 8-line spectrum. From these it is easy to derive the I.S.[3,4] and the values of the excited state energies E_i. The magnitude of the magnetic field H and the quadrupole splitting are then given by:[4,5]

$$g_0 \mu_N H = \Delta$$

$$|2\varepsilon| = \sqrt{\sum_{i=0}^{4} E_i^2 - 5\left(\frac{g_1}{g_0}\Delta\right)^2} \qquad (2)$$

where g_0 and g_1 are the g-factors of ground- and excited state, respectively: $g_0 = 0.18048$[6] and $g_1/g_0 = -0.5716$.[7]

Table I lists the values for I.S. and Q.S. obtained at 4.2, 13.2, 14.1 and 14.9 K. Within statistical accuracy the quadrupole splitting and isomer shift are the same for the three spectra. This is in accord with the Mössbauer spectra observed in the paramagnetic (trigonal) phase of Fe–Br–boracite, where only one Q.S. is observed.[8]

As explained elsewhere[4,5] η, ϑ and φ cannot be determined in a unique way. The values of η, ϑ and φ

TABLE I
Quadrupole splitting and isomer shift for three component spectra A, B and C at various temperatures. Values are averages of results of the two spectra measured simultaneously, and given in mm/sec. Isomer shift is with respect to a Rh(^{57}Co)-source at 4.2 K.

T		4.2 K	13.2 K	14.1 K	14.9 K	19 K
2ε (±0.008)	A	3.105	3.074	3.064	3.084	
	B	3.105	3.094	3.086	3.069	3.065
	C	3.100	3.088	3.086	3.077	
I.S. (±0.003)	A	1.035	1.020	1.027	1.030	
	B	1.033	1.030	1.031	1.031	1.028
	C	1.030	1.032	1.029	1.027	

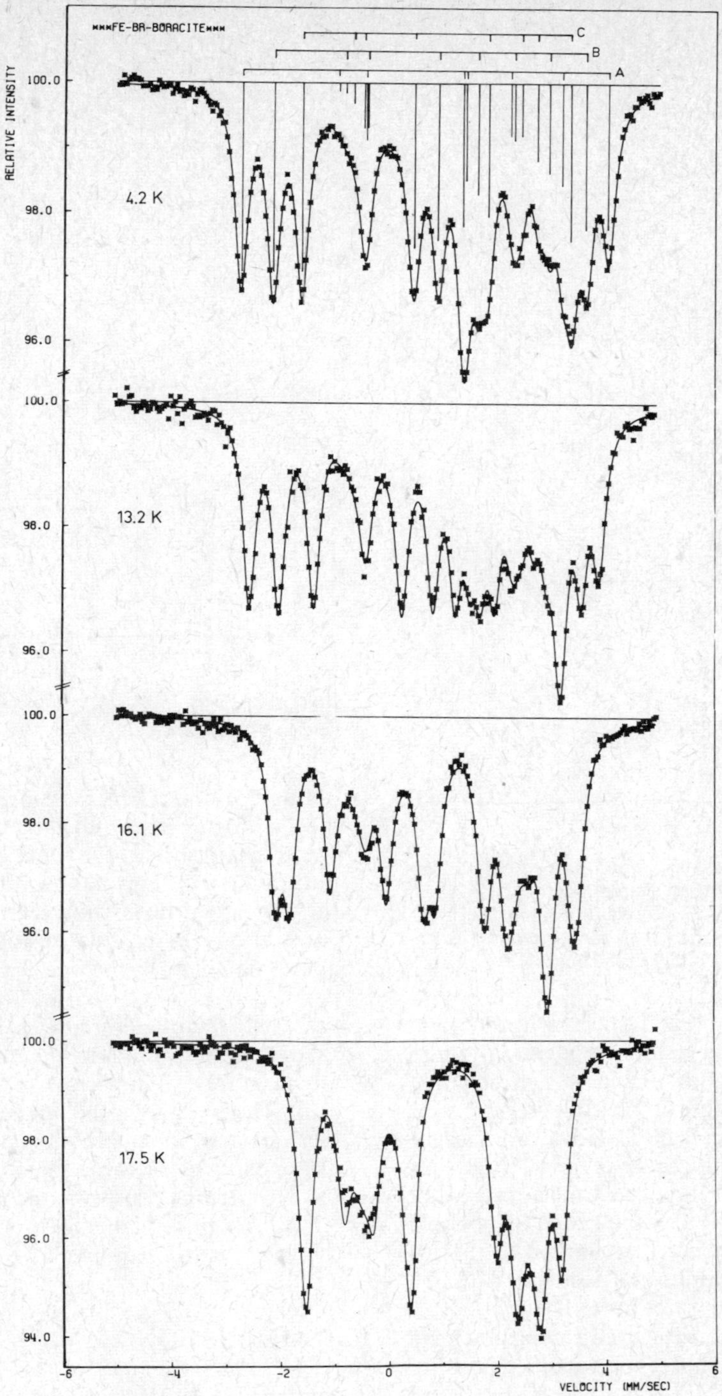

FIGURE 2 Spectra of Fe–Br–boracite below the Néel temperature. Curves drawn through spectra of 4.2 K and 13.1 K are calculated using parameters listed in Table II. Curves drawn through spectra of 16.1 K and 17.5 K are least square fits using constraints described in text. Line width is 0.28 mm/sec.

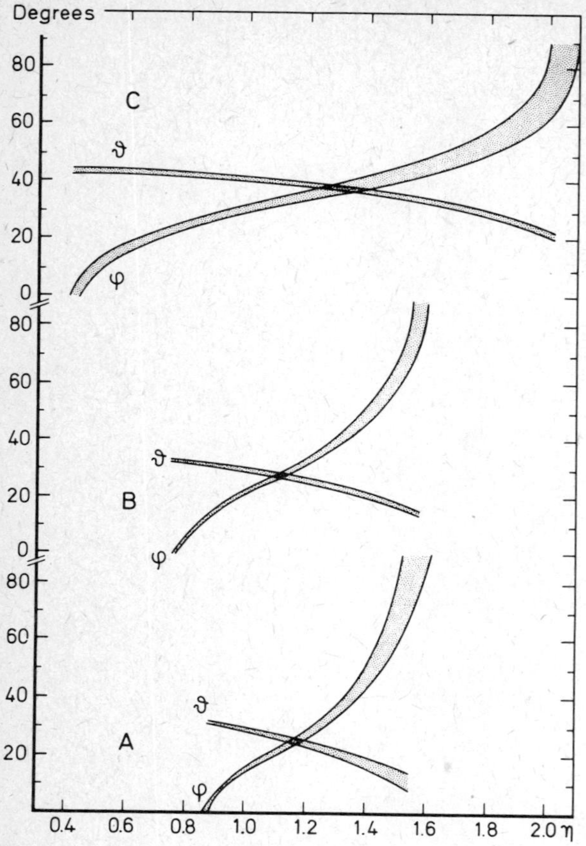

FIGURE 3 Possible values of η, ϑ and φ for the three sites A, B and C as derived from the measured spectra at $T = 4.2$ K, 13.1 K, 14.1 K and 14.9 K. V_{zz} (or 2ε) was taken to be negative, in this case the limits of η are given by φ becoming 0° or 90°. Bands indicate estimated accuracy in the angles.

compatible with experimental spectra were calculated analytically.[4,5] The results are given in Figure 3, where the polar angles ϑ and φ are given as function of η. The range of possible solutions is particularly large for spectrum C. However, because the quadrupole splitting is the same for all three sites it is highly probable that the electric field gradient of all three sites is identical and the range of possible solutions for all sites is therefore limited to that of spectrum A, with the additional restriction that η should be the same for all three spectra. In the calculation of η, ϑ and φ the measured intensities have not been taken into account. This is justified because no additional information is obtained by including them in the analysis.[5] However, the positions and the intensities of the three component spectra are correlated in the least square fitting program. As a final check on the validity of the results the measured spectra were compared with the spectra calculated with the values of H, Q.S., I.S., η, ϑ and φ derived as described above. Generally, excellent agreement was found. This is illustrated in Figure 2 for the spectra taken at 4.2 K and 13.1 K.

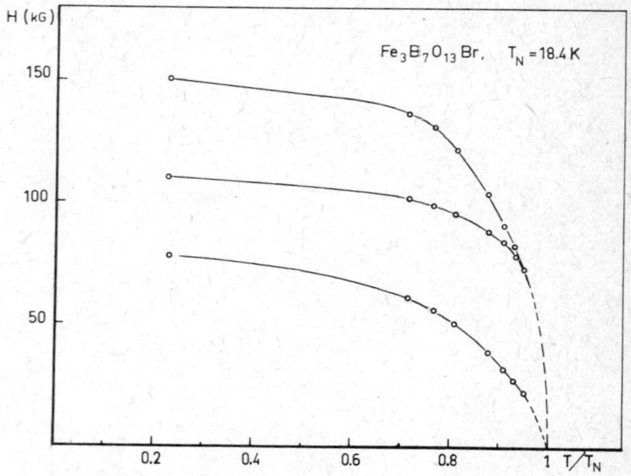

FIGURE 4 Temperature dependence of internal magnetic field in Fe–Br–boracite.

For spectra taken at temperatures close to the Néel temperature (typified by the 16.1 K and 17.5 K spectra in Figure 2) the number of parameters used in fitting the spectra had to be reduced. To this end the intensity ratios within each 8-line spectrum were fixed at values calculated for η, ϑ, φ and H as extrapolated from the spectra taken at lower temperatures. In this way we could obtain reliable values of $|H|$ for all three sites. No attempt was made to determine η, ϑ and φ independently at these temperatures. Numerical results are listed in Table II. In Figure 4 the temperature dependence of the internal magnetic field is given. From the disappearance of the magnetic hyperfine splitting the Neël temperature was determined to be 18.4 ± 0.1 K, which is higher than that determined from susceptibility measurements by Quezel and Schmid.[9]

4 DISCUSSION

The crystal structure of $M_3B_7O_{13}Cl$ in cubic and orthorhombic phase was determined by Ito *et al.*[10] Mössbauer measurements on the iron boracites above the Néel temperature were discussed in an

TABLE II

Values of H, ϑ, φ and relative intensities (Int) at various temperatures, derived for negative 2ε and $\eta = 1.0$. Values are given separately for the two spectra measured simultaneously. Accuracy of H is $<\pm 2$ kG.

T	A				B				C			
(K)	H(kG)	ϑ	φ	Int	H(kG)	ϑ	φ	Int	H(kG)	ϑ	φ	Int
4.2	150.0	31	17	1.0	110.8	32	22	1.01	78.1	40	29	1.06
	149.2	29	17	1.0	110.6	30	22	1.02	78.1	40	29	1.02
13.2	137.5	30	13	1.0	102.1	30	23	1.06	61.4	40	32	1.08
	137.0	28	13	1.0	102.0	28	24	1.05	60.7	38	34	1.04
14.1	131.5	28	14	1.0	99.6	30	21	1.02	56.6	40	35	1.04
	132.5	29	0	1.0	99.0	28	21	1.03	55.1	35	40	1.00
14.9	121.9	27	16	1.0	95.6	30	20	1.02	51.5	43	32	1.03
	122.2	26	0	1.0	95.6	30	14	1.04	49.7	37	36	1.01
16.1	105.5	—	—	1.0	88.5	—	—	1.0	38.7	—	—	1.0
	103.8	—	—	1.0	88.1	—	—	1.0	38.3	—	—	1.0
16.7	91.2	—	—	1.0	84.2	—	—	1.0	32.3	—	—	1.0
	91.1	—	—	1.0	84.0	—	—	1.0	32.1	—	—	1.0
17.1	82.6	—	—	1.0	78.5	—	—	1.0	27.6	—	—	1.0
	82.6	—	—	1.0	78.7	—	—	1.0	27.0	—	—	1.0
17.5	69.0	—	—	1.0	71.6	—	—	1.0	22.9	—	—	1.0
	68.0	—	—	1.0	72.2	—	—	1.0	22.1	—	—	1.0

earlier paper.[8] Fe–Br–boracite was shown to undergo a transition to trigonal symmetry at 403 K. The conclusions drawn therein regarding the crystal structure in the trigonal phase were substantiated by a recent structure determination by Dowty and Clark.[11] In the paramagnetic trigonal phase there are four crystallographically inequivalent sites. Nevertheless only one quadrupole splitting was observed. The present results show that below the Néel temperature the EFG remains the same for all Fe^{2+}-ions. This means that the electronic environments are identical. On the other hand three internal magnetic fields are observed, differing in magnitude and orientation with respect to the principal axis of the EFG. These three fields may be expected to arise from the Fe^{2+}-ions situated along the mutually perpendicular (cubic) axes. It is reasonable to place the z-axis of the ionic contribution to the EFG as well of the zero-field-splitting tensor along the halogen–metal–halogen axis, as these are predominantly determined by the local symmetry. The lattice contribution to the EFG on the other hand will reflect the total crystal symmetry. Mössbauer spectra above the Néel temperature and a point charge calculation showed that the lattice contribution to the EFG is considerable.[8] The principal axes of the resulting total EFG will be different from those determined by the local symmetry. We now assume that $S_z = 0$ is the ground state of the effective spin Hamiltonian so that the plane perpendicular to the Br–Fe–Br axis is the easy plane for the magnetization of the Fe^{2+}-ions. Different orientations of the spin within this plane will result in different values of both ϑ and φ with respect to the principal axes of the total EFG. Similarly because of deviation of axial symmetry, different spin orientation within the XOY plane will result in different contributions of the dipolar field to the internal magnetic field at the nucleus.

Lack of detailed knowledge on the magnetic hyperfine interaction tensor as well as of the total EFG-tensor preclude a determination of the spin directions from the measured internal magnetic fields. Measurements of Fe–Cl–boracite and ^{57}Fe in Co–Cl–boracite are in progress. For Fe–Cl–boracite the results obtained up to now resemble closely those of Fe–Br–boracite. These results, together with a more detailed discussion, will be presented in a forthcoming paper.

ACKNOWLEDGEMENTS

The boracite crystals were kindly supplied by Dr. H. Schmid. An enlightening discussion was held with Drs. R. Link and W. Wurtinger who also gave us results on their Mössbauer measurements on Fe–Cl–boracite. H. van Dongen-Torman assisted in the analysis of the spectra.

REFERENCES

1. A. Abragam, *Principles of Nuclear Magnetism* (Oxford University Press, London, 1961), chapters VI and VII.

2. R. Fritz and D. Schulze, *Nucl. Instr. and Meth.* **62**, 317 (1968).
3. G. K. Wertheim, *Physics Letters* **30A**, 237 (1969).
4. S. V. Karyagin, *Fiz. Tver. Tela.* **8**, 493 (1966); Engl. Transl. *Sov. Phys. Sol. State* **8**, 391 (1966).
5. L. Dabrowski, J. Piekoszewski and J. Suwalski, *Nucl. Instr. Methods* **103**, 545 (1972).
6. P. R. Locher and S. Geschwind, *Phys. Rev.* **139**, A991 (1965).
7. R. S. Preston, S. S. Hanna and J. Heberle, *Phys. Rev.* **128**, 2207 (1962).
8. J. M. Trooster, *Phys. Stat. Sol.* **32**, 179 (1969).
9. G. Quezel and H. Schmid, *Solid State Comm.* **6**, 447 (1968).
10. I. Ito, N. Morimoto and S. Sadanaga, *Acta Cryst.* **4**, 310 (1951).
11. Eric Dowty and Joan R. Clark, *Solid State Comm.* **6**, 447 (1972).

Discussion

V. L. FOLEN Do you see significant differences in the hyperfine fields when comparing the Cl, Br and I boracite compounds?

J. M. TROOSTER Fe–Cl and Fe–Br boracites are very similar. The iodine compound, however, is quite different, possibly six hyperfine fields will be needed to explain the spectra.

ELECTRO-MAGNETO-STRICTION AND MAGNETICALLY INDUCED PSEUDO-PIEZOELECTRIC EFFECTS

KEI-ICHIRO EHYSHIMA and TOMOYA OGAWA

Department of Physics, Gakushuin University, Mejiro, Tokyo 171, Japan

An unusually large deformation coefficient is obtained in a conducting Mn–Zn ferrite crystal when an electric field is applied to it. If the deformation is regarded as electrostrictive, the coefficient (χ) defined by $S = \chi E^2/2$ (S: strain, E: electric field) is about 10^{-10} $(m/V)^2$. This is about 10^{11} times larger than the electrostrictive coefficient of $BaTiO_3$, since χ in $BaTiO_3$ is about 10^{-21} $(m/V)^2$ in both ferroelectric and paraelectric phases. However, the deformation is regarded as a sort of magnetostriction, because the electric current passing through the crystal induces magnetic flux which is closed around the current and has no free pole to make any demagnetization field.

When the crystal is biased by a static magnetic field or a d.c. current, it acts as a pseudo-piezoelectric crystal. Its pesudo-piezoelectric constant is about 10^{-9} C/N at a suitable bias and it has the same order as the piezoelectric constant of a PZT ceramic element. The sign of the deformation is changed by the inversion of the direction of the biasing magnetic field or the biasing d.c. current.

1 INTRODUCTION

Induced reversible deformations, linearly proportional to an applied electric or magnetic field are known as "piezoelectric" and "piezomagnetic" respectively. Piezoelectricity is allowed in all non-centrosymmetric crystals. The Shubnikov groups permitting piezomagnetism are also well known (see e.g. Ref. 1, pp 141/142); all ferromagnetic groups are necessarily piezomagnetic. If the deformation is proportional to the square of the electric and magnetic field, we speak of "electrostriction"[2] and "magnetostriction"[3] (which can also be called the quadratic piezoelectric[4] and piezomagnetic effects[4]) respectively. For the case of ferroelectrics and ferromagnetics, the pure electrostrictive and magnetostrictive effects (pure in the above-mentioned sense) may be superposed on and even hidden by field-induced deformations due to domain reorientations at fields below saturation; furthermore, respective piezoelectric and piezomagnetic contributions to the total deformation may occur concomitantly.

The deformation which is observed in a conducting Mn–Zn ferrite crystal[5] due to an electric field is somewhat different from those mentioned above, because an electric current passes through the crystal, inducing a magnetic field around the current.

If the deformation is regarded as electrostrictive, the coefficient (χ) which is defined by $S = \chi E^2/2$ (S: strain, E: strength of electric field) is unusually large, i.e., about 10^{-10} $(m/V)^2$. This is about 10^{11} times larger than the electrostrictive coefficient of $BaTiO_3$, since χ in $BaTiO_3$ is about 10^{-21} $(m/V)^2$ in both ferroelectric and paraelectric phases.[6] If the deformation is evaluated as a magnetostrictive strain given by $S = \alpha H^2/2$ (α: magnetostrictive coefficient, H: strength of magnetic field), the coefficient α is a reasonable value as a magnetostrictive coefficient in Mn–Zn ferrite crystals.[7]

When a static magnetic field was applied to the crystal, the strain was proportional to an applied electric field. The sign of the deformation was changed by the inversion of the direction of the magnetic field. When mechanical stresses were applied to the ferrite crystal biased by a magnetic field, electric signals were observed as a result of the inverse process of the phenomenon. That is, the crystal acted as a piezoelectric one. When a d.c. current flows through the crystal as bias, the strain[8] is linearly proportional to an a.c. current which is superposed on the d.c. current as a signal. And an electric signal is also generated by a mechanical stress.

All these phenomena may seem to be piezomagnetoelectric (PME) ones.[4,9] However, because the current flowing through the crystal induces a

---

The main part of this paper was presented at the Symposium on Magnetoelectric Interaction Phenomena in Crystals, Seattle, May 1973.

magnetic field and causes magnetostrictive strain in the crystal, the phenomena should be called "Electro-magneto-strictive (EMS) effect."[5]

In this paper, the interactions among electric, magnetic and elastic systems, especially elastic deformations due to an electric and/or a magnetic fields, in a conducting Mn–Zn ferrite crystal are discussed from the viewpoints of the PME and EMS effects.

2 EXPERIMENTAL PROCEDURES

Figure 1 is a schematic diagram of the apparatus used for measuring the strain induced in a Mn–Zn ferrite crystal. The ferrite crystal (*F*), *x*-cut quartz plate (*Q*) and a PZT ceramic element (*P*) were clamped in order to detect the deformation of the ferrite or the quartz crystal. A d.c. electric and/or a magnetic fields were applied to the ferrite crystal to get an appropriate bias. The deformation of the ferrite and the quartz crystals caused by an a.c. signal was measured by piezoelectric voltages on the PZT (bimorph) element with a lock-in amplifier and/or a spectrum analyser. The strain and/or pseudo-piezoelectric coefficient of the ferrite crystal were

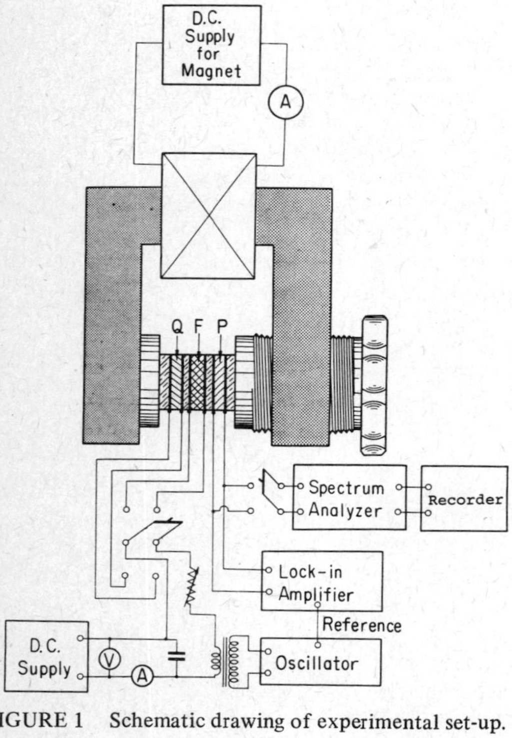

FIGURE 1 Schematic drawing of experimental set-up.

obtained numerically in comparison with those of the *x*-cut quartz plate.

A biasing magnetic field was applied perpendicularly to the a.c. electric field in the crystal by using another apparatus constructed on the same principle as the system mentioned above.

When an electric field was applied to the PZT element, the emf signals were generated in the ferrite and the quartz crystals by the stress from the element and they were also detected by this system using the amplifier and/or the analyser.

The ferrite crystals used here were prepared by Fuji Electrochemical Co., and their composition, density, lattice constant, Curie temperature, easy axis of magnetization, magnetic anisotropy constant and specific resistivity are $Mn_{0.6}Zn_{0.3}Fe^{2+}_{0.1}Fe^{3+}_{2}O_4$, 5.1 g/cm^3, 8.49 Å, about 200°C, ⟨111⟩, -10^3 J/m^3 = -10^4 erg/cm^3 and 1 ohm-cm, respectively.

Three specimens were prepared from an ingot. They were shaped in rectangular parallelepipeds with dimensions of 1 × (0.8 ~ 1) × 0.4 cm^3. Their largest surfaces were respectively (111), (110) and (100), on which electrodes of indium amalgam were mounted after the specimens had been etched by conc. HCl for about 30 min. in order to remove the strains due to cutting and polishing.[10]

3 EXPERIMENTAL RESULTS

3.1 *Without Bias*

The frequency spectrum of the strain which was induced in a Mn–Zn ferrite crystal by an a.c. current was detected by the PZT element and analysed by a spectrum analyser (Hewlett-Packard: 8556A-8852B). The spectrum is shown in Figure 2. The strain has a good many harmonics, up to the 20th ones.

3.2 *Biased by a d.c. Current*

When a d.c. current was superimposed on the a.c. current as bias, the fundamental and odd harmonic components in the strain increased while even harmonics diminished with increasing bias d.c. current as shown in Figure 3. The dependence of the fundamental and harmonic components on biasing d.c. current showed a little hysteresis.[8] The induced strain are plotted against the a.c. current in Figure 4, where the biasing d.c. current is taken as a parameter.

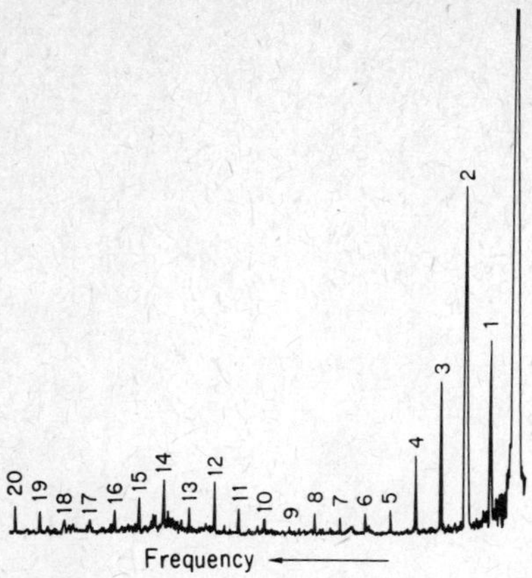

FIGURE 2 Frequency spectrum of a.c.-current induced strain without bias. The arabic numerals indicate the orders of harmonics. The ordinate is the strain (arbitrary units in dB). A.c. current: 34 mA/0.8 × 1.0 cm² = 42.5 mA/cm² (1 kHz). A.c. electric field ‖[111]. Strain measured ‖[111].

When mechanical stresses were applied to the specimen biased by a d.c. current, the electric signals were observed as a result of the inverse process of the effect.

3.3 *Biased by an External Magnetic Field*

When the crystal was biased by an external magnetic field, the following relation was found to hold between the induced strain (S) and the applied electric field (E) in the crystal:

$$S = d \cdot E.$$

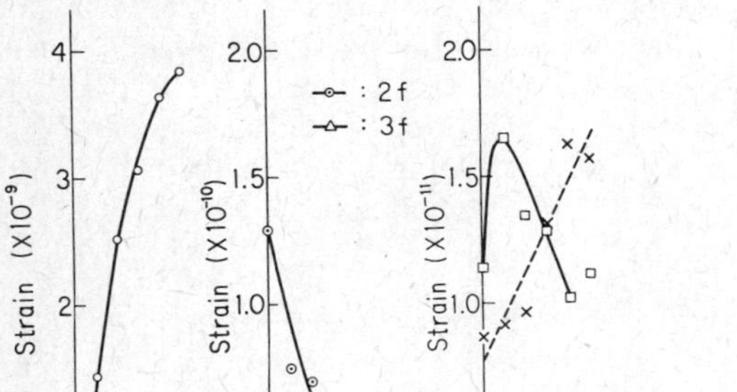

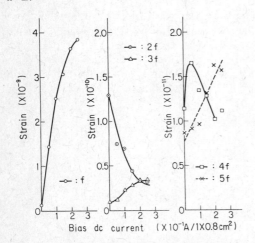

FIGURE 3 A.c.-current induced strain versus d.c.-bias current density. The fundamental, second, third, fourth and fifth harmonic components in the strain are indicated by the symbols of f, $2f$, $3f$, $4f$ and $5f$, respectively. A.c. current: 11 mA/0.8 × 1.0 cm² = 14.2 mA/cm² (1 kHz). A.c. and d.c. electric fields ‖ [111]. Strain measured ‖ [111].

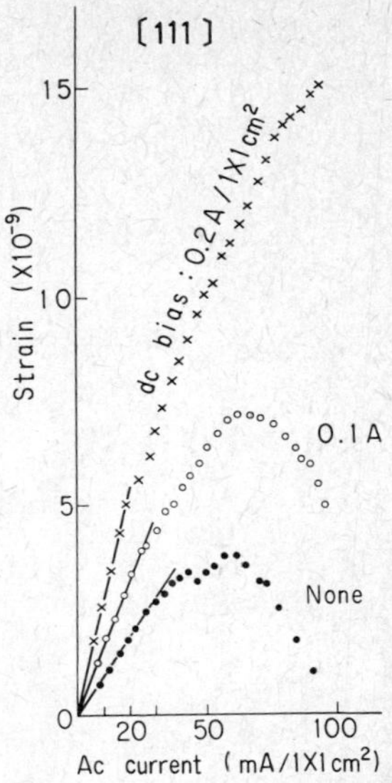

FIGURE 4 A.c.-current induced strain component at fundamental frequency (20 kHz) versus a.c.-current density, at 0, 0.1 and 0.2 A/cm² d.c.-bias; a.c. and d.c. currents parallel [111].

This is the relation that holds in piezoelectric materials, that is, the ferrite crystal biased by a magnetic field acted as a piezoelectric crystal as shown in Figure 5. The pseudo-piezoelectric d-coefficient is shown as a function of the biasing magnetic field H in Figure 6. The sign of the deformation was changed by the inversion of the direction of H. Below saturation, d-coefficient was scattered and each value often showed an aging phenomenon even when the crystal was biased by a constant magnetic field.

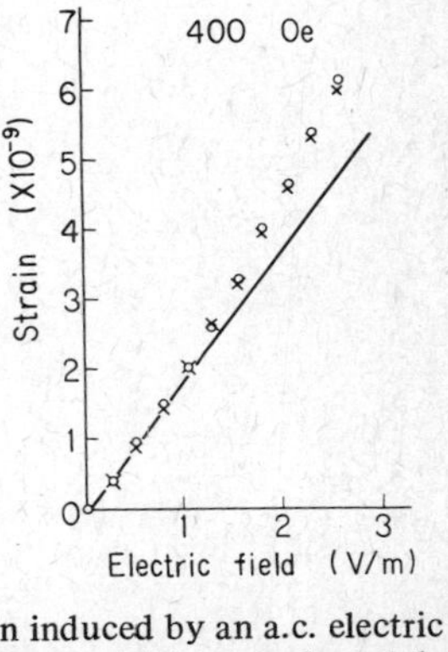

FIGURE 5 Strain induced by an a.c. electric field (1 kHz). Resistivity and resistance of the sample: ~1.1 ohm–cm and 0.6 ohm. A.c. electric field ‖[111]. Magnetic field ⊥ [111]. Strain measured ‖[111].

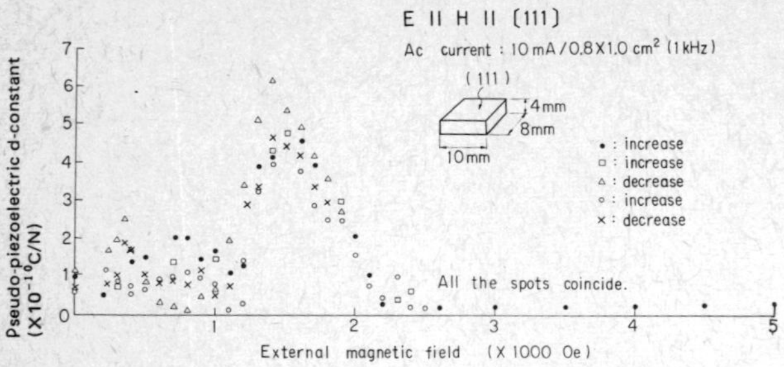

FIGURE 6 Magnetic field induced pseudo-piezoelectric coefficient versus inducing magnetic field. The derivatives of the strain with respect to the applied electric field are plotted as a function of the external magnetic field. A.c. current: 10 mA/0.8 x 1.0 cm^2 = 12.5 mA/cm^2 (1 kHz). A.c. electric field ǁ[111]. Magnetic field ǁ[111]. Strain measured ǁ[111].

The fundamental and second harmonic components in the strain were observed and shown in Figure 7 as a function of an external magnetic field applied to the crystal. The other higher components could not be measured. The emf signal due to an elastic stress was also observed in the sample biased by a magnetic field.

After the magnetization was saturated, the signal on the PZT element was proportional linearly to the strength of the magnetic and electric fields applied to the sample. However, the magnitude of the signal on the PZT element coincides quite well with that due to the force on a current (I) in a magnetic field (H), i.e., $I \times H$. Therefore, even if there exists the PME effect, it is very difficult to separate it from the signal due to the force of $I \times H$.

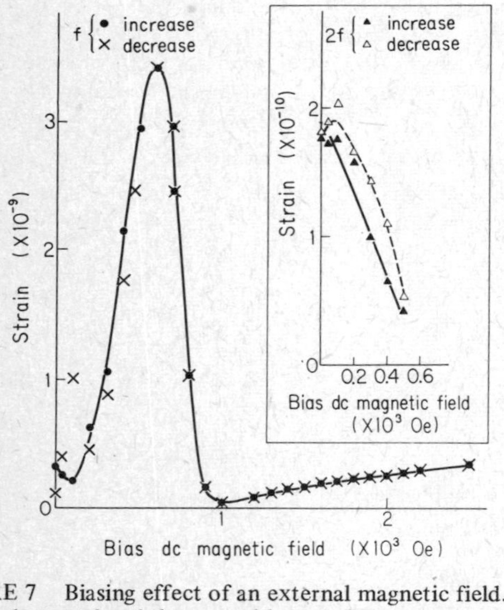

FIGURE 7 Biasing effect of an external magnetic field. The fundamental and the second harmonic components are plotted as a function of a bias magnetic field. Other higher components could not be measured. The magnetization is saturated at about 1000 Oe of external field. A.c. current: 14 mA/cm^2 (1 kHz). A.c. electric field ǁ[111]. Magnetic field ⊥[111]. Strain measured ǁ[111].

4 THEORIES AND DISCUSSIONS

4.1 *Electro-Magneto-Striction*

The strain (S) due to the magneto-striction effect is given by

$$S = \beta M^2/2 \tag{1}$$

where M is magnetization and β is a magnetostrictive coefficient. The magnetization M is expanded in a series of odd powers as

$$M = \gamma_1 H - \gamma_3 H^3 + \gamma_5 H^5 - \ldots \tag{2}$$

where H is magnetic field. Then the strain is given by

$$S = \alpha_2 H^2 - \alpha_4 H^4 + \alpha_6 H^6 - \alpha_8 H^8 + \ldots. \tag{3}$$

Assuming that the magnetic field in the crystal is evaluated by the relation $J = \int H dl = H l_0$ (J = total current through the crystal, l_0 = mean length of the magnetic flux) and that the magnetic field H is the sum of d.c. and a.c. components:

$$H = H_0 + h \sin \omega t, \tag{4}$$

where H_0 is a d.c. (static or bias) magnetic field and h is the amplitude of magnetic field with frequency ω, and substituting Eq. (4) into Eq. (3), the strain induced by a magnetic field includes all the harmonic components. For example, the fundamental, second and third harmonic components are respectively given by

$$S(\omega) = h \sin \omega t \times \{(2\alpha_2 - 3\alpha_4 h^2)H_0 - (4\alpha_4 - 15\alpha_6 h^2){H_0}^3 + 6\alpha_6 {H_0}^5 + \tfrac{15}{4}\alpha_6 H_0 h^4 + \ldots\}, \tag{5a}$$

$$S(2\omega) = -\tfrac{1}{2}h^2 \cos 2\omega t \times \{\alpha_2 - \alpha_4 h^2 - (6\alpha_4 - 15\alpha_6 h^2){H_0}^2 + 15\alpha_6 {H_0}^4 + \tfrac{15}{16}\alpha_6 h^4 + \ldots\}, \tag{5b}$$

$$S(3\omega) = -h^3 \sin 3\omega t \times \{-\alpha_4 H_0 + 5\alpha_6 {H_0}^3 + \tfrac{15}{8}\alpha_6 H_0 h^2 + \ldots\}. \tag{5c}$$

The dependence of the components upon a magnetic field H_0, which is mainly the magnetic field caused by the biasing d.c. current and partly a residual one (H_r) in the crystal, as shown in Figure 3, is correctly represented by Eqs. (5a), (5b) and (5c). Here the values of α's and residual magnetic field are respectively evaluated as

$\alpha_2 \sim 1.7 \times 10^{-9}$ (m/A)^2,
$\alpha_4 \sim 0.8 \times 10^{-10}$ (m/A)^4,
$\alpha_6 \sim 3.4 \times 10^{-12}$ (m/A)^6, and
$H_r \sim 0.1$ A/m.

The value of α_2 coincides with the magnetostrictive coefficient of 9×10^{-9} (m/A)^2 obtained on the assumption that $\lambda = \frac{1}{2}\alpha(M_s/\mu)^2$, where the magnetostriction constant[5] (λ) is 1.4×10^{-5}, the saturation magnetization (M_s) is 0.35 Wb/m^2, the magnetic permeability (μ) is 5000 μ_0 and $\mu_0 = 4\pi \times 10^{-7}$ H/m.

The fundamental component was increased and then decreased with increasing amplitude of the magnetic field (h) as represented by Eq. (5a) when a residual magnetic field served as H_0, which is correctly shown in Figure 4.

The induced strain, when the specimen is biased by a d.c. current, is therefore caused by the magnetostriction due to the magnetic field around the current passing through the crystal.

4.2 *Magnetically Induced Pseudo-Piezoelectric Effect*

When the crystal is biased by a static magnetic field, the observed phenomena will be discussed according to the following two items:

1) a sort of magnetostriction phenomena,

2) phenomena which is caused by the "piezo-magneto-electric (PME) effect."[9]

In the first case it is considered that the magnetic field induced by an a.c. current moves magnetic domains and causes strain as magnetostriction. In this case the d-coefficients are given using Eq. (1) by

$$d = \frac{\partial S}{\partial E} = \beta M \left(\frac{\partial M}{\partial H}\right)\left(\frac{\partial H}{\partial E}\right). \tag{6}$$

If the magnetoresistance effect[11] can be neglected, the magnetic field due to the current flowing through the crystal does not depend upon the biasing magnetic field, that is, $\partial H/\partial E$ is independent of the biasing magnetic field.† The fact that the d-coefficient increases and then decreases with increasing magnetic field corresponds with the fact that $M(\partial M/\partial H)$ increases and then decreases with the biasing magnetic field H as the magnetization approaches the saturation. Below saturation the d-coefficients changed slightly with time and they were not reproducible, in a strict sense, against the biasing magnetic field. The reason why the d-coefficients were scattered is that they are dependent upon the domain structure of the crystal.

The observed behavior of the d-coefficients in Figure 6 suggests the above mentioned mechanism.

Next, the possibility of the PME effect will be discussed.

Strain (S), dielectric displacement (D) and magnetic flux density (B) are respectively expanded in Taylor series:

$$S = \frac{\partial S}{\partial T}T + \frac{\partial S}{\partial E}E + \frac{1}{2}\frac{\partial^2 S}{\partial E \partial E}EE + \frac{1}{2}\frac{\partial^2 S}{\partial E \partial H}EH + \frac{\partial S}{\partial H}H + \frac{1}{2}\frac{\partial^2 S}{\partial H \partial H}HH, \tag{7a}$$

$$D = \frac{\partial D}{\partial E}E + \frac{1}{2}\frac{\partial^2 D}{\partial H \partial T}HT + \frac{\partial D}{\partial T}T + \frac{\partial D}{\partial H}H \tag{7b}$$

$$B = \frac{\partial B}{\partial H}H + \frac{1}{2}\frac{\partial^2 B}{\partial E \partial T}ET + \frac{\partial B}{\partial T}T + \frac{\partial B}{\partial E}E \tag{7c}$$

where T, E and H are respectively stress, electric and magnetic fields.

The PME interaction constant Ξ is an axial fourth rank tensor, because

$$\Xi = \frac{\partial^2 S}{\partial H \partial E} = \frac{\partial^2 D}{\partial H \partial T} = \frac{\partial^2 B}{\partial E \partial T} = -\frac{\partial^3 G}{\partial E \partial H \partial T}$$

(G = Gibbs' function). Considering that the Shubnikov point group of the paramagnetic high-temperature phase of Mn–Zn ferrite is m3ml′ (see e.g. Ref. (12)) and that the easy axis of magnetization lies along a cubic ⟨111⟩ direction, the Shubnikov point group of the ferromagnetic phase must be $\bar{3}$m′ because it is the

† The sign reversal observed in the EMS phenomena is explained as follows. The pseudo-piezoelectric coefficient (d) is defined by $d = \beta M(\partial M/\partial H) \cdot (\partial H/\partial E)$ as mentioned in Eq. (6). Here the magnetic field (H) includes the static biasing magnetic field (H_0) and the field (h) due to the current ($J = \sigma E$) passing through the crystal. Then $\partial H/\partial E$ is given as:

$\partial H/\partial E = \partial H_0/\partial E + \partial h/\partial E$.

On the assumption that the magnetoresistance effect is neglected, $\partial h/\partial E$ is a constant value determined by the law of Ampère. The derivative $\partial M/\partial H$ is positive whether the magnetization (M) is positive or negative. Therefore, the sign of pseudo-piezoelectric d-coefficient should be changed by the reversal of M.

only maximal M-subgroup of m3ml′ having the magnetization along the pseudo-cubic ⟨111⟩ direction (see e.g. Ref. 13, p. 467, Item 2). For reasons of symmetry, point group $\bar{3}m'$ permits the piezomagnetic effect[4], but not the piezoelectric[4] and PME effects. As a consequence, the two latter effects can be excluded as potential interpretation of the observed interaction phenomena.

5 SUMMARY

The induced deformation included a good many harmonics when the a.c. current passing through the crystal had fairly large amplitude. When d.c. current was superimposed on the a.c. current as bias, the fundamental and odd harmonic components in the deformation increased while even harmonics diminished with increasing bias d.c. current.

When a static magnetic field was applied to the crystal, it acted as a piezoelectric crystal. Its piezoelectric constant abruptly increased and then decreased with increasing magnetic field. The sign of the deformation was changed by the inversion of the direction of the magnetic field. The magnitude of the pseudo-piezoelectric coefficient was about 10^{-9} C/N at a suitable biasing magnetic field[5] and it had the same order as the constant of a PZT ceramic element.

When mechanical stresses were applied to the Mn–Zn ferrite crystal biased by a magnetic field or a d.c. current, the electric signals were observed as a result of the inverse process of the above phenomena.

These phenomena were observed in any conducting Mn–Zn ferrite crystal. The induced deformation responded to the frequency of the applied electric field up to 100 kHz. This frequency is not the upper limit of the phenomena but the limit of the measuring system used here.

The observed phenomena will, therefore, be mainly caused by the magnetostriction effect due to the magnetic flux induced by the current passing through the crystal until its magnetization is saturated.

ACKNOWLEDGEMENTS

The authors wish to express their cordial thanks to Dr. A. Takeda, Dr. I. Tsuboya and Mr. K. Hoshikawa of The Electrical Communication Laboratory, Nippon Telegraph and Telephone Public Corporation for their useful suggestions and discussions.

REFERENCES

1. R. R. Birss, *Symmetry and Magnetism* (North Holland Pub. Co., Amsterdam, 1966) p. 57.
2. W. Känzig, "Ferroelectrics and antiferroelectrics", *Solid State Physics* (Academic Press, New York, 1957) Vol. **4**, pp. 68–88.
3. D. A. Berlincourt, D. R. Curran and H. Jaffe, "Piezoelectric and piezomagnetic materials and their function in transducers", *Physical Acoustics* (Academic Press, New York, 1964) Vol. **1**–Part A; pp. 257–270.
4. H. Schmid, Abstract Booklet of "Symposium on Magneto-electric Interaction Phenomena in Crystals", May 1973, Seattle, Table I, p. 3f/6.
5. T. Ogawa and K. Ehyshima, *Oyo Buturi* (Monthly Journal of the Japan Society of Applied Physics), **41**, 831 (1972).
6. S. Iida, *et al.*, Tables of Constants for Physical Measurements and Experiments (Asakura Pub. Co., Tokyo, 1969) p. 154.
7. R. M. Bozorth, E. E. Tilen and A. J. Williams, *Phys. Rev.* **99**, 1788 (1955).
8. T. Ogawa and K. Ehyshima, *Proc. IEEE.* **62**, No. 1, 140 (1974).
9. G. T. Rado and V. J. Folen, *J. Appl. Phys.* **33**, 1126 (1962); G. T. Rado, *Phys. Rev.* **128**, 2564 (1962).
10. Y. Kawai, Private Communication.
11. E. Tatsumoto, *Phys. Rev.* **109**, 658 (1958).
12. R. W. G. Wyckoff, *Crystal Structures* (Interscience Pub. Co., New York, 1965) 2nd ed., Vol. **3**, Chap. VIII, p. 76.
13. E. Ascher, *Helv. Phys. Acta.* **39**, 466 (1966).

Appendix

THE ORIGIN OF THE OBSERVED DEFORMATION

Deformation Due to the Lorentz Force

The deformation in ferrite crystals provoked by the Lorentz force can be neglected below saturation because it is very small (at external field 1 KOe: the strain about 10^{-11}) as compared with the deformation due to magnetostriction effect (at 500 Oe: the strain nearly 10^{-9}). Above saturation, however, deformation due to Lorentz force becomes of comparable order with the deformation due to magnetostriction effect because of the decrease of the latter, and we can not discriminate these two deformations experimentally.

Deformation of the Piezomagnetic Nature

If the deformation were caused by piezomagnetic effect, the pseudo-piezoelectric coefficient (d') would be defined as: $d' \propto (\partial M/\partial H)(\partial H/\partial E)$. The $(\partial H/\partial E)$ being a constant value and the $(\partial M/\partial H)$ positive for either direction of the magnetic field, the sign of d'-coefficient is not reversed by the reversal of the magnetic field. Therefore, the deformation in which the sign-reversal was provoked by the reversal of magnetic field is not of piezomagnetic nature.

As for the small deformation observed at zero of the external static magnetic field, it could be said to be due to piezomagnetic effect if piezomagnetic effect can be defined under residual magnetization. But this deformation can also be classified as the one caused by magnetostriction effect, because the residual magnetization works as a biasing magnetic field. Therefore, we think it is scarcely possible to distinguish the purely piezomagnetic deformation from the one due to magnetostriction biased by residual magnetic field.

MEASUREMENT OF THE MAGNETOELECTRIC EFFECT IN Ni-Cl BORACITE†

J.-P. RIVERA,‡ H. SCHMID,§ J. M. MORET‡ and H. BILL‡

‡ *Department of Physical Chemistry, University of Geneva, 1205 Geneva, Switzerland*
§ *Battelle, Geneva Research Center, 1227 Carouge, Geneva, Switzerland*

By using three different crystallographic cuts of ferroelectric orthorhombic single domain samples of Ni–Cl boracite, seven components out of the linear magnetoelectric tensor (α_{ij}) have been subjected to measurement. Among these seven components, two non-diagonal ones have been found non-zero though with different behaviour below and above a critical temperature of 9°K. The ME_H effect disappears entirely at about 25°K.

At 4.5°K the magnetoelectric coefficients are

$$|\alpha_{23}| \approx |\alpha_{32}| = (3.3 \pm 0.4) \cdot 10^{-12} [\mathrm{s/m}] \text{ or } (1.00 \pm 0.12) \cdot 10^{-3}$$

in Gaussian units. At 8.5°K, α_{32} reaches a peak with twice that value. The measurements suggest the occurrence of magnetic phase transitions at 9°K and 25°K. These results complete knowledge of the sequence of phases in Ni–Cl boracite:

	$m'm2'$	(weakly ferromagnetic/ferroelectric)
9°K	$\updownarrow$	
	$mm2$	(antiferromagnetic/ferroelectric)
25°K	$\updownarrow$	
	$mm21'$	(paramagnetic/ferroelectric)
610°K	$\updownarrow$	
	$\bar{4}3m1'$	(paramagnetic/paraelectric)

1 INTRODUCTION

In this paper, evidence will be given for the existence of the linear magnetoelectric effect in Ni–Cl boracite, measured via the magnetic field induced polarization $P_i = \alpha_{ij} H_j$. The existence of the magnetoelectric effect in boracites has been demonstrated already on the Ni–I[1,2] and Co–Cl[3] compositions. The common feature of most boracites with paramagnetic transition metal ion is the simultaneous occurrence of ferroelectricity and weak ferromagnetism in their magnetoelectric phase (magnetoelectric Type FM I/FE I[3]).

Let us recall that boracites form a large crystal family named after the mineral boracite, $Mg_3B_7O_{13}Cl$. In their general formula, $Me_3B_7O_{13}X$, Me stands for a bivalent metal ion such as Mg, Cr, Mn, Fe, Co, Ni, Cu, Zn, Cd, and X for Cl, Br, I[4] and OH.[5,6] Only with a slight change of the formula, the halogen may even be replaced by S, Se, and Te,[5-7] and the bivalent metal ion by Li^+.[8] Most of the compositions have been synthesized by chemical vapour transport.[4] The cubic high-temperature phase of all boracites has space group $F\bar{4}3c$, as repeatedly confirmed by x-ray measurements.[9,10] Thus the Shubnikov point group is $\bar{4}3m1'$ of that prototype phase. In the case of Ni–Cl boracite, the $\bar{4}3m1'$ phase transforms at 610°K into a paramagnetic ferroelectric phase of orthorhombic symmetry $mm21'$. This ferroelectric phase has been known to become (weakly) ferromagnetic below 15°K.[11] Because Ni–I boracite has point symmetry $m'm2'$ below 60°K down to at least 15°K,[1] it was plausible to assume by analogy that Ni–Cl boracite would have the same magnetic point group in its ferromagnetic phase. The experiments described in this paper were so devised as to permit examination of this assumption.

2 GENERAL ASPECTS

Let us assume a function g, describing the density of stored free enthalpy of a crystal at zero mechanical stress:

$$g = g(T, \vec{E}, \vec{H})$$

† In part presented at the Symposium on Magnetoelectric Interaction Phenomena in Crystals, Seattle, Washington, May 21–24, 1973.

Working in the frame of a research program supported by the Swiss National Foundation for Scientific Research‡ and Battelle, Geneva Research Center §.

where T is the temperature, $\vec{E}$ the electric field, and $\vec{H}$ the magnetic field. At constant temperature, this function has the following form:[12,3]

$$-g = \ldots + \kappa_i{}^0 E_i + \chi_i{}^0 H_i + \alpha_{ik} E_i H_k + \tfrac{1}{2}\kappa_{ik} E_i E_k + \tfrac{1}{2}\chi_{ik} H_i H_k + \tfrac{1}{2}\alpha_{ijk} H_i E_j E_k + \tfrac{1}{2}\beta_{ijk} E_i H_j H_k + \ldots$$

By differentiating with respect to E_i, we obtain the total polarization:

$$P_i = -\frac{\partial g}{\partial E_i} = \ldots + \kappa_i{}^0 + \alpha_{ik} H_k + \kappa_{ik} E_k + \alpha_{jki} H_j E_k + \tfrac{1}{2}\beta_{ijk} H_j H_k + \ldots$$

With a view to measuring the magnetic field induced magnetoelectric polarization (ME_H effect[13]), we have to consider only the coefficients α_{ik} and β_{ijk} of the linear and second order magnetoelectric effect respectively.

The symmetry of the magnetic point group determines which of the elements of the tensors are permitted. Let us note that the tensor (β_{ijk}) has the same form as the piezoelectric tensor.[14] For the magnetic point group $mm2$ and $m'm2'$, the tensor of the linear effect has the same form.[15,16] By choosing the coordinate system as shown in Figure 1, we obtain:

$$\begin{pmatrix} P_1 \\ P_2 \\ P_3 \end{pmatrix} = \begin{pmatrix} 0 & 0 & 0 \\ 0 & 0 & \alpha_{23} \\ 0 & \alpha_{32} & 0 \end{pmatrix} \cdot \begin{pmatrix} H_1 \\ H_2 \\ H_3 \end{pmatrix}$$

For the symmetries $mm21'$, $mm2$, and $m'm2'$, the tensor (β_{ijk}) of the second order effect has the same form as the piezoelectric tensor (d_{ijk}) for point group $mm21'$.

As $d_{ijk} = d_{ikj}$, we can put $d_{ijk} = d_{il}$ (if $j = k$) and $d_{ijk} = \frac{1}{2}d_{il}$ (if $j \neq k$) with i, j, k running from 1 to 3, and l from 1 to 6 for jk = 11, 22, 33, 23, 13 and 12. As the only non-zero coefficients are $d_{31}, d_{32}, d_{33}, d_{24}$, and d_{15} for $mm21'$,[17] one obtains, for the magnetic field induced polarizations, owing to contributions of the first *and* second order magnetoelectric effect:

$$P_1 = \beta_{113} H_1 H_3$$
$$P_2 = \alpha_{23} H_3 + \beta_{223} H_2 H_3$$
$$P_3 = \alpha_{32} H_2 + \tfrac{1}{2}(\beta_{311} H_1{}^2 + \beta_{322} H_2{}^2 + \beta_{333} H_3{}^2)$$

For measurement of P_1, P_2, and P_3, three platelets are used (Figure 1). The orientation of the principal refractive indices (definition: $n_\alpha < n_\beta < n_\gamma$) with respect to the crystallographic axes a_0, b_0, c_0 (definition: $a_0 < b_0 < c_0$) is also indicated. From ferro-

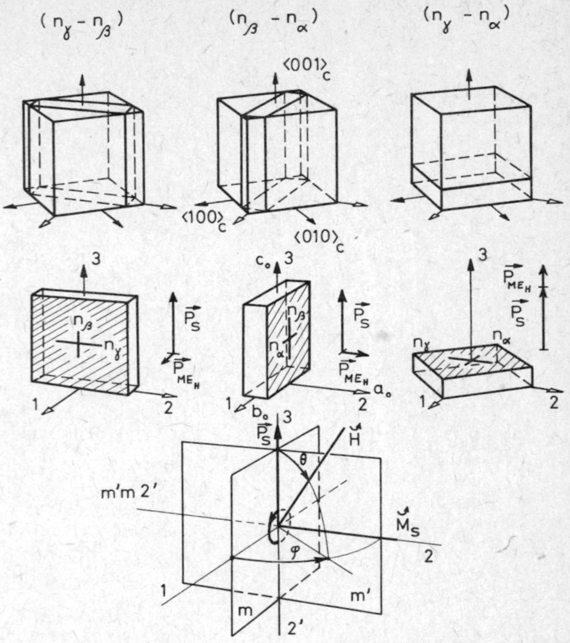

FIGURE 1 Mutual orientation of crystallographic axes and optical indicatrix for the three platelets of Ni–Cl boracite used for ME_H measurements. The orthorhombic axes are $1(b_0)$, $2(a_0)$, and $3\ (c_0 // \vec{P}_S)$ for the definitions $a_0 < b_0 < c_0$ and $n_\alpha < n_\beta < n_\gamma$. Bottom: orientation of spontaneous polarization $\vec{P}_S$, spontaneous magnetization $\vec{M}_S$, and magnetic field $\vec{H}(\theta, \varphi)$ with respect to the crystallographic axes and to the symmetry elements of Shubnikov group $m'm2'$.

elastic experiments, it is known[18] that $n_\alpha // b_0, n_\gamma // a_0$, and $n_\beta // c_0 // \vec{P}_S$. In the chosen coordinate system $b_0 // 1, a_0 // 2$, and $c_0 // 3$.[19]

Since there exist several ways of defining the tensor of the linear magnetoelectric effect according to the system of units chosen, we shall discuss this point in some more detail.

Experimentally one measures the change of a voltage over a capacitor, the charge of which is induced by the crystal under the influence of an applied magnetic field. We use the rationalized MKSA system. One of Maxwell's equations is div $\vec{D} = \rho$, or in integrated form

$$\oint_\Sigma \vec{D} \cdot d\vec{\Sigma} = \int_V \rho\, dV = Q$$

where Q is the charge inside volume V, limited by the surface Σ, which contains one of the crystal's electrodes.

For a platelet of surface S_i perpendicular to the axis i:

$$P_i = D_i = \frac{Q}{S_i} = \frac{CU}{S_i}$$

For the case that all components of the tensor (β_{ijk}) are zero,

$$\alpha_{23} = \frac{CU}{S_2 H_3} \quad \text{and} \quad \alpha_{32} = \frac{CU}{S_3 H_2}$$

Here, Q is measured in Coulomb (= [As]), S in [m^2] and H in [A/m] (1 [A/m] = $4\pi \cdot 10^{-3}$ [Oe]), whereas α_{ik} has the dimension of the inverse of a velocity: [s/m].

The charge can be measured independently of the capacity of the system by means of a "Coulomb-meter," indicating the voltage over an internal capacity of known value, or, alternately, it can be measured with an "electrometer" if the total capacity of the system (crystal, coaxial cable, etc.) can be calculated or measured.

By using the "mixed Gaussian" system (noted by a prime) and defining the thermodynamic potential in the following form:[20]

$$-\Phi' = \ldots + \frac{\alpha'_{ik}}{4\pi} E'_i H'_k + \ldots$$

and since

$$D'_i = -4\pi \frac{\partial \Phi'}{\partial E'_i}$$

we have

$$D'_i = \alpha'_{ik} H'_k \quad \text{or} \quad 4\pi P'_i = \alpha'_{ik} H'_k$$

In the Gaussian system div $\vec{D}' = 4\pi\rho'$ or

$$\oint_{\Sigma'} \vec{D}' \cdot d\vec{\Sigma}' = 4\pi \int_{V'} \rho' \, dV' = 4\pi Q'$$

hence

$$D'_i = \frac{4\pi Q'}{S'_i} \quad \text{and} \quad \alpha'_{ik} = \frac{D'_i}{H'_k} = \frac{4\pi Q'}{S'_i H'_k}$$

Here, Q' is given in statcoulomb (= [cm$^{3/2}$ g$^{1/2}$ s^{-1}]), S' in [cm^2], H' in Oersted (= [cm$^{-1/2}$ g$^{1/2}$ s^{-1}]) thus α'_{ik} is dimensionless.

For passing from the value α_{ik} in rationalized MKSA units to the value of α'_{ik} in units of the Gaussian system, α_{ik} has simply to be multiplied by the velocity of light c (expressed in MKSA):

$$\alpha'_{ik} = c \cdot \alpha_{ik} \simeq 3 \cdot 10^8 \text{ [m/s]} \cdot \alpha_{ik} \text{ [s/m]}$$

3 EXPERIMENTAL

3.1 *Preparation of Samples*

The single crystal of Ni–Cl boracite were prepared by gas phase transport[4] and had a size of about 1 to 3 mm. By means of wire saw, oriented platelets (Figure 1) were cut out of various crystals. In case of sufficient area of single domain crystals, the thickness was ground down to 80 μm, sometimes even to 30 μm, by means of carborundum paper. Subsequent polishing was done with diamond paste (grain size 1 μm). The properties of the optical indicatrix,[2] and the correlation between crystallographic axes and principal refractive indices[18] being known, it was easy to determine the orientation and thickness of the crystal platelet by means of a polarizing microscope.

Electrodes were vacuum evaporated on to the platelets. The platelet called $(n_\gamma - n_\alpha)$ in Figure 1, was covered with circular aluminium electrodes with a surface of $S = 0.98$ mm^2, this surface being smaller than the entire single domain. Prior to electroding definitely, the crystal was poled in between mobile transparent tin oxide electrodes, and cleaned in HCl and NaOH consecutively. It is not possible to pole electrically the platelets $(n_\beta - n_\alpha)$ and $(n_\gamma - n_\beta)$ of Ni–Cl boracite in which case the spontaneous polarization lies within the plane of the platelet. This difficulty arises because the coercive field is too high (10–20 kV/cm) to achieve poling by applying the field on the electroded narrow edges of the crystals as was done for samples of the $mm21'$ phase of Fe–I boracite.[21] However, by applying a uniaxial pressure (e.g., of 1 Kp/mm^2) along a $\langle 1\bar{1}0 \rangle$ direction in the cubic phase and by cooling under that pressure from above the Curie temperature, it was possible to obtain a $(n_\beta - n_\alpha)$ single domain. By this operation the smallest orthorhombic axis, a_0, becomes parallelly oriented to the direction of the stress applied. At the end of the operation the crystals often fractured. In Figure 2 we show such a platelet before and after the treatment.

The crystal $(n_\beta - n_\alpha)$ which was used for measuring α_{23} is represented in Figure 3. Transparent gold electrodes were deposited under high vacuum (10^{-7} torr) on both the entire large surfaces. In order to achieve better adherence of the gold, a thin chromium layer was deposited first. The deposition of the layers was preceded by outgasing at about 350°C. Because the ferroelectric Curie point of Ni–Cl boracite is 335°C, the crystal transformed into the cubic phase. After this heat treatment and subsequent cooling, the initially polydomain crystal had incidentally become entirely single domain with $(n_\beta - n_\alpha)$ within the plane of the platelet. In order to obtain isolating side facets, the edges were ground off with the saw or slightly polished on fine grained carborundum paper.

A platelet $(n_\gamma - n_\beta)$ was also singled out and electroded on both its entire large surfaces.

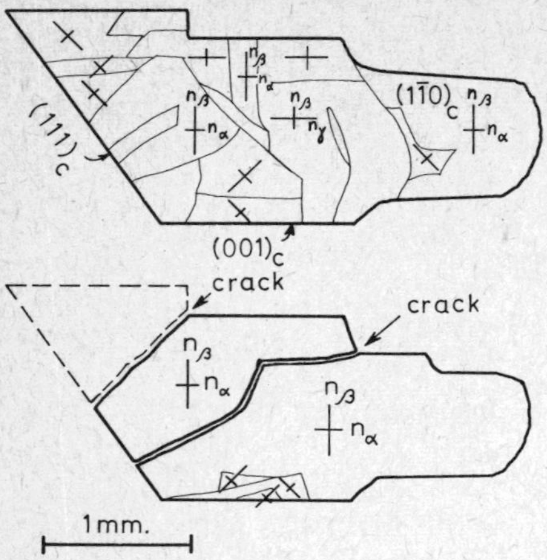

FIGURE 2 Platelet cut parallel $(1\bar{1}0)_{cubic}$ (thickness: 33 μm, surface: 3.4 mm^2) at room temperature, before and after cooling under uniaxial pressure (1 Kp/mm^2 perpendicular to plane of platelet) from the cubic phase.

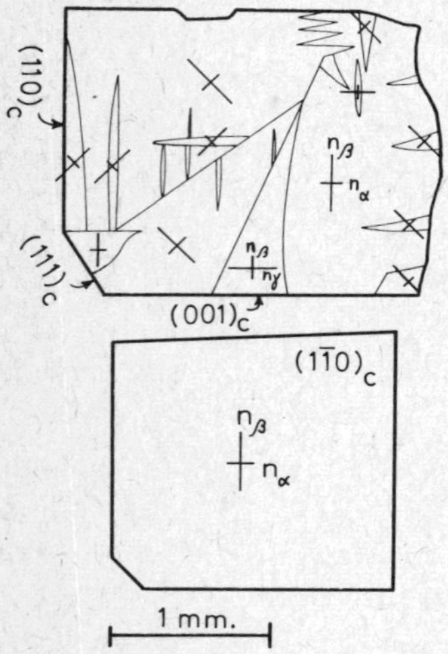

FIGURE 3 Crystal platelet used for measurement of α_{23}, before and after deposition of transparent gold electrodes and cutting of the edges (thickness: 25 μm, surface: S_2 = 2.95 mm^2) Bottom: the platelet has accidentally become a 100% single domain after being cooled from above the Curie temperature.

The electric contacts on the electrodes were realized by means of thin gold wires (∅ = 0.04 mm) which were fixed with a mixture of silver powder and epoxy resin, necessitating polymerization at 150°C for one hour.

Because the spontaneous polarization of Ni–Cl boracite single domains is very stable at room temperature and below, it was not necessary to apply an electric field during cooling to the magnetically ordered phase. However, to be sure that a ferromagnetic single domain would form at low temperature, the samples were always cooled to 4°K in a magnetic field of 6 kOe, applied parallelly to the direction of the refractive index n_γ.

3.2 *Apparatus and Measuring Technique*

For quasistatic measurements of the ME_H effect, we use an Inox liquid helium dewar and a Varian 9 inch magnet, the whole system being usually employed for EPR work.[22] The wave guide for EPR work in the dewar is interchangeably replaced by an Inox tube (∅ 16 mm, with thickness 0.5 mm), in the centre of which an Inox wire (∅ 0.8 mm) is stretched over a length of 115 cm. This tube-wire assembly serves as a low capacity co-axial cable (45 pF). The exit of the coaxis is realized by means of a high resistance (~10^{14} Ω) ceramic-metal element. In the interior of the tube, teflon disks are used for centering of the wire. At the lower extremity of the wire, the crystal is fixed on an interchangeable copper support by means of small pieces of "Scotch" tape. A HF coaxial cable of 40 cm length links the system to a vibrating capacity electrometer ("Keithley 640") of 10^{16} Ω input resistance and internal capacity of 20.0 picofarad (±0.25%). By operating in the integrating (Coulombmeter) regime, $Q = CU = 20.0$ pF $\cdot U$. The voltage U over the capacitor is read on the instrument scale, or else it is amplified and put on the Y amplifier of an X-Y recorder by using the ±1 volt output of the electrometer. The magnetic field is measured by means of a Hall effect Gaussmeter which may be linked direct to the X input of the recorder.

By slowly increasing the magnetic field at constant temperature (e.g. 6000 Oe/mn) by means of the "Varian system," the slope of the recorded curve gives us α_{ik}.

The temperature is measured with a germanium resistor fixed to the copper support. The thermal contact between support and liquid helium reservoir is realized by means of helium exchange gas and beryllium bronze springs. By pumping over the He (which is later recycled), 1.4°K can be reached. A resistance heater permits the temperature of the sample holder to be raised.

On one of the Ni-Cl boracite crystals with $(001)_{cub.}$ cut and $(n_\gamma - n_\alpha)$ orientation, i.e. the spontaneous polarization ($\vec{P}_s$) being perpendicular to the electrodes, a dc voltage—hence a charge—was observed already at room temperature and in absence of a magnetic field. After lifting the short-circuit over the crystal electrodes, this charge increased steadily and stabilized after about

10 minutes. After cooling the crystal to liquid helium, the final charge obtained after lifting the short circuit (at constant temperature) had strongly increased. At that temperature it took several hours for the voltage to increase and stabilize. The final voltage appeared to depend on the ferroelectric switching history of the crystal. It was observed on several single domain $(n_\gamma - n_\alpha)$ platelets, having $\vec{P}_s$ perpendicular to the electrodes. Since the electric resistance of the Ni–Cl boracite samples was very high ($\sim 10^{12}$ Ω) at room temperature and increased even more during cooling, the phenomenon can certainly be ascribed to an electret effect. This interpretation is supported by the fact that the effect could hardly be observed on the orthorhombic (100) and (010) cuts, i.e. $(n_\gamma - n_\beta)$ and $(n_\beta - n_\alpha)$, respectively, both containing the polarization within the plane of the platelet.

In order to compensate for the undesirable drift, the electric zero of the electrometer had to be displaced. Then the signal was measured for one minute at zero or constant magnetic field. Subsequently, the

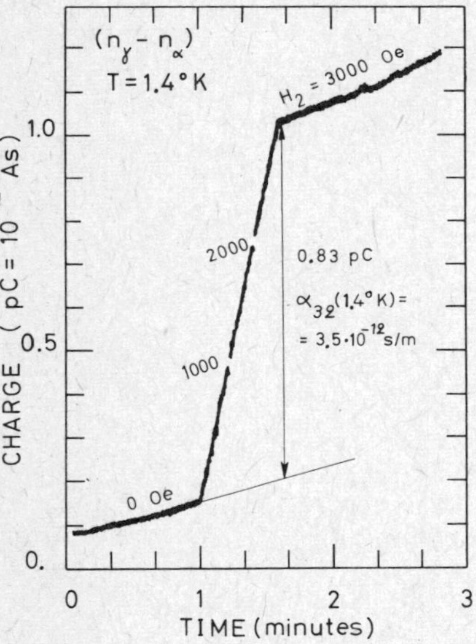

FIGURE 4 Graph showing the measurement of charge (ME_H induced) on crystal $(n_\gamma - n_\alpha)$ as a function of magnetic field and time, and for the case $\vec{P}_{ME_H} // \vec{P}_s$. The drift is due to the electret effect.

magnetic field was slowly increased and again held constant after it had reached a certain value, the recording having been made as a function of time. Figure 4 shows such a recording at $T = 1.4$°K with $\vec{H} // 2$ and $\vec{P}_{ME_H} // \vec{P}_s // 3$, hence α_{32} (1.4°K).

4 EXPERIMENTAL RESULTS

4.1 *Crystal* $(n_\gamma - n_\beta)$

Thickness 75 μm, surface of the electrode $S_1 = 1.22$ mm^2.

The spontaneous electric polarization $\vec{P}_s$ and the spontaneous magnetization $\vec{M}_s$ both lie within the plane of the platelet (Figure 1).

The coefficients α_{11} and α_{12} are found to be zero within the accuracy of the measuring apparatus. It was not attempted to measure the coefficient α_{13}.

4.2 *Crystal* $(n_\beta - n_\alpha)$

25 μm, $S_2 = 2.95$ mm^2.

The spontaneous polarization lies within the plane of the platelet and the spontaneous magnetization perpendicular to it (Figure 1).

Figure 5 shows the angular dependence $P_2/(\alpha_{23} \cdot H) \cong \cos\theta$ for $\varphi = 90°$ and $T = 5.5$°K. A sudden change of sign of the $\vec{M}E_H$ signal occurs when the magnetic field becomes parallel to the spontaneous polarization $\vec{P}_s$ and hence perpendicular to the spontaneous magnetization $\vec{M}_s$. Figure 6 shows the change of sign after an increase of the positioning angle of the magnetic field by one degree, and this without changing the sense of the magnetic field. One observes (at $\theta = 181°$) that part of the magnetic domains returns suddenly, but with a certain lag. It is also noteworthy to

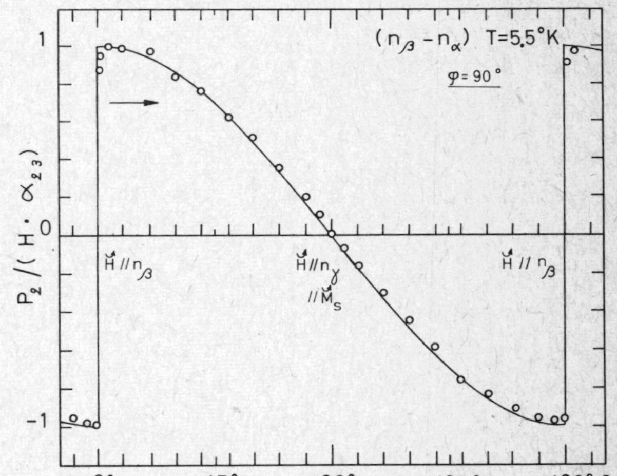

FIGURE 5 Angular dependence of the normalized ME_H signal (circles) compared with the half-period of cos θ (full line) for platelet $(n_\beta - n_\alpha)$ at $T = 5.5$°K. An abrupt change of sign occurs when $\vec{H}$ becomes parallel to $\vec{P}_s$, hence perpendicular to $\vec{M}_s$. The arrow indicates (as in some of the subsequent figures) increases in time.

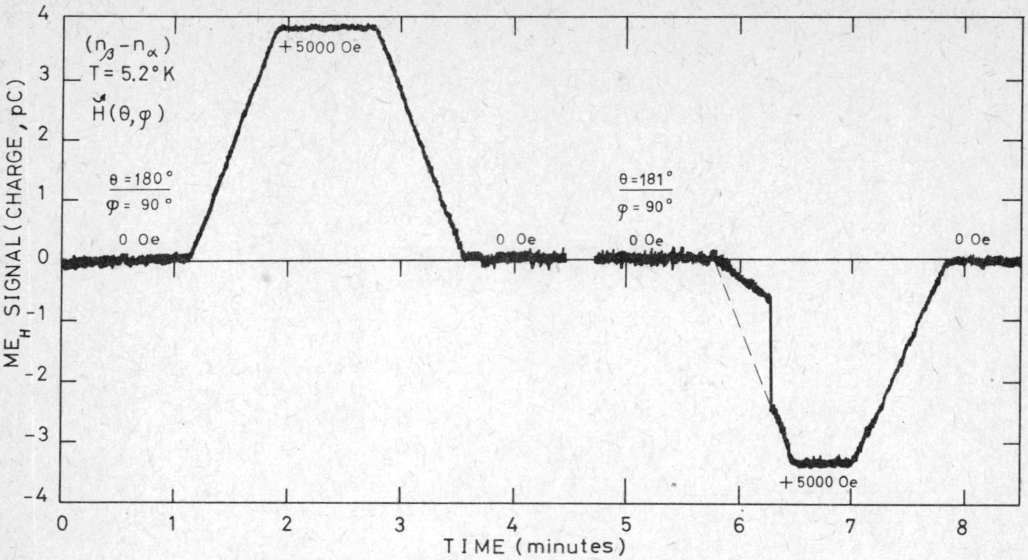

FIGURE 6 ME_H signal for platelet $(n_\beta - n_\alpha)$, ($T = 5.2°K$, $\vec{H} // \vec{P}_s$). When the angle θ of $\vec{H}$ is increased by one degree in the plane $(\vec{P}_s - \vec{M}_s)$, a sudden reversal of the magnetic domains occurs for $H_c(// \vec{M}_s) = H \cdot \sin 1° \cong 70$ Oe.

mention the absence of drift of the ME_H signal for the case of P_2 being perpendicular to $\vec{P}_s$.

Figure 7 shows the linearity of the effect as measured close to the position of sign reversal at $T = 5.3°K$, i.e. $\alpha_{23}(5.3°K) \cdot \cos 5°$. Finally, Figure 8 shows—for the same orientation of the magnetic field—the temperature dependence of α_{23}. The coefficient slowly decreases from 1.4°K to 6°K and tends rapidly towards zero at 9°K. It remains zero above that temperature. Its value at 4.5°K is

$$|\alpha_{23}(4.5°K)| = (3.3 \pm 0.4) \cdot 10^{-12} \text{ [s/m]}$$

i.e.

$$|\alpha'_{23}(4.5°K)| = (10.0 \pm 1.2) \cdot 10^{-4}$$

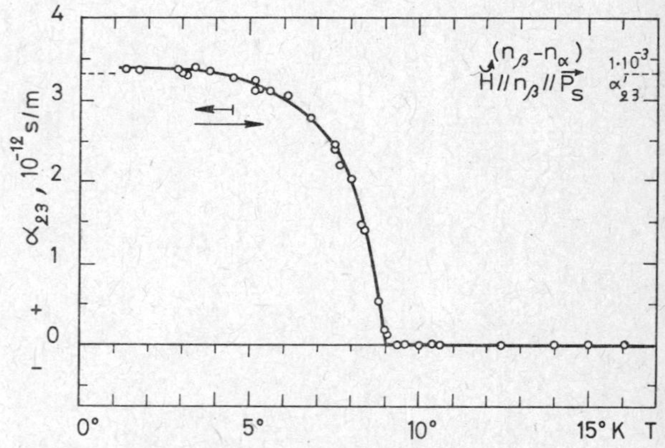

FIGURE 8 Temperature dependence of coefficient α_{23}, measured on crystal $(n_\beta - n_\alpha)$ with $\vec{H}$ at $\theta = 5°$ inclination from $\vec{P}_s$ in plane $(\vec{P}_s - \vec{M}_s)$.

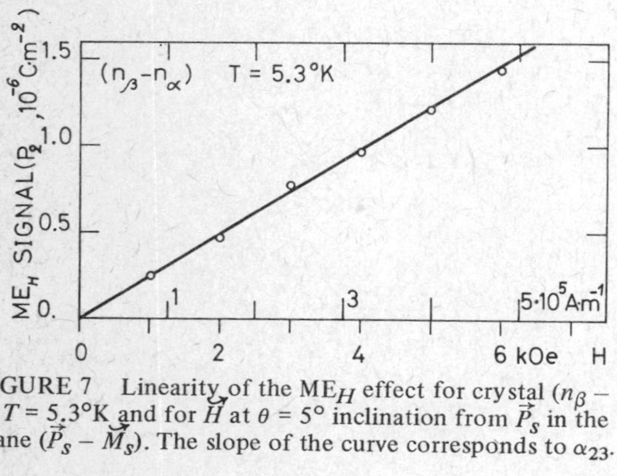

FIGURE 7 Linearity of the ME_H effect for crystal $(n_\beta - n_\alpha)$ at $T = 5.3°K$ and for $\vec{H}$ at $\theta = 5°$ inclination from $\vec{P}_s$ in the plane $(\vec{P}_s - \vec{M}_s)$. The slope of the curve corresponds to α_{23}.

The element α_{22} of the magnetoelectric tensor was found to be zero as shown in Figure 5. No attempt was made to measure α_{21}.

4.3 *Crystal* $(n_\gamma - n_\alpha)$

80 μm, $S_3 = 0.98$ mm^2.

The spontaneous polarization $\vec{P}_s$ lies perpendicular to the plane of the platelet and the spontaneous magnetization $\vec{M}_s$ within (Figure 1) and parallel to n_γ.

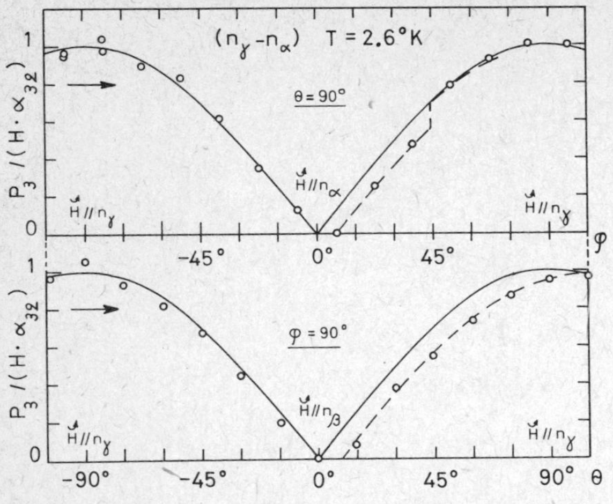

FIGURE 9 Angular dependence of normalized ME_H signal of crystal $(n_\gamma - n_\alpha)$, compared with $|\cos\varphi|$ (full line) for $\vec{H}$: (a) in the plane at $\theta = 90°$ and compared with $|\cos\theta|$ for $\vec{H}$: (b) in the plane at $\varphi = 90°$.
A lag in switching is observed owing to the finite coercive field.

Figure 9 shows the angular dependence of the magnetic field induced polarization for the case of the magnetic field being rotated within the plane of the platelet:

$$P_3/(\alpha_{32} \cdot H) \cong |\sin\varphi|$$

or for the case of rotating in a plane perpendicular to the plane of the platelet:

$$P_3/(\alpha_{32} \cdot H) \cong |\sin\theta|$$

The linearity of the effect was measured for several temperatures with $\vec{H}//\vec{M}_s//2//n_\gamma$ (Figure 10). In

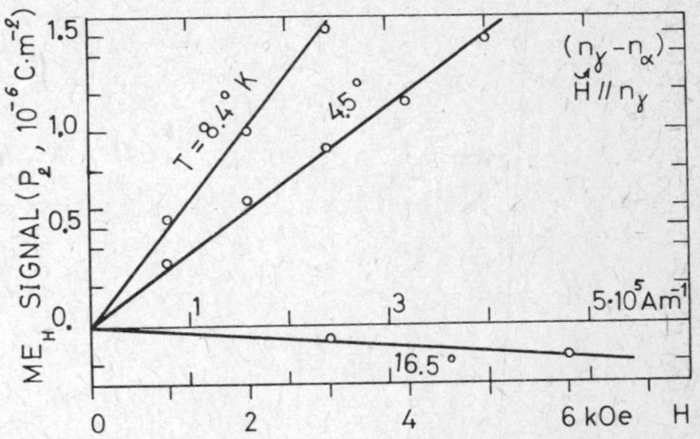

FIGURE 10 Linearity of ME_H signal for crystal $(n_\gamma - n_\alpha)$ at $T = 4.5°K$, $T = 8.4°K$, and $T = 16.5°K$ $(\vec{H}//n_\gamma//\vec{M}_s)$, the slopes yield $\alpha_{32}(T)$.

Figure 11 the temperature dependence of α_{32} is represented, showing a peak at 8.5°K and a change of sign at 10.5°K. Above that temperature the signal is small, and it tends to disappear at about 25°K. Figure 12 makes it possible to see the magnetic coercive field

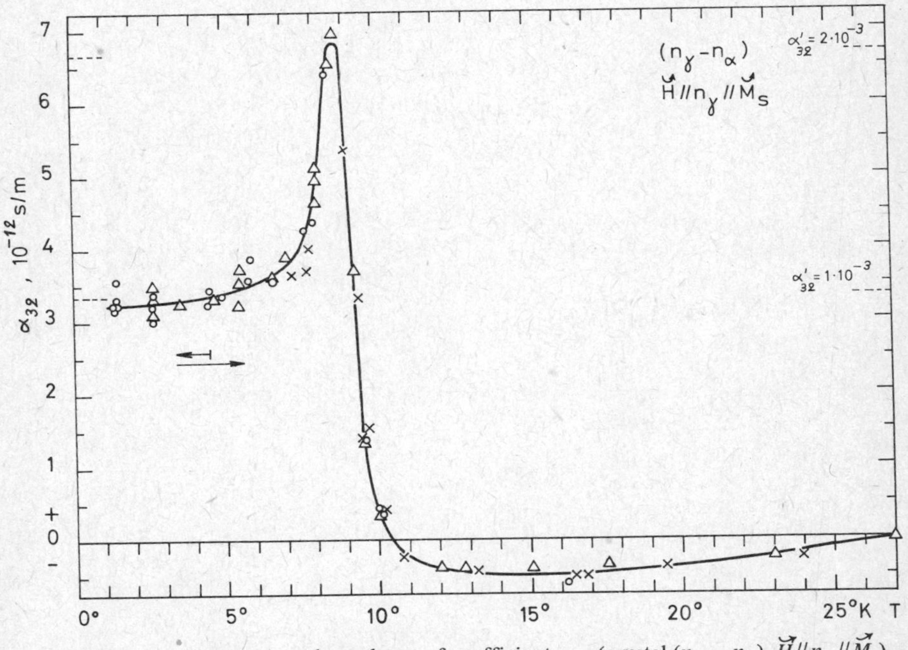

FIGURE 11 Temperature dependence of coefficient α_{32} (crystal $(n_\gamma - n_\alpha)$, $\vec{H}//n_\gamma//\vec{M}_s$).

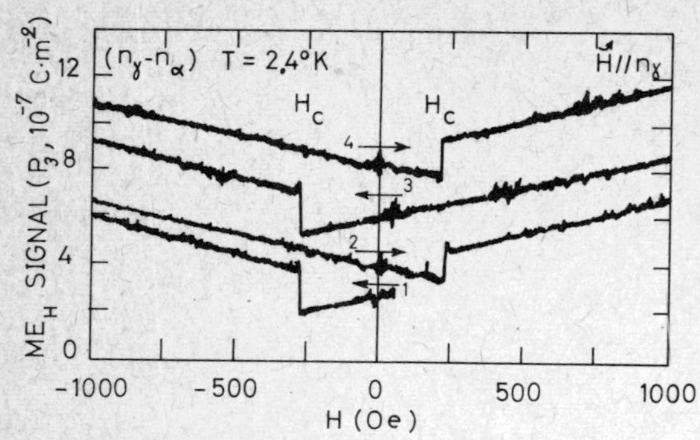

FIGURE 12 Observation of the ME_H signal (not corrected for the drift) on crystal ($n_\gamma - n_\alpha$), obtained by alternatively changing the direction of $\vec{H}$ (1, 2, 3, 4, . . .). This figure is the analogue of the "butterfly loops" observed for Ni–I boracite.[1,2]

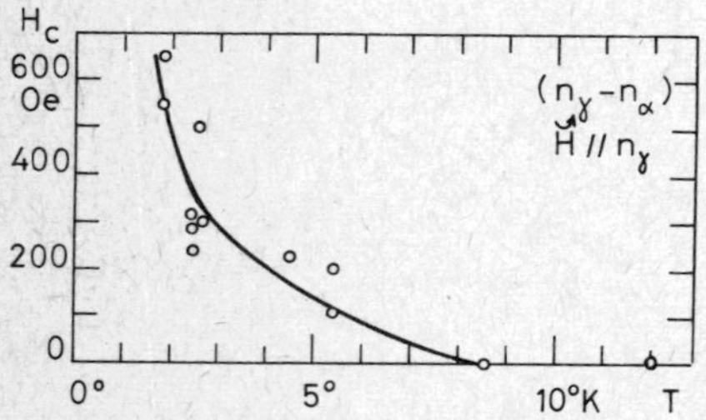

FIGURE 13 Magnetic coercive field H_c of Ni–Cl boracite versus temperature. The steep increase of H_c for decreasing temperature is typical of weak ferromagnets.

H_c. It is of the order of 250 Oe at 4.5°K. This confirms the value of 300 Oe measured at 4°K on powder.[11] The coercive field decreases with rising temperature and tends to zero at the critical temperature of 9°K to 10°K (Figure 13). No coercive field was detectable above 10°K.

The value of α_{32} at $T = 4.5$°K is

$$|\alpha_{32}(4.5°\mathrm{K})| = (3.3 \pm 0.5) \cdot 10^{-12}\ [\mathrm{s/m}]$$

i.e.

$$|\alpha'_{32}(4.5°\mathrm{K})| = (10.0 \pm 1.5) \cdot 10^{-4}\ \text{(Gaussian system)}$$

(half the value at 8.4°K).

The elements α_{31} and α_{33} are found to be zero, as shown for example in Figure 9. Above 10.5°K, the coefficient α_{32} changes sign, but—within the limits of measuring accuracy—the effect appears to remain still linear as a function of the magnetic field (Figure 10).

5 DISCUSSION OF RESULTS

5.1 *Phase Transitions*

5.1.1 Various independent measurements (see especially Figures 5, 6, 9, 12) clearly show that the spontaneous magnetization in the phase below 9°K lies parallel to n_γ, i.e. parallel to the orthorhombic a_0-axis (2-direction).

5.1.2 Seven components out of the nine potential ones of the magnetoelectric tensor have been submitted to measurement. From 1.4°K to 9°K or 10°K, α_{23} and α_{32} are non-zero. All diagonal elements are zero within the accuracy of measurement.

5.1.3 The coefficient α_{32} shows a pronounced peak value at 8.5°K, and a change of sign at 10.5°K. The signal gently disappears at about 25°K. The coefficient α_{23} is zero above 9°K within the accuracy of measurement.

5.1.4 The sharp rise of the magnetic coercive field H_c is typical of weak ferromagnets. The weak value of H_c seems to be typical of ferromagnetoelectric chlorine-boracite.[11,3]

As the results of the preceding items and the fact that magnetic susceptibility measurements on powders clearly showed antiferromagnetic coupling and a weak ferromagnetic moment,[11,24] it is most probable that *two* magnetic phase transitions take place at low temperature. Thus we arrive at the following sequence of phases as a function of temperature:

Temperature	Phase			Magneto-electric type[3]
	$\bar{4}3m1'$	cubic,	paramagnetic	AA V
$T_{C(\mathrm{Fe})}$ = 610°K	↕			
	$mm21'$	orthorhombic,	paramagnetic	FE IV
25°K	↕			
	$mm2$	orthorhombic,	antiferromagnetic	FE II
9°K	↕			
	$m'm2'$	orthorhombic,	weakly ferromagnetic ferromagnetoelectric	FE I/FM I

Whereas point group $m'm2'$ is practically certain below 9°K, point group $mm2$ between 9°K and 25°K is assumed on the basis of observed linearity up to 6 kOe of the ME_H effect in that range and the disappearance of the coercive field at 10.5°K. However, in order to confirm this interpretation, measurements at still higher magnetic fields would be desirable in order to see whether the observed signals might not be due to a higher order ("paramagnetoelectric"[25]) effect.

In order to verify these results, measurements of the spontaneous magnetization and Faraday effect versus temperature are currently underway.

5.2 *Relative Sign of ME_H Coefficients, Accuracy of Measurements, Comparison with Ni–I boracite*

The relative and absolute signs of α_{23} and α_{32} are not known. The relative accuracy of the angles is ± 0.3°, and the absolute accuracy is estimated to be ± 5°. The relative errors are: 5% for the electrode surface, 4% for the magnetic field, and 3% (crystal $n_\beta - n_\alpha$) or 7% (crystal $n_\gamma - n_\alpha$) for the measurement of the charge. Hence the relative total error for α_{23} and α_{32} is 12% and 16%, respectively.

In the case of Ni–I boracite, the maximum value found for $\alpha_{zy} = P_z/H_y$ (at 15°K),[1] and calculated in a mixed Gaussian system, is $\alpha_{zy} = 3.3 \cdot 10^{-4}$. Because $\alpha_{zy} = P'_z/H'_y = \alpha'_{32}/4\pi$, one obtains in the Gaussian system: $\alpha'_{32} = 4\pi \cdot \alpha_{zy} = 4.1 \cdot 10^{-3}$. Hence $\alpha_{32} = \alpha'_{32}/c = 1.4 \cdot 10^{-11}$ [s/m] in rationalized MKSA units, and one finds that the maximal value of α_{32} measured for Ni–I boracite is twice that of the peak value of Ni–Cl boracite.

The negative peak of α_{32} observed by various authors (quasistatic ME_H measurements;[1,26] dynamic ME_H and ME_E measurements[27]) is possibly of analogous origin as the negative part of α_{32} of Ni–Cl boracite, and not necessarily due to spurious domain switching as initially[2] believed.

5.3 *"Pulse" Measurements*

In this paragraph we shall discuss some peculiarities of quasistatic ME_H measurements with "rapid" variation of the magnetic field (e.g. 6000 Oe per second).

During our first experiments on Ni–Cl boracite, we applied such "rapid" pulses as had been done in the case of Ni–I boracite.[1,2] In the case of Ni–Cl boracite, this meant that a signal as a function of time was recorded as that shown in Figure 14 for $T_0 = 7.5$°K. The variation with time of the magnetic field–measured with the Hall probe–is also shown. With the magnetic field finally being kept constant, the amplitude of the ME_H signal decreases. It is also noteworthy that the sign of the signal is opposite to that measured at "slow" variation of the magnetic field. For all Ni–Cl boracite crystals, such signals were obtained in the temperature ranges above 2.5°K. Contrary to the situation above 2.5°K, the negative pulse like response at $T_0 = 1.6$°K is very short (Figure 14) and the signal stabilizes rapidly at the same value as in the case of "slow" variation of field. Because, on the other hand, a sudden increase of temperature (several degrees), followed by a return to the initial temperature, was observed by means of the Ge-probe in the case of "rapid" variation of the magnetic field, the thermal origin of the anomalous behaviour became evident. For example, rapid change of magnetic field can induce eddy currents in the copper support that heat the system. This may have been aggravated by the fact that the sample was mounted in a chamber having a low helium gas pressure. The different behaviour observed at $T_0 = 1.6$°K could be explained by the fact that the He gas, condensing on the copper support, is superfluid (He II below 2.2°K!) and any weak variation of temperature is propagated at the velocity of second sound.[28] In this way, the copper support of the crystal rapidly restores its equilibrium temperature, which is not the case above 2.2°K.

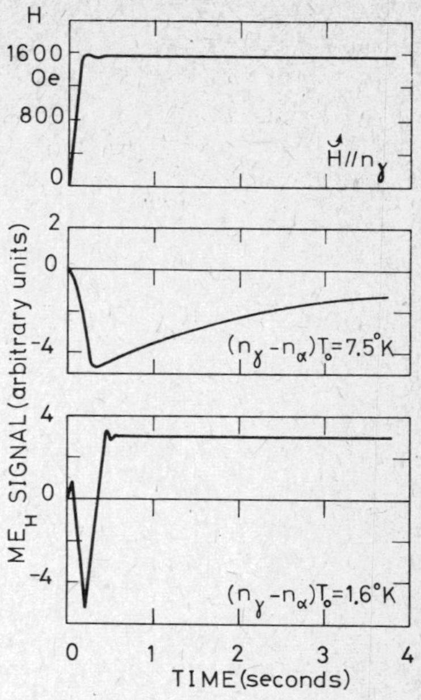

FIGURE 14 Measurement of the ME_H effect by "rapid" variation of H (and subsequent maintenance of H at a constant value) for $T_0 = 7.5$°K and $T_0 = 1.6$°K (< 2.2°K, i.e. in He II).

These observations show that particular care is necessary for ME_H measurements at low temperature with rapid variation of the magnetic field.

In the case of similar measurements on Ni–I boracite,[1,2] the sample was cooled by a flow of He gas at 1 atm, and the lowest temperature attained was 15°K. No difficulties of the above-described nature were encountered in that case.

ACKNOWLEDGEMENTS

The authors wish to thank Dr. E. Ascher for helpful discussions, Mr. H. Tippmann (Battelle, Geneva) for the very able assistance in preparing the crystals, and Mr. Ch. Husler (Department of Physical Chemistry) for his helpfulness in matters mechanical.

REFERENCES AND NOTES

1. E. Ascher, H. Rieder, H. Schmid and H. Stössel, *J. Appl. Phys.* **37**, 1404 (1966).
2. H. Schmid, *Growth of Crystals* 7, 25 (1969), Consultants Bureau, New York–London; translated from *Rost Kristallov* 7, 32 (1967).
3. H. Schmid, *Int. J. Mag.* (this Symposium).
4. H. Schmid, *J. Phys. Chem. Solids* **26**, 973 (1965).
5. J.-C. Joubert, J. Muller, M. Pernet and B. Ferraud, *Bull. Soc. fr. Minéral. Cristallogr.* **95**, 68 (1972).
6. A. Levasseur, Thesis, University of Bordeaux I, 1973.
7. B. Rouby, Thesis, University of Bordeaux I, 1973.
8. A. Levasseur, C. Fouassier and P. Hagenmuller, *Mat. Res. Bull.* **6**, 15 (1971).
9. I. Ito, N. Morimoto and R. Sadanaga, *Acta Cryst.* **4**, 310 (1951).
10. S. Sueno, J. R. Clark, J. J. Papike and J. A. Konnert, *Amer. Min.* **58**, 691 (1973).
11. G. Quèzel and H. Schmid, *Solid State Comm.* **6**, 447 (1968).
12. E. Ascher, *J. Phys. Soc. Jap.* **28** (suppl.), 7 (1970).
13. G. T. Rado and V. J. Folen, *Phys. Rev. Letters* **7**, 310 (1961).
14. E. Ascher, *Phil. Mag.* **17**, 149 (1968).
15. S. Bhagavantam, *Crystal Symmetry and Physical Properties* (Academic Press, London–New York, 1966).
16. T. H. O'Dell, *The Electrodynamics of Magneto-Electric Media* (North-Holland Publishing Co., Amsterdam, London, 1970).
17. J. F. Nye, *Physical Properties of Crystals* (Clarendon Press, Oxford, 1957).
18. H. Schmid, H. Tippmann and J. Kobayashi, unpublished.
19. We have maintained the same system of axes (1, 2, 3) as in the case of Ni–I boracite (x, y, z),[1,2] however, it must be noted that for the latter case, it is not known so far whether $\vec{M}_S // a_0$ or $\vec{M}_S // b_0$.
20. I. E. Dzyaloshinskii, *J. Exptl. Theor. Phys.* (*USSR*) **37**, 881 (1959); translation: *Soviet Phys. JETP* **10**, 628 (1960).
21. J. Kobayashi, H. Schmid and E. Ascher, *Phys. Status Solidi* **26**, 277 (1968).
22. EPR measurements[23] on Mn^{2+} doped Mg–Cl, Zn–Cl, Zn–Br, and Zn–I boracites show a very high total Zero Field Splitting, ranging from 1.6 cm^{-1} for Zn–Cl to 3.8 cm^{-1} for Zn–I boracite.
23. J.-P. Rivera, H. Bill, J. Weber, R. Lacroix, G. Hochstrasser and H. Schmid, *Solid State Comm.* **14**, 21 (1974).
24. H. Schmid, H. Rieder and E. Ascher, *Solid State Comm.* **3**, 327 (1965).
25. S. L. Hou and N. Bloembergen, *Phys. Rev.* **138A**, 1218 (1965).
26. W. von Wartburg, *Magnetische Struktur von Nickel–Jod-Borazit $Ni_3B_7O_{13}I$*, Eidg. Institut für Reaktorforschung, Würenlingen, Schweiz (1973).
27. M. Mercier (private communication).
28. M. W. Zemansky, *Heat and Thermodynamics* (McGraw-Hill Book Company, Inc., New York, 1957).

Note added in proof

The magnetic Curie temperature and the Néel temperature of Ni–Cl Boracite, as determined in this work are consistent with results obtained by magnetic torque measurements by K. Kohn, Waseda University, Tokyo (to be published shortly).

Part IV

Applications and Miscellaneous

POSSIBLE APPLICATIONS FOR MAGNETOELECTRIC MATERIALS†

VAN E. WOOD and A. E. AUSTIN

Battelle-Columbus Laboratories, 505 King Avenue, Columbus, Ohio 43201

The applications which have been proposed for materials displaying magnetoelectric effects are reviewed and a realistic appraisal of the probability of some of these potential applications being demonstrated within a few years is attempted. Both

1) materials exhibiting a magnetoelectric effect in an applied electric or magnetic field, and
2) materials possessing simultaneously magnetic and ferroelectric ordering

are considered. Only technological applications are discussed, more fundamental applications, such as determination of magnetic space groups, being omitted.

The principal operating assumption is that materials with dramatically increased magnetoelectric coefficients will *not* be found in the near future, and the question we try to answer is whether any worthwhile applications are likely to be found anyway. We do assume that some improvements may take place in other important materials characteristics, improvements such as better optical transparency, growth in the form of sizable single crystals, lowered electrical conductivity, etc.

A simple classification of the ways magnetoelectric effects may be used in devices is described, and then possible magnetoelectric devices are classified on the basis of function, operating range, effect used, and other characteristics required. A tabulation, and classification according to the scheme just described, of all magnetoelectric devices we have been able to find in the literature or think up is included, as is a critical parameter analysis intended to determine which possible applications may be eliminated out of hand, at least on the basis of the ground rules given above.

While there does not seem to be at present any device need which can be filled uniquely or most effectively by devices using particular properties of known magnetoelectrics, novel and in some ways superior devices of the "signal-processing" type operating at frequencies from microwave to visible are possible in principle, although considerable further work will be necessary before it can be shown that effective specific devices can be designed. Magnetoelectric memory devices do not at present seem very feasible. The following areas seem promising enough to merit further investigation:

1) Magnetic field switching or modulation of electric polarization, in turn affecting electrooptic properties, eliminating need for thin plates and high applied electric fields.
2) Exploitation of high dielectric constant in high-frequency low-loss microwave Faraday rotators.
3) Efficient generation, modulation, or modification of spin waves or hybrid spin-electromagnetic waves using the magnetoelectric effect.
4) Use of irreversible light propagation in a sensitive interference sensor.

The last two areas must be considered speculative since the fundamental properties to be exploited have not been demonstrated experimentally. There is no reason to believe that these effects are particularly difficult to observe, but no serious attempts to find them have been made. Finally, we present some suggestions concerning the lines future research might best take if magnetoelectric materials are to find device applications.

INTRODUCTION

The purpose of this paper is to attempt a realistic appraisal of the potential for application of magnetoelectric materials. The discussion will be limited, however, to what is known or can be legitimately assumed about materials already prepared and their close relatives, and the possibility of any dramatically superior materials being discovered will be discounted. By "magnetoelectric materials" (abbreviated "ME") we mean all those materials which in some temperature range either

1) exhibit a magnetoelectric effect in an applied electric or magnetic field, or

† Work supported by Advanced Research Projects Agency, U.S. Department of Defense, and monitored by U.S. Army Missile Command under Contract No. DAAH01-70-C-1076.

2) possess both magnetic and ferroelectric ordering.†

It has been pointed out many times by numerous authors that both these effects could lead to certainly unique and possibly useful devices were the effects sufficiently strong, present in a useful temperature range, and present in materials with other desirable characteristics (e.g., optical transparency). This has not proved to be the case so far, although there do not seem to be any fundamental reasons much better materials might not occur. In the present paper we try to assess whether there are any promising areas for application without counting on such superior ME's being found.

There are several factors which make such an assessment difficult.

1) Even in the most thoroughly investigated ME's, not all the relevant parameters for device operation have been measured. In some cases, reasonable estimates can be made, but these verge all too easily into guesses that cannot be substantiated.

2) As we will discuss below, many of the most attractive opportunities for applications of ME devices lie in the regions of millimeter and submillimeter wavelengths. There are relatively few devices operating in these regions at present, but since interest in these regions is growing rapidly, it can be anticipated that this situation will change over the next few years. A comparative analysis of ME vs. "other" devices is largely precluded by not knowing what the "other" will be.

† We repeat here for convenience our definitions[1] of the more important terms met with in this area. Ferroelectric materials which possess some sort of magnetic ordering in the temperature range in which there exists a spontaneous polarization are called *magnetoferroelectrics*. If there is any kind of net spontaneous magnetic moment (that is, if the order is not purely antiferromagnetic) the material is called *ferromagnetoferroelectric*. As it happens, all known examples of these are *weak ferromagnets*; that is, the order is nearly antiferromagnetic, but a slight canting of the spins leads to a net moment. These materials and other magnetically ordered materials may (or may not, depending on the magnetic point group symmetry[2]) exhibit "the" *magnetoelectric effect*, which is a dependence of the dielectric polarization on an applied magnetic field and a dependence of the magnetization on an applied electric field. Ordinarily we refer to the "linear" effect (bilinear, if one is talking about the free energy) in which the additional polarization or magnetization is proportional to the strength of the applied field; non-linear effects depending on particularly strong applied fields, simultaneous application of electric and magnetic fields, or application of stress, are also possible.

3) There are very few classes of materials (the boracites may be an exception) in which sufficient materials preparation research has been carried out that one can tell what the optimum parameters may be. This is particularly so in the areas of reducing electrical conductivity, improving dielectric, magnetic, or magnetoelectric properties by solid solution, improving optical quality, and obtaining larger crystals. Thus limiting ourselves to those extrapolations that have some basis in experiment may sometimes be too stringent a policy.

So although a critical comparative analysis cannot presently be made, one may at least point out where the possibilities and, within the ground rules already discussed, impossibilities lie.

2 BASES FOR MATERIALS SELECTION

The device designer is frequently faced with a choice among several physical effects for the accomplishment of some desired device function, and with a choice among many materials showing a given effect. If he attempts to rationalize his choice, he will often do so on the basis of a figure of merit (perhaps several) indicating the relative effectiveness of the phenomenon and material under consideration in performing the desired function. Such figures of merit are commonly chosen to be dimensionless where possible, and they are defined such that "big" means "good." Generally speaking, the "function" of any of the devices considered here is the modification (change of amplitude, frequency, phase, polarization, etc.) of an electromagnetic, acoustic, or hybrid (e.g., magnetoelastic) wave. For some devices, specification of a principal figure of merit is straightforward.[3] Thus for an *isolator*, the natural figure of merit is the ratio of reverse to forward attenuation; for a *Faraday* (plane-polarization) *rotator*, it is the angle of rotation per unit length divided by the average attenuation in this length; for a *phase shifter* it is the phase shift per unit average attenuation. For switching and memory devices, specification of a figure of merit is not as simple, but a useful figure for the specific case of magnetooptical memories has recently been developed by Cohen and Mezrich.[4]

Even in the simpler cases, there is a certain amount of arbitrariness in the definition of a figure of merit which is hard to justify on grounds other than simplicity. For instance, if there is a threshold below which power transmitted in the reverse direction through an isolator will not be detected, an increase in reverse attenuation beyond that necessary to reduce the power below this level is of no value and ought not be counted. It should also be borne in mind that there is almost always some time constant —switching or response time—requirement which the material must also meet. And there are many other technological factors that must be considered, such as temperature stability or fabricability, for instance, to say nothing of important additional factors such as familiarity, availability, and cost. Still, the figure of merit provides the simplest criterion for materials comparison and we use it where possible. Because of the paucity of experimental data, there are few cases where complete calculations can be carried out, but quantitative discussions of the effects of magnetoelectric and other parameters on the figure of merit can sometimes be given.

3 CLASSIFICATION OF POSSIBLE ME DEVICES

Any ME device may be specified in a fairly general, if somewhat abstract way by giving

1) its function,
2) its operating range,
3) the magnetoelectric effect used, and
4) any other required characteristics.

Some types of functions of interest have been mentioned in the last paragraph. More specific operations are given in Table I, which is organized predominantly by function. Under the general term "switches" we may include proposed memory devices, which amount to switches with non-destructive sensors of their polarity.

By "operating range," we mean predominantly frequency range of operation, but also mean to include temperature requirements, voltage and power requirements and limitations, etc., as appropriate.

There are a number of ways one can classify devices by the ME effect used. One way is simply to identify the term or terms in the expansion of the free energy in powers of the applied fields important to the device operation. Such an expansion has been given by Schmid.[5] This is useful for reminding oneself of the effects which may occur, but is a little complicated for our purposes, and does not too easily distinguish effects based on the presence of spontaneous magnetization or polarization from induced effects. So we adopt the following simple scheme, which is used in the fourth column of Table I:

a) Devices requiring presence of both spontaneous polarization P_{sp} and spontaneous magnetization M_{sp}, but not requiring any interaction between them. (This doesn't seem to be a very promising area.)

b) Devices requiring interaction between P_{sp} and M_{sp}. Such an interaction might come about through the magnetoelectric tensor, but might also occur via magneto- or electro-striction or other higher-order effects. One might also include here interaction between the polarization and the sublattice magnetization in antiferromagnets. Such interaction effects may be expected to be largest near the ferroelectric Curie temperature when that temperature is below the magnetic ordering temperature. This situation is seldom met with, though.

c) Devices based on the interaction between electric and magnetic properties brought about by the non-zero components of the magnetoelectric tensor, whether or not there is a P_{sp} or M_{sp}. (There is some overlap between this category and the previous one.)

d) Devices using ferroelectricity simply to get a high dielectric constant, these devices being conventional ones for modulation of microwaves otherwise. (No converse devices—utilizing ferrimagnetism just to get a large magnetic permeability in a ferroelectric—have been suggested, and indeed it is difficult to see what the use of any such thing might be.)

e) Devices based on higher-order ME effects.

f) Devices in which ME effects are not used at all. These are not listed in Table I, but the likelihood that some of the materials studied so far mainly for their magnetoelectric interest will be found to have other useful attributes should not be underestimated. Among the types of applications that might be mentioned are electrooptic or magnetooptic modulators (for instance, iron iodine boracite appears to have electrooptic coefficients[6] comparable to those found in $LiNbO_3$), piezoelectric transducers, pyroelectric detectors, magnetic or ferroelectric semiconductors (both of which might show interesting negative differential resistance effects,[7,8] although this has never been demonstrated), high-electric-field switches,[9] and so forth. Many of the materials so far investigated are of the type in which the ferroelectric transition is accompanied by, or even induced by, a spontaneous deformation which is divertible (can be switched between two (or more) different states by an applied stress). These so-called *ferroelastic* properties may eventually be exploited in devices.

The "other required characteristics" mentioned above (item 4) include such things as preparation in single-crystal form, optical transparency, semi-conductivity, etc. Such things are mentioned in the column headed "Comments" in Table I.

4 TYPES OF ME MATERIALS

We do not present here any list of classes of materials which have been found to have interesting ME properties, but simply refer the reader to the recent compilations of Schmid,[5] Hornreich,[10] and Bertaut.[11] In the discussion below, we refer frequently to boracites and to rare-earth-substituted bismuth orthoferrites as examples of ferromagneto-ferroelectric materials with which we are most familiar. The former materials, of formula $(3d)_3B_7O_{13}$(Hal), where (3*d*) represents any 3*d* ion and (Hal) any halogen, have the spontaneous polarization and magnetization perpendicular; while in the latter type of materials, of formula $Bi_{1-x}(RE)_xFeO_3$ with little enough rare earth (RE) to retain the rhombohedral $BiFeO_3$ structure, the polarization and magnetization are parallel. Only a few of the boracites are FMFE, and these only at low temperatures. Those properties of other types of ME materials of interest for particular applications are described where the applications are discussed; more details are given in the accompanying article of Schmid.[5]

5 CRITICAL PARAMETERS FOR DEVICE APPLICATIONS

The essential factors in utilization of magnetoferro-electrics are principally those of importance in ferrimagnets and ferroelectrics separately. These are:

1) Magnetization and reversing magnetic field.

2) Polarization and reversing electric field.

3) Optical birefringence (natural and induced) and absorption.

4) Resonance frequencies of the spin system and their dependence on the applied magnetic field.

In addition, the magnitude of the magneto-electric coefficients and the size of the effective fields they lead to will be of importance in those applications depending on this effect. Application in devices involves either detection of a polarization or magnetization state or dynamic interaction with fields. The major area of interest for static states is that of data storage (memories) while that for dynamic interaction is control of microwaves, particularly millimeter to submillimeter waves. The analysis of critical parameters has been made with a view toward possible utilization in these fields.

5.1 *Memory Requirements*

There have been several recent reviews[12,13] on memory systems which have emphasized the hierarchy of the systems with trade-offs among access time, capacity and cost per bit. Typical requirements for fast-access memories are a density of 2×10^5 to 2×10^6 bits/cm^2 with access time of 10^{-6} sec. These values are used in the following two sections.

5.2 *Magnetization*

The sensing of the magnetic state of a bit, i.e., its direction of magnetization, involves the detection of its magnetic flux where $\varphi =$ induction $\times$ area $\leqslant \pi^2 d^2 M_s$ (gaussian units), $d =$ bit diameter, and M_s is magnetization per unit volume. An optimistic estimate of minimum flux[14] detectable without using superconducting flux meters is 2×10^{-5} maxwells (2×10^{-13} webers) while present conventional devices require an order of magnitude greater. For an area of 5×10^{-6} cm^2, an M_s of 0.3 emu/cm^3 or an induction of 4 gauss is required. The weakly ferromagnetic orthoferrites have M_s values of around 3.6 emu/cm^3 or inductions of 45 gauss for detectable domains of this size, 25 μ diameter, while ferrimagnetic garnets with much greater M_s have inductions of 200 gauss for smaller domains of 6 μ diameter. Typical bits, 5×10^{-6} cm^2 area, of γ-Fe_2O_3 may have inductions up to 350 gauss. The known magnetoferroelectrics have very weak ferromagnetism. Our data on $Bi_{0.9}Nd_{0.1}FeO_3$ gave M_s values of 0.14 to 0.45 emu/cm^3 which is just about the estimated minimum detectable value for bits of 5×10^{-6} cm^2. Any increase in bit density, which would be desirable, would require higher magnetizations. This could probably be achieved by partial substitution of Al, Ga, or Cr for Fe. No magnetization data are available for FMFE nickel–iodine boracite, $Ni_3B_7O_{13}I$. However, the value of M_s is expected to be comparably low, since its transition is from an antiferromagnetic state to a weak ferromagnetic state. For $Ni_3B_7O_{13}Br$ below the Néel temperature of 40 K the magnetization[15] is of order 10^{-2} emu/cm^3.

5.3 *Polarization*

The use of ferroelectrics for digital data storage requires both storage of the charge from polarization and the detection of its sign and magnitude. According to recent reviews[13, 16] the characteristics of ferroelectrics are not nearly as good as magnetic materials for digital data storage and retrieval. Leakage limits use for long-time storage, and the electrical detection of charge involves some dissipation with resultant reduction on each access. The digital bit size is limited by the minimum stable domain. In sandwiches of thin ferroelectrics/photoconductors, the switching and detection of charge has been demonstrated on 25-μ diameter spots for typical polarization of 8 μC/cm^2 [17] and switching voltages of 20 V across 4-μ thick films, i.e., fields of 50 kV/cm. The need for a photoconductor adds further leakage of charge and also limits cycle time to about 250 μsec.

The known magnetoferroelectrics should have polarizations comparable to other oxide ferroelectrics, 5–20 μC/cm^2. $BiFeO_3$ itself has too high a coercive field for switching or even for measurement of hysteresis at room temperature.[18] The solid solutions such as $Bi_{0.9}Nd_{0.1}FeO_3$ may have lower coercivity, although this is not definitely known. It appears that the conductivity can be sufficiently low, $<10^{-8}$ Ω^{-1} cm^{-1}. Nickel–iodine boracite switches with a field of 5 kV/cm, which is typically that of most usable ferroelectrics.

As noted above, ferroelectrics will not find widespread application in digital memory and display. The potential utilization of the ferroelectric property of magnetoferroelectrics is basically the same as for ferroelectrics not having magnetic ordering. The applications of interest depend upon the large changes in birefringence produced by switching the polarization. Possible applications are shutter array as a page composer in a holographic memory system and light valves in slow-scan displays.

5.4 *Optical Properties*

Pertinent optical properties are absorption, linear birefringence, and Faraday rotation (magnetic circular birefringence). The absorption is most critical for transmitted signals. One can assume the linear birefringence change with electric polarization reversal in the magnetoferroelectric is of the order of 0.5 to 1% (i.e., around 10^5 °/cm birefringent phase retardation) as for good ferroelectrics. Then for rotations of 90° or 180° of transmitted light in the visible range, as desirable for phase shifters in optical processing,[13] one needs thicknesses of the order of 50 to 100 μ. The absorption coefficient must be $\leqslant 10^2$ cm^{-1} to permit 10 to 50% transmission for these thicknesses. Many ferroelectric materials have much lower absorptions so that for half-wave retardation they are virtually lossless. There are insufficient absorption data on most magnetoferroelectrics. The boracites have strong absorption bands with coefficients from about 5×10^2 to 3×10^3 cm^{-1} [19] in the visible. Other materials also have 3*d* transition metal cations, such as Fe^{3+}, Ni^{2+}, Cr^{3+} and Mn^{3+} producing absorption bands in the visible. so one expects low transmission of visible and uv light with absorption coefficient on the order 10^3 to 10^4 cm^{-1}. For instance the ferrimagnetic garnets have absorption values $\sim 10^3$ cm^{-1}. Impurities and nonstoichiometry, particularly of oxygen, can increase the absorption of these oxidic materials. Thus it is questionable that the absorption coefficients of magnetoferroelectrics for the visible range could be decreased from $>10^3$ cm^{-1} to $<10^2$ cm^{-1} as would be desirable for use in place of good ferroelectrics alone. However it is likely that the magnetoferroelectrics containing transition metals may have much lower absorption coefficients in the near-to-medium ir range as do the garnets and orthoferrites.

We must also note the distinction between uniaxial and biaxial crystals for use in optical processing. The uniaxial crystal with 180° domain switching, which is the more common, leaves transmitted light unaltered. The more useful effect requires a biaxial crystal where the optical index ellipsoid is rotated with polarization switching.[13] Of the known magnetoferroelectrics, we may cite $Bi_{0.9}Nd_{0.1}FeO_3$ as an example of a uniaxial and nickel–iodine boracite of a biaxial crystal.

The Faraday rotation F is the magnetooptical effect of major interest. For transmission applications, the parameter of concern is the ratio of Faraday rotation to absorption coefficient, F/α (deg) while for a magnetooptical memory it is F^2/α (deg^2/cm).[4] Rotations of $\pi/4$ radians are desired for transmitted signal processing (rotators, isolators, modulators, etc.). To attain this with acceptable loss (50–95%) an F/α parameter of 15 to 60° is needed. For magnetooptic memories the F^2/α value should be $>10^5$ and preferably $\approx 10^6$ deg^2/cm to be comparable to the intermetallic ferromagnets.

The pertinent data are not available for many known magnetoferroelectrics; however, estimates may be made on the basis of other known weak ferromagnets—orthoferrites, $FeBO_3$, FeF_3—and the ferrimagnetic garnets. A reasonable estimate of the maximum likely spontaneous Faraday rotation is 10^3 deg/cm for which F/α is about 1°. For nickel–iodine boracite[19,20] at 0.633 μ, F/α is around 4°. The corresponding F^2/α value is 10^3 deg^2/cm assuming α is 10^3 cm^{-1}. As noted above, decreased absorption in the near-ir could possibly increase the F/α parameter to the desirable range, assuming F did not change. Raising the F^2/α value to 10^5 deg^2/cm does not appear likely, however.

There is also a problem of interference of the linear birefringence with the Faraday rotation in magnetoferroelectrics. The interference depends upon the relative orientation of the magnetization, M, and polarization, P. Thus for $M \perp P$ the light propagating along the direction of M undergoes birefringent phase retardation. This results in an oscillating spatial variation of the plane of light polarization with period equal to the wavelength of the light and markedly reduces the maximum rotation. This is the case for naturally birefringent $FeBO_3$ and FeF_3 where rotations are limited to 0.7 and 0.3°.[21] According to the estimates above, the linear birefringence is much greater than the Faraday rotation. Nickel–iodine boracite would be expected to show such interference since it has $M \perp P$. For $M \| P$ one would have only the Faraday rotation for light transmitted along the magnetization direction. $Bi_{0.9}Nd_{0.1}FeO_3$ apparently satisfies this latter condition. Thus for utilization of the magnetooptical Faraday effect one would prefer *a priori* uniaxial crystals with $M \| P$. Favorable situations with $M \perp P$ may occur, however. Miyashita and Murakami[20] found, for instance, that a 0.05 cm thick NIB platelet with light propagation in the (110) direction showed a good spontaneous Faraday rotation of 200–350 deg/cm. They did not report the degree of ellipticity of the emerging light, though.

5.5 *Magnetoelectric Coefficient*

The linear magnetoelectric coefficient α (actually, the largest component of the linear magnetoelectric tensor) is the critical parameter for interaction of magnetic and electric fields in applications such as memories or control devices. The higher-order coefficients as well, however, enter into possible applications involving dynamic phenomena. A compilation[11] of measured coefficients gives a maximum linear coefficient of 10^{-2} (c.g.s. units) with 10^{-4}–10^{-3} being more typical. Thus even in strong electric fields (say 2×10^4 V/cm), the induced magnetization, $(4\pi)^{-1}\,\alpha E$, is unlikely to be more than 0.01–0.02 emu/cm^3. For the quadratic effects the likely maximum is a little smaller yet. For a magnetic field of 10 kOe an induced polarization of $\sim 10^{-9}$ C/cm^2 can be estimated similarly. The magnetization is comparable to that of very weak ferromagnets. Thus it is borderline for potential applications involving its use alone. The polarization is several orders of magnitude less than the spontaneous polarization of ferroelectrics. It is too low for use by itself.

The gyrator[22] can be taken as an example of the need for greater magnetoelectric coefficient for purely ME devices. For efficient use, the figure of merit $\alpha^2/\chi_e\chi_m$ should be ≈ 1 where χ_e and χ_m are normal electric and magnetic susceptibilities. The quantity $\alpha^2/\chi_e\chi_m$ for nickel–iodine boracite is about 5×10^{-3} which is very much less than required.

We also need to consider the effects of the magnetoelectric interaction on the optical properties. These can be separated into static effects in magnetoferroelectrics and induced dynamic effects. The principal optical parameters are, as described above, absorption, rotation and linear birefringence. The magnetoferroelectrics can have domain switching by either applied magnetic or electric field. Again there are the two cases of $M \| P$ and $M \perp P$ to consider. In the first, switching of either magnetic or electric field can produce reversal of dielectric polarization and thus of linear birefringence for light propagated transverse to the unique axis and reversal of Faraday rotation for transmission parallel to the axis. The second case involves a 90° change of magnetization direction with reversal of dielectric polarization. Here one can still have reversal of the linear birefringence but the Faraday rotation will show interference with the linear birefringence. The relative magnitudes of the two optical effects point toward utilization mainly of the linear birefringence. Other parameters of importance are the electric and magnetic coercivities and domain size. The small magnetoelectric coefficients lead to only a small contribution to the fields which separately are needed for domain switching of either the ferroelectric or weak ferromagnet. These fields are of order 5 kV/cm or 5 kOe. The necessity of switching domains limits the minimum size of bits for memory applications to a few microns in diameter.

The linear magnetoelectric effect in magnetoelectric materials causes plane-polarized light to become, in general, elliptically polarized, the orientation of the major axis depending on the path difference.[23, 24] The linear ME effect also under certain conditions leads to irreversible light propagation; that is, light travelling in one direction the material may move at a slightly different velocity from that travelling in the opposite direction. It can be shown that the change in index of refraction on reversing direction is given approximately by $|\Delta n| = 2|\alpha|$, or on the order of 10^{-3} [25] if we assume that the magnetoelectric coefficient is not much changed at optical frequencies from its low frequency value. While this is not a large index change, it should be easily measurable, and should lead to interesting interference phenomena between forward- and backward-travelling waves in materials where the absorption is not too high. It is possible that such interference effects might have useful applications, such as the precise control of electric or magnetic fields, or of temperature or pressure. It seems unlikely that this phenomenon can be achieved *any other way* than through the ME effect. The effect might be much larger at frequencies near resonances, where the effective permeability is large.[26]

The effects of higher-order magnetoelectric terms on the optical properties have been discussed by Ascher.[27] The interaction of a linearly polarized electromagnetic wave with a material of the proper magnetic structure through the quadratic magnetoelectric effect can yield second-harmonic generation and optical rectification. The effect again is proportional to the magnitude of the quadratic magnetoelectric coefficients. Too little is known about the magnitudes of these higher-order terms to justify further speculation.

5.6 *Collective Magnetic Resonances*

Known magnetoferroelectrics are antiferromagnets with some exhibiting weak ferromagnetism. Therefore they have natural resonance frequencies falling in the millimeter and submillimeter wavelength range.[28] Neglecting losses, the resonance frequencies, ω, for uniaxial two-sublattice antiferromagnets are

$$\omega^{\pm}/\gamma = [2\lambda K + (\beta H/2)^2]^{1/2} \pm H(1 - \beta/2) \tag{1}$$

where γ is the gyromagnetic factor, λ is the molecular-field exchange constant, K is the anisotropy constant, and β is the susceptibility ratio $\chi_{\parallel}/\chi_{\perp}$. For λK we may write $H_e H_a$ where H_e is the exchange field, λM, and H_a is assumed $\ll H_e$. Or the term $2\lambda K$ can be taken as the square of the "critical" field H_c which can range from 60 kOe to >100 kOe. One must operate at applied fields, H, less than H_c to avoid "spin flop" to the state with the M's perpendicular to the applied field. Either resonance mode can be used in devices. The resonance linewidth ΔH can vary in a given material, depending on its purity, from hundreds to thousands of oersteds. The figures of merit are $F = (4H/\Delta H)^2$ for waveguide resonance isolators and $F = 4H/\Delta H$ for phase shifters.[29] Therefore the pertinent parameters are the critical field H_c and linewidth ΔH. For antiferromagnetic ferroelectrics such as $BiFeO_3$ or $YbMnO_3$, the critical field H_c is expected to be about 100 kOe. Assuming a linewidth of 1 kOe or more, we find the applied fields would need to be of magnitude 5 to 50 kOe for reasonable figures of merit.

An applied electric field can affect the antiferromagnetic resonances through the magnetoelectric coupling. In a strongly anisotropic uniaxial antiferromagnet, magnetoelectric coupling can modify the zero-field resonance frequency[30] to

$$\omega_0'/\gamma = H_c \pm H_{me} \tag{2}$$

where $H_{me} = \alpha E$, α is the magnetoelectric effect coefficient, and E is the applied electric field. For an α of 10^{-4}–10^{-3}, E would need to be very high, $\sim 10^8$ V/cm, to be an appreciable fraction of H_c. Shavrov[30] suggested that a smaller electric field could produce a significant resonance shift in those antiferromagnets showing the magnetoelectric effect in which the minimum excitation energy results from magnetostriction effects only. This should be an observable effect; the resonance, however, will lie in the low microwave region, which does not seem to be of great interest at present. No experimental studies have been made.

For ferroelectric antiferromagnets, an oscillating electric field is equivalent to an added component to the applied magnetic field given by[31]

$$H \rightarrow H + 2\xi M \tag{3}$$

where

$$\xi = \frac{\gamma \eta^2 M(E(\omega_e))^2 \varphi}{8\omega_e}$$

where ω_e is frequency of the electric field and φ is a polarization factor which may be as large as unity. η is the magnetoelectric coefficient if the effect is expressed in terms of cross product of the sublattice magnetizations. According to Akhiezer,[31] η is $\sim 10^{-3}$.

TABLE I
Proposed applications of ME materials

Device	Frequency range[a]	Functions	Phenomenon used[b]	Comments	References
1. Magnetoelectric gyrator	AF–RF	Isolation, phase-shifting, amplification	c	Requires large ME susceptibility, impractical (see Section 5.5)	22, 26, 39
2. Magnetoelectric memory	—	Bit-by-bit read-write memory	c	Requires both *E* and *H* field to switch magnetoelectric domain, thus writing more similar to coincidence writing in ferrite cores than proposed optical-read memories; writing speed possibly slow unless very high fields used; reading by sensing of magnetization change resulting from a.c. electric field may be very fast. Overall present prospects definitely unpromising.[40] Bit density fairly low	11, 26, 40, 41
3. Optical-rectification IR detector	FIR, SMM probably	Sensitive broad-band detector in difficult region	e	The terms $\frac{1}{2}\alpha_{ijk} H_i E_j E_k$ and $\frac{1}{2}\beta_{ijk} E_i H_j H_k$ in the free energy[5] lead to changes in polarization and magnetization, respectively, (the inverse magnetoelectric Pockels and "Mockels" effects) when an electromagnetic wave passes through the material. The conventional wisdom that higher-order optical-property coefficients are frequently less frequency-dependent than linear ones leads one to conjecture that such devices might be useful in the FIR or SMM region. Components of the α and β tensors have been measured in a few materials, but neither of these specific inverse, rectification effects has been demonstrated experimentally	42
4. Electric-field-controlled ferromagnetic resonance device	MW, possibly MM	Switch or isolator, most likely—possibly temperature-stabilized (by *E* field) FMR device	c?	Rather large changes in the *K*-band ferrimagnetic resonance spectrum of $Ga_{0.85}Fe_{1.15}O_3$ were found by Petrov and coworkers[35] on application of 2–3 kV/cm electric fields. A shift of line position and a reduction of linewidth of kOe magnitude and a simplification of line shape were observed. Folen and Rado[36] pointed out that the observed effects were too large to be directly attributable to the magnetoelectric effect, and suggested that they might be due to heating of the samples by the applied *E* field, producing changes in the anisotropy energy. Experimental work of Folen and Rado under slightly different conditions also did not show any large effects. While this seems to be the most likely explanation, there is perhaps room for further experimental work, since there might be an increase in the effective magnetoelectric effect near the resonance,[26] and since the unusual line shapes are not really accounted for. For a material with a tolerably narrow resonance line, which also has a reasonably low d.c. electrical conductivity to minimize Joule heating effects, it should be possible to use an applied *E* field to offset the drift in the resonance caused by temperature change due to microwave absorption. This would greatly simplify the circuitry for the applied variable *H* field being used for phase shifting. The present prospects for this are not good, though, since a satisfactory material is not known. Good single crystals are required	35, 36
5. Electric-field-controlled anti-ferromagnetic resonance device	MM, SMM, MW?	Switch, isolator, phase shifter	c	Does not seem promising at present. (See discussion, Section 5.6.) Almost no experimental work has been done	28–30, 43

TABLE I—*continued*

Device	Frequency range[a]	Functions	Phenomenon used[b]	Comments	References
6. Magnetically switched electro-optic device	V, IR	Optical switch, isolator, amplitude modulator	b	To obtain an electrooptic shutter operating by ferroelectric domain reversal (using spontaneous birefringence) it is necessary to apply high electric fields to the sample, thus usually requiring the use of thin samples and either transparent electrodes or overlying fine wire grids. In certain magnetoferroelectrics, the spontaneous polarization can be reversed by applying a magnetic field (see Section 3, 5.5) of not more than a few kOe. This eliminates the need for thin plates (if absorption not too high) and high electric fields, which could be troublesome in certain environments. Something is presumably sacrificed in frequency response. A modulator working along the same lines would probably work best if it involved a change in domain pattern, rather than relying on the magnetoelectric effect. The kind of coupling required has been demonstrated in the boracites.[44] For the present application, uniaxial materials might be somewhat simpler to work with. Any device of this sort of course has to prove its superiority to straightforward magnetooptical devices operating at the same frequency	44
7. Improved microwave phase shifter or polarization rotator	MW, MM ?	Phase shifter, Faraday rotator	d	At frequencies well above the ferromagnetic resonance frequency, the figure of merit for a microwave Faraday rotator is proportional to $\varepsilon' M_z/\varepsilon''$, where M_z is the magnetization component in the direction of propagation and ε' and ε'' are the real and imaginary parts of the dielectric constant at the frequency of operation. Even at[47, 48] microwave frequencies, the dielectric constant in certain lead-based ME's, such as $PbFe_{0.5}Ta_{0.5}O_3$ may be 2 or more orders of magnitude greater than in comparable conventional ferrites. So the lower magnetization in weak ferromagnets may not be too severe a problem. Additional study of these lead-based materials would be a good idea. Known materials require cooling. Also possibly worth consideration are frankly two-phase ferroelectric ferromagnetic materials.[49] No similar advantage is found for isolators	3, 45–49
8. Multiple-state memory element	—	Element for nonbinary memory	a	It has been frequently suggested that the presence of more than two discernible polarization states in ME materials could be used for storing more than one bit of information in a single domain. But the packing density still would not be very high, and the associated electronics rather complicated, and the advantages nebulous at best; so such devices do not seem promising	
9. Electric-field-modulated visible Faraday rotator	V, IR	Amplitude modulator	a	Suppose a polarizer, Faraday rotator, and analyzer are set so that no light is transmitted. An electric field applied to the rotator will cause the light being transmitted to become elliptically polarized leading to a transmitted signal whose strength depends on the applied electric field. This will occur in any material, but probably most strongly in one which is ferroelectric. But there does not seem to be any particular advantage to this over any of the many other methods of accomplishing the same result. If any ME material were known with stripe domains, an electric-field modulated stripe-domain deflector[50] might be an interesting thing to try	

TABLE I—*continued*

Device	Frequency range[a]	Functions	Phenomenon used[b]	Comments	References
10. Magnetically (electrically)-modulated piezoelectricity (piezomagnetism) device	AF, RF	Variable transducer	e	The terms $s'_{ijklm}T_{ij}T_{kl}E_m$ and $s''_{ijklm}T_{ij}T_{kl}H_m$ in the free energy[5] lead to electric and magnetic field modifications, respectively, of the effective elastic moduli, which might be used to vary the frequency of a transducer somewhat. To our knowledge, no experimental work has been done	51
11. Irreversible-light-propagation device	V, IR	Sensor, optical isolator?	c	Under some circumstances, the velocity of light propagation in ME materials is indicated by theory to depend on the direction of transmission (i.e., "forward" and "backward" are not the same). The corresponding effective change in index of refraction is about 10^{-3}. While this is not a large change, observable interference effects should occur when forward and backward beams are combined. Other electro- and magnetooptical effects occurring simultaneously might obscure these effects without proper experimental arrangement, including applied fields. The interference pattern should be very sensitive to applied field changes and thus might serve as a sensor. In very transparent materials, this effect might be used for optical isolation, or (with applied field tuning) beam deflection. It is very unfortunate that no observation of this effect has been made	23, 26, 32
12. Magnetoelectric non-linear-optical device	V, IR, SMM	Frequency doubler, parametric oscillator	e	The non-linear terms responsible for rectification (item 3) could also be used, in conjunction with the usual non-linear susceptibility, to produce second-harmonic generators or optical parametric oscillators which could be controlled by applied fields. For doubling the antiferromagnetic resonance frequency, the magnetoelectric non-linear effects could outweigh the usual ones. One might also hope that precise temperature control of OPO's, now necessitating voluminous and complex ancillary equipment, could be replaced by control with a moderate electric field. Again, the basic effects (that is, properties of the relevant tensors in the visible–IR range) have not been observed, and device design is complicated by the variety of effects all occuring at once	11,42
13. Magnetoferro-electric semiconductor spin-wave generator	—	Generate spin waves	c	Akhiezer[52] suggests that in an antiferromagnetic ferroelectric semiconductor, it should be easier to generate a spin wave than in an ordinary magnetic semiconductor, because of the direct coupling between the electric field and the spins; furthermore spin waves should be generated by a strong enough applied d.c. electric field much in the way ultrasonic waves are amplified in piezoelectric semiconductors. No experiments have been attempted	52
14. Oscillating-electric-field magnetoelectric spin-wave amplifier	—	Amplify spin waves	c	A high-frequency electric field acting on an antiferromagnetic ME material should also lead to spin wave amplification through the effective change in magnetic field discussed in Section 5.6. A strong applied uniform magnetic field and a good-sized sublattice magnetization are also necessary. No experiments reported	31

TABLE I—*continued*

Device	Frequency range[a]	Functions	Phenomenon used[b]	Comments	References
15. Coupled-wave devices	—	Produce new types of hybrid waves	c	In analogy with conversion of spin waves to magnetoelastic waves through magnon–phonon coupling, a number of suggestions have appeared in the Russian literature concerning the production of novel hybrid spin-electromagnetic or spin-acoustic waves through the magnetoelectric effect. The possible uses of these hybrid excitations do not seem to have been considered, though	53

[a] Of the detected electromagnetic field—not of modulating fields or of other types of waves. *Notation:* AF—audio frequency; RF—radio frequency; MW—microwave region; MM—millimeter wave region; SMM—sub-millimeter wave or very-far-infrared region (0.1–1 mm wavelength); FIR—far infrared (20–100 μ); IR—infrared (0.8–20 μ); V—visible.

[b] See Section 3 for explanation of symbols.

So the contribution depends on the quantity $(\eta M)^2$, and may become appreciable for strong sublattice magnetizations, that is, for $\eta M \sim 10$ or more. Also it should be remembered that the magnetoelectric coefficient depends on frequency and may show resonant behavior at antiferromagnetic resonance frequencies.[32] Unfortunately, no experimental investigations of these effects have been attempted.

Next, we turn to ferromagnetic (or more properly ferrimagnetic) resonance in weak ferromagnets and conventional ferrimagnets. For the low-frequency mode for which the net magnetic moment vectors of each magnetic unit cell remain parallel to each other and to the applied field, it is adequate to describe the system as if it were a simple ferromagnet characterized by a magnetization and spin that are just the vector sums of those due to the individual sublattices.[28]

Ferromagnetic resonance has been observed in rather few weak ferromagnets. This is because *good single crystals* are required for its observation. This is in contrast to the situation in conventional ferrimagnets, where FMR is observed, and used, in polycrystals and sintered materials. The reason for this is just that, in a polycrystalline weak ferromagnet, the distribution of magnetization directions from crystallite to crystallite produces a wide range of effective resonance fields, and thus an extremely broad line. In a garnet, say, with its considerably larger total magnetization, the interaction of the magnetizations of adjoining crystallites tends to produce a more uniform magnetization, and thus a much narrower line.[33] To our knowledge, FMR has not been observed in any weak ferromagnets in which linear magnetoelectric coefficients have been measured or in which ferroelectricity has been demonstrated; it has, however, been observed in a number of rather similar materials, and its detection in ME weak ferromagnets presumably awaits only the preparation of single crystals. Since the total magnetization is small, the location of the resonance is determined almost entirely by anisotropy properties, the specimen shape making little difference. The resonance frequency, for not too small applied fields, and for crystals of not too low symmetry, is given approximately by[34]

$$(\omega/\gamma)^2 = (H + C_1 H_a)(H + C_2 H_a),\quad -1 \leqslant C_1, C_2 \leqslant 1 \qquad (4)$$

where the C's are factors depending on the crystal orientation and type of anisotropy. (For a uniaxial crystal magnetized along the axis, $C_1 = C_2 = 1$, for instance.) Thus depending on these factors and the magnetization, the resonance may be found anywhere from very low microwave to near millimeter-wave frequencies in reasonable applied fields.

Among conventional ferrimagnets, the only ones known to display linear ME effects are the gallium ferrites. (They are not ferroelectric.) Some unusual and somewhat controversial electric-field effects on ferromagnetic resonance lines have been observed in these materials.[35, 36] These effects are by no means small, and do not seem to be attributable directly to effects of the magnetoelectric interaction. This is discussed further in Table I. Resonance linewidths are 400 Oe or more.

Improvements in material purity, stoichiometry, and so forth, leading to improved linewidths are important not only for microwave, but also for optical, applications. For high-speed optical modulation, rf losses must be low.[37] The FMR linewidth, proportional to the magnetic damping, provides a good measure of this. Known ME's have a long way to go to catch up with YIG (linewidth as narrow as 0.5 Oe) in magnetic switching speed.

6 CONCLUSIONS AND RECOMMENDATIONS

From the foregoing discussion and from Table I, it appears that there are four applications areas for ME materials which, on paper, are decidedly promising and merit further investigation. These are:

1) Magnetic field switching or modulation of electric polarization, in turn affecting electrooptic properties, avoiding the need for thin plates and high applied electric fields (the magnetic fields required are moderate—a few kOe at most).

2) Exploitation of high dielectric constant in high-frequency, low-loss microwave Faraday rotators.

3) Efficient generation and/or modulation or modification of spin waves or hybrid spin-electromagnetic waves utilizing the magnetoelectric effect.

4) Use of the phenomenon of irreversible light propagation in one or another type of highly sensitive interference sensors.

Some other reasonably good possibilities are given in Table I. There does not seem at present to be any device need which can be filled uniquely or most effectively by devices utilizing the particular properties of known ME's. This may change as systems operating in millimeter-wave and ir regions become more common and as some prototype ME devices are demonstrated. Incidentally, several of the devices described in Table I require good-sized (1–15 kOe) static magnetic fields for their operation. High fields (over small regions) are particularly necessary for millimeter-wave devices. A natural application of rare-earth-cobalt permanent magnets is for the production of such fields.

Each of the suggested uses above has both advantages and disadvantages or impediments. The descriptions above point out the obvious advantages; the drawbacks are summarized very briefly here. For magnetic-field control of electrooptic properties, the chief possible disadvantages are high absorption, nullifying the advantage of thicker samples, and effects of natural and magnetic birefringence. The first difficulty may not be too severe in good crystals in some region in the near ir, and the second can be to some extent designed around.[21] For efficient high-frequency Faraday rotators or phase shifters, the principal difficulties are that there are insufficient experimental data to establish clearly the superiority of any particular magnetoferroelectric (compared to some non-ferroelectric material, such as iron borate,[38] for example) and that there is still some controversy over the properties of the highest-dielectric-constant materials. As for the last two proposed applications, the overwhelming drawback is that the phenomena involved have never been observed experimentally. (To our knowledge, no attempts to observe them have been made.) Until this is done, detailed discussion of applications appears otiose.

If the unique properties of ME materials are to be exploited in devices, the following research should be done, at a minimum:

1) Further experimental work should be done on some of the better-known ME materials, so that reasonably complete and reliable sets of data are available for at least a few materials. Some work towards better crystals should be included, aiming particularly at higher transparency, lower conductivity, narrower resonance lines.

2) Further studies should be conducted of antiferromagnetic ferroelectrics (such as $BiFeO_3$ and hexagonal rare-earth manganites), which are probably just as useful as weakly ferromagnetic ferroelectrics for many millimeter-wave and far-ir applications. Work on crystal growth and electrical properties is of primary importance.

3) Lingering questions concerning properties of lead-based materials such as $PbFe_{2/3}W_{1/3}O_3$ and $PbFe_{0.5}Ta_{0.5}O_3$ should be answered by applying modern materials characterization techniques to samples of these materials.

4) Attempts should be made to demonstrate magnetoelectric coupling of spin waves to external electric fields, to demonstrate irreversible light propagation, and to measure the components of the magnetoelectric tensor at microwave and higher frequencies. Further theoretical work along these lines is also desirable.

5) The search for improved materials should continue, although it seems to us that dramatic improvements should not be counted on, and expectable improvements within classes of materials already studied are more likely to lead to early application.

These recommendations are not to be considered as fixed; they are intended in part to stimulate discussion. Those involved in work on magnetoelectric effects ought also to maintain an interest in all aspects of the development of the as yet little-utilized regions of the electromagnetic spectrum.

REFERENCES

1. V. E. Wood *et al.*, Annual Technical Report to ARPA on Contract DAAH01-70-C-1076, June 30, 1971 (AD 726201).
2. R. R. Birss, *Symmetry and Magnetism* (North-Holland, Amsterdam, 1966).
3. B. Lax, *Proc. IRE* **44**, 1368 (1956).
4. R. Cohen and R. S. Mezrich, *RCA Review* **33**, 55 (1972).
5. H. Schmid, *Int'l. J. Magnetism*, accompanying paper (1973); see also G. T. Rado and V. J. Folen, *J. Appl. Phys.* **33**, 1126 (1962).
6. J. Kobayashi *et al.*, *J. Phys. Soc. Japan* **28**, Suppl., 67 (1970).
7. C. Haas, *IEEE Trans. Mag.* **5**, 487 (1969).
8. E. V. Chenskii, *Sov. Phys.-Solid State* **11**, 534 (1969); M. S. Shur, *Bull. Acad. Sci. USSR*, Phys. Ser. **33**, 187 (1969).
9. P. J. Freud, *Phys. Rev. Letters* **29**, 1156 (1972); I. G. Austin and R. Gamble, *Eilat Conference on Conduction in Low-Mobility Materials* (Taylor and Francis, London, 1971), p. 1.
10. R. M. Hornreich, *Solid State Communs.* **7**, 1081 (1969).
11. E. F. Bertaut and M. Mercier, *Mat. Res. Bull.* **6**, 907 (1971).
12. A. H. Eschenfelder, *J. Appl. Phys.* **41**, 1372 (1970); E. W. Pugh, *Proc. 17th Annual Conference on Magnetism and Magnetic Materials* (AIP Conf. Proc. No. 5, 1972).
13. L. K. Anderson, *Ferroelectrics* **3**, 69 (1972).
14. W. Strauss, *J. Appl. Phys.* **42**, 1251 (1971).
15. G. Quézel and H. Schmid, *Solid State Communs.* **6**, 447 (1968).
16. M. H. Francombe, *Ferroelectrics* **3**, 199 (1972).
17. R. R. Mehta, *J. Appl. Phys.* **42**, 1842 (1971).
18. J. R. Teague *et al.*, *Solid State Communs.* **8**, 1073 (1970).
19. R. V. Pisarev *et al.*, *Soviet Phys.-Solid State* **11**, 766 (1969); *Phys. Stat. Sol.* **35**, 145 (1969); **40**, 503 (1970).
20. T. Miyashita and T. Murakami, *J. Phys. Soc. Japan* **29**, 1092 (1970).
21. R. Wolfe, A. J. Kurtzig and R. C. LeCraw, *J. Appl. Phys.* **41**, 1218 (1970).
22. B. D. H. Tellegen, *Philips Research Reports* **3**, 81 (1948).
23. R. R. Birss and R. G. Shrubsall, *Phil. Mag.* **15**, 687 (1967).
24. V. M. Lyubimov, *Sov. Phys.-Doklady* **13**, 739 (1969).
25. V. M. Lyubimov, *Sov. Phys.-Crystallog.* **13**, 877 (1969).
26. T. H. O'Dell, *The Electrodynamics of Magneto-Electric Media* (North-Holland, Amsterdam, 1970), and references therein.
27. E. Ascher, *Phil. Mag.* **17**, 149 (1968).
28. S. Foner, in *Magnetism* (edited by G. T. Rado and H. Suhl) (Academic Press, New York, 1963). Vol. I, Chap 9.
29. G. S. Heller, J. J. Stickler and J. B. Thaxter, *J. Appl. Phys.* **32**, 3075 (1961).
30. V. G. Shavrov, *Sov. Phys.-Solid State* **7**, 265 (1965).
31. I. A. Akhiezer and L. N. Davydov, *Sov. Phys.-Solid State* **13**, 1499 (1971).
32. R. A. Fuchs, *Phil. Mag.* **11**, 647 (1965).
33. A. M. Clogston, *J. Appl. Phys.* **29**, 334 (1958).
34, C. Kittel, *Phys. Rev.* **73**, 155 (1948).
35. M. P. Petrov, S. A. Kizhaev and G. A. Smolenskii, *Phys. Stat. Sol.* **30**, 871 (1968); *JETP Letters* **6**, 306 (1967).
36. V. J. Folen and G. Rado, *Solid State Communs.* **7**, 433 (1969); J. Dweck, *Phys. Rev.* **168**, 602 (1968).
37. R. C. LeCraw, *Intermag. Conf. Presentation, Stuttgart, 1966* (unpublished).
38. R. C. LeCraw, R. Wolfe and J. W. Nielsen, *Appl. Phys. Letters* **14**, 352 (1969).
39. J. W. Miles, *J. Acoust. Soc. Am.* **19**, 910 (1947).
40. T. J. Martin, *Phys. Letters* **17**, 83 (1965).
41. T. H. O'Dell (private communication).
42. E. Ascher, *Helv. Phys. Acta* **39**, 446 (1966); L. N. Bulaevskii and V. M. Fain, *JETP Letters* **8**, 165 (1969).
43. T. Penney, P. Berger and K. Kritayakirana, *J. Appl. Phys.* **40**, 1234 (1969); K. Kritayakirana, P. Berger and R. V. Jones, *Optics Communs.* **1**, 95 (1969).
44. E. Ascher, H. Rieder, H. Schmid and H. Stossel, *J. Appl. Phys.* **37**, 1404 (1966); H. Schmid, *Rost Kristallov* **7**, 32 (1967) (translation: *Growth of Crystals* **7**, 25 (1969), Consultant's Bureau, New York).
45. C. L. Hogan, *Bell Syst. Tech. J.* **31**, 1 (1952).
46. G. A. Smolenskii *et al.*, *Ferroelectrics and Antiferroelectrics* (Izdatelstvo "Nauk", Leningrad, 1971) (translation: U.S. Army Foreign Science and Technology Center, 1972, AD 741037); Chap. 18.
47. S. Nomura, H. Takabayashi and T. Nakagawa, *Japan J. Appl. Phys.* **7**, 600 (1968).
48. D. N. Astrov *et al.*, *Sov. Phys.-JETP* **28**, 1123 (1969).
49. K. Leibler, V. A. Isupov and H. Bielska-Landowska, *Acta Phys. Polon.* **A40**, 815 (1971).
50. D. S. Lo, D. I. Norman and E. J. Torok, *J Appl. Phys.* **41**, 1342 (1970); T. R. Johansen, D. I. Norman and E. J. Torok, *J Appl. Phys.* **42**, 1715 (1971).
51. E. Ascher and H. Schmid (private communication).
52. A. I. Akhiezer and I. A. Akhiezer, *Sov. Phys.-JETP* **32**, 549 (1971).
53. A. I. Akhiezer and L. N. Davydov, *Soviet Phys.-Solid State* **12**, 2563 (1971); V. G. Baryakhtar and I. E. Chupis, *Sov. Phys.-Solid State* **10**, 2818 (1969); **11**, 2628 (1970); I. F. Ioffe and A. L. Kazakov, *Sov. Phys.-Solid State* **13**, 1806 (1972).

Discussion

G. T. RADO *Comment*: In unpublished work, which I shall reproduce on the blackboard, I have shown that the usual reciprocity theorem of electromagnetic theory [see, for example, L. D. Landau and E. M. Lifshitz, *Electrodynamics of Continuous Media* (Addison-Wesley, Reading, Mass., 1960, p. 288)] is invalid in a magnetoelectric medium. Note inserted in proof: In a recent paper [G. T. Rado, *Phys. Rev.* **B8**, 5239 (1973)], I proposed and derived a modified reciprocity theorem which is valid under nonequilibrium as well as equilibrium conditions and which states, in effect, that the usual reciprocity theorem does remain valid in a magnetoelectric antiferromagnetic medium provided the medium is time-reversed whenever an electromagnetic source and a detection apparatus are interchanged. This shows that the basic reason for the nonvalidity of the usual reciprocity theorem in a magnetoelectric medium is the noninvariance of the magnetoelectric susceptibility under time reversal.

S. SHTRIKMAN In your talk you seem to stress nonreciprocity as of central importance in the applications discussed. I hate to be the devil's advocate but the Faraday effect is also nonreciprocal and a much larger effect. So what would be the advantage of using magnetoelectrics?

VAN E. WOOD Two possible advantages are:

1) Greater sensitivity of destruction of interference nodes to changes in applied fields through interferences between waves of slightly different wave length.

2) Sensitivity to applied electric as well as magnetic fields.

E. J. POST

i) Is there a rotation of the plane of polarization in the situation you depicted?

ii) Remark: the discussion of these matters (nonreciprocal birefringence) becomes much less involved if one uses the complete constitutive tensor.

VAN E. WOOD

i) No, not in this case.

ii) This is indeed frequently advantageous.

SYMMETRY CHANGES AT PHASE TRANSITIONS ACCORDING TO THE LANDAU THEORY†

S. GOSHEN

Department of Physics, Nuclear Research Centre-Negev, P.O.B. 9001 Beer Sheva, Israel

D. MUKAMEL

Department of Electronics, Weizmann Institute of Science, Rehovot, Israel and Department of Physics, Nuclear Research Centre-Negev, P.O.B. 9001 Beer Sheva, Israel

and

S. SHTRIKMAN‡

Department of Electronics, Weizmann Institute of Science, Rehovot, Israel

We derive the second condition of the Landau theory of second-order phase transitions, whereby the antisymmetric part of T^2 must have no representation in common with the vector representation, following a method proposed by Dzyaloshinskii. The physical interpretation of this condition is given for a simple example of a ferromagnetic transition in a non-centrosymmetric tetragonal crystal.

1 INTRODUCTION

Landau's theory[1] of phase transitions imposes restrictions on the possible symmetry G that can appear after a second-order phase transition in a crystal with initial symmetry G_0. These restrictions are:

1) The group G is such that the charge or spin density which is associated with the transition belongs to only one irreducible representation of G_0 below the transition.

2) T cannot be any representation:[2] the allowed T representations are only those for which the antisymmetric part of T^2 (denoted $\{T^2\}$) does not have any representation in common with the vector representation. This condition insures that the crystal remains homogeneous after the transition.

3) The symmetric part of T^3 (denoted $[T^3]$) must not contain the unit representation. Otherwise the transition is first order.

While the first and third conditions of this theory are generally accepted, the validity of the second condition was extensively argued in the literature from theoretical and experimental points of view. Using the second condition, Lifshitz[3] has shown that a crystal may, after transition, increase its period by an integer factor which is four, at most. This means that representations with k whose components[4] are different from $0, \frac{1}{2}, \frac{1}{3}, \frac{1}{4}$ do not satisfy the second condition, and therefore cannot appear below the transition. This result is in contradiction with both the experimentally observed helicoidal and related structures in many materials[5] and with model calculations not based on the Landau theory which predict the appearance of such structures[6–8] after second-order transitions. Furthermore, Dimmock[9] and later on Kaplan[10] pointed out a deficiency in the derivation of the second condition and therefore claimed that this condition must be omitted from the theory.

The contradiction with model calculations and with the experimental results mentioned above suggest that condition 2 with Lifshitz deduction is incorrect as it stands, in agreement with Dimmock and Kaplan.

Dzyaloshinskii[11] tried to overcome these difficulties by developing a theory based on that of

† This paper was presented at the Symposium on Magneto-Electric Interaction Phenomena in Crystals, May 21, 1973, at Battelle Seattle Research Center, Seattle, Washington".

‡ Supported in part by the Air Force Materials Laboratory (AFSC), United States Air Force under Grant AFOSR 72-2327 through the European Office of Aerospace Research.

Landau. In view of his work one may replace the second condition by the following:

2′) The representation T may be either due to symmetry or not ("due to symmetry" will be defined below). In the former case the only allowed T representations are those given by Landau's condition 2. In the latter case this condition gives no restrictions on the allowed representations and any representation may appear after the transition.

The classification of T as being either due or not due to symmetry is convenient because it shows under what assumptions the original 2 is valid. As pointed out by Dzyaloshinskii the magnetic transitions which lead to helicoidal structures are not due to symmetry and, therefore, are not forbidden by this condition. Dzyaloshinskii demonstrated a method[12] different from that of Landau to derive the second condition in its corrected form. However, his derivation leads to Lifshitz's result for representations T which are due to symmetry, without explicitly obtaining Landau's second condition; namely, that the antisymmetric part of T^2 must have no representation in common with the vector representation.

In this work we use the method proposed by Dzyaloshinskii to show that Landau's second condition holds when T is due to symmetry. The physical interpretation of this condition is illustrated on a simple model of ferromagnetic transitions in a non-centrosymmetric tetragonal crystal.

2 THE DERIVATION OF THE SECOND CONDITION

Let $\rho(x, y, z)$ be the "density" function of the crystal which changes its symmetry at the transition point. It may be a real atomic density in a structural transition, a magnetic moment density (a pseudo-vector density function) in magnetic transitions, etc. The density $\rho(x, y, z)$ can be decomposed near the transition point into two parts:

$$\rho(x, y, z) = \rho^0(x, y, z) + \delta\rho(x, y, z)$$

where ρ^0 is that part of ρ which has the symmetry of G_0 of the crystal above the transition, while $\delta\rho$ is the part with lower symmetry which determines the symmetry of the crystal below the transition. It is clear that $\delta\rho$ is zero above the transition and non-zero below. The density $\delta\rho$ can in turn be decomposed into a linear combination of functions which transform according to the irreducible representations of G_0:[14]

$$\delta\rho = \sum_l{}' \sum_i C_{li}\phi_{li}(r)$$

where l enumerates the irreducible representations of G_0, i is an index which specifies the functions within any representation, C_{li} are numerical coefficients and Σ' means that the sum does not include the unit representation (the unit representation which is excluded in $\delta\rho$ appears in ρ_0). Since $\delta\rho$ is a real quantity, any non-real function ϕ_{li} must appear in the expansion together with its complex conjugate ϕ_{li}^*. We therefore combine the functions ϕ_{li} of the representation T_l and the functions ϕ_{li}^* of the representation T_l^* to form a basis of a single real representation $T_l + T_l^*$, which is called a physically irreducible representation.[15] The representations l which appear in the expansion are therefore either real irreducible representations or the physically irreducible representations.

The thermodynamic potential Φ of a crystal with density function ρ is a function of the temperature T, the pressure P and $\delta\rho$ (i.e. the coefficients C_{li}). We expand Φ in a power series with respect to C_{li} near the transition point T_c. The expansion is meaningful whenever $\delta\rho$ is small in the vicinity of the transition point. This is the case for second-order phase transitions, where $\delta\rho \to 0$ as $T \to T_c$, or even for some first-order transitions, if the discontinuity of $\delta\rho$ at $T = T_c$ is sufficiently small.[16] The thermodynamic potential must be invariant under G_0. Hence, the expansion up to second order in C_{li} takes the form:

$$\Phi(T, P, C_{li}) = \Phi_0(T, P) + \sum_l{}' A_l(T, P) \sum_i C_{li}^2$$

A term linear in C_{li} cannot be present since the unit representation of G_0 is not included in the expansion of $\delta\rho$. Only one second-order invariant exists for each representation l, and this is $\Sigma_i C_{li}^2$. For $T > T_c$, $\delta\rho = 0$ or $C_{li} = 0$ for every l and i. This structure minimizes the thermodynamic potential only if $A_l(T, P) > 0$ for every l. The occurrence of a transition is conditioned by a change of sign of at least one of the A_l's. Landau assumed that only one of the A_l's, hereafter denoted by A_{l_0}, changes its sign at $T = T_c$. In this case the density $\delta\rho$ belongs to the representation l_0 below the transition, which means that it can be written

$$\delta\rho = \sum_i C_{l_0 i}\phi_{l_0 i}$$

where the sum includes functions which belong only to l_0.[19]

In order to determine the restriction on the possible representations which are associated with the transition, let us consider the A_l's. All A_l's are positive above the transition. One of them, A_{l_0}, changes its sign at the transition point, while all the others remain positive. It is obvious therefore that the l_0 which appears during transition must be the one for which $A_l(T, P)$ has a minimum as a function of l at the transition point. This may happen in two ways:

1) $A_l(T_c, P_c)$(as a function of l) has a minimum at $l = l_0$ not associated with symmetry. Such a minimum may occur for any l, and it bears no restriction on the possible representations which may appear below the transition. Furthermore, the l_0 at which $A_l(T_c, P_c)$ has a minimum may vary on the transition curve $T_c(P_c)$(since it is not imposed by symmetry).[20] As pointed out by Dzyaloshinskii the magnetic transitions which lead to helicoidal structures are of this type.

2) A_l has an extremum at $l = l_0$ which is associated with symmetry.

Let us consider the second possibility and determine for which value l_0 can the quantity A_l have an extremum which is associated with symmetry. To do this we make use of the irreducible representations of the space groups.[4] They are classified according to two parameters: one of them, $\bar{f}$, is continuous in the first Brillouin zone of the crystal, while the other, n, is discrete, and enumerates the small representations of the vector $\bar{f}$. The functions which form a basis of the representation $l \equiv (\bar{f}, n)$ have the form:

$$\phi^j_{(\bar{f}',n)} = u^j_{\bar{f}'n}(r)\, e^{i\bar{f}',\bar{r}}$$

where $u^{j'}_{\bar{f}n}(r)$ are periodic functions with the period of the crystal, $\bar{f}'$ can be any one of the vectors of the star of $\bar{f}$, and j enumerates the functions which belong to any one of these vectors.

We assume that the energy bands $A_{\bar{f},n}(T, P)$ are continuous functions of the continuous parameter $\bar{f}$ (a schematic form of $A_{\bar{f},n}$ is shown in Figure 1). The condition of an extremum of $A_{\bar{f},n}$ at $\bar{f} = \bar{f}_0$, is the absence of the linear term in $\bar{k}$ in the expansion of $A_{\bar{f}_0+\bar{k},n}$ near $\bar{k} = 0$.

Let us check first the possibility of the appearance of the linear term for $f_0 = 0$, using symmetry considerations. The basis functions of the representation $(0, n)$ are:

$$\phi_1 \equiv u^1_{0n}, \ldots, \phi_N \equiv u^N_{0n}$$

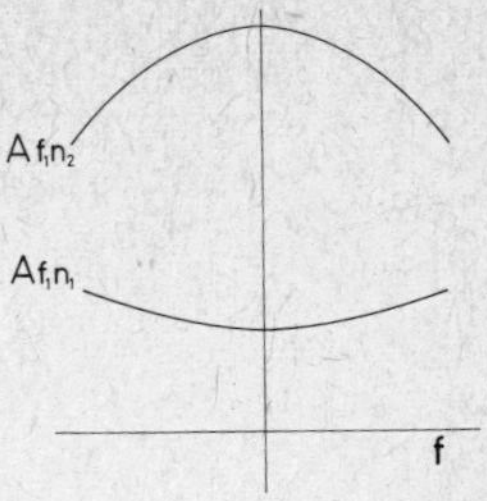

FIGURE 1 A schematic band structure.

where N is the order of the representation. We can choose the functions ϕ_j to be real, since the representations we are dealing with are real. Now let us consider the functions which belong to the vector $\bar{k}$:

$$\phi_1\, e^{i\bar{k}\cdot\bar{r}}, \ldots, \phi_N\, e^{i\bar{k}\cdot\bar{r}} \tag{1}$$

and check whether the degeneracy between them and the functions $\phi_1, \ldots, \phi_n$ is removed in the first order in $\bar{k}$. Since the density $\delta\rho$ is real, it follows that $\delta\rho$ which belong to the representation $(\bar{k}, n)$[21] cannot be expressed as a linear combination of the functions (1) alone, but rather by the functions (1) together with their complex conjugates. Thus we have to look at the set of $2N$ functions:

$$\phi_1\, e^{i\bar{k}\cdot\bar{r}}, \ldots, \phi_N\, e^{i\bar{k}\cdot\bar{r}}; \qquad \phi_1\, e^{-i\bar{k}\cdot\bar{r}}, \ldots, \phi_N\, e^{-i\bar{k}\cdot\bar{r}}$$

or at an equivalent set of real functions:

$$\phi_1 \cos\bar{k}\cdot\bar{r}, \ldots, \phi_N \cos\bar{k}\cdot\bar{r};$$
$$\phi_1 \sin\bar{k}\cdot\bar{r}, \ldots, \phi_N \sin\bar{k}\cdot\bar{r}$$

The density $\delta\rho$ associated with the vector $\bar{k}$ can be written as:

$$\delta\rho = \sum_{j=1}^{N} (a_j\phi_j \cos\bar{k}\cdot\bar{r} + b_j\phi_j \sin\bar{k}\cdot\bar{r}) \tag{2}$$

where a_j and b_j are real numbers. Applying the symmetry operations of G_0 on $\delta\rho$, and regarding the transformations as if acting on the numerical coefficients (a_j, b_j) rather than on the functions ϕ_j themselves, one concludes that the sets $\{a_j\}$, $\{b_j\}$ transform as two bases of the representation $(0, n)$. The vector $\bar{k}$ transforms as the vector representation of G_0.

The thermodynamic potential of a crystal for which $\delta\rho$ is given by Eq. (2) is a function of T, P, a_j, b_j and $\bar{k}$. The expansion of this potential to the second order in a_j, b_j and first order in $\bar{k}$ takes the form:

$$\Phi = A \sum_1^N a_j^2 + B \sum_1^N b_j^2 + C \sum_1^N a_j b_j + \sum_{l=1}^3 D_l k_l$$
$$+ \sum_{il} E_{il} a_i k_l + \sum_{il} F_{il} b_i k_l + \sum_{ijl} \alpha_{ijl} a_i a_l k_l$$
$$+ \sum_{ijl} \beta_{ijl} b_i b_j k_l + \sum_{ijl} \gamma_{ijl} a_i b_j k_l$$

where k_l ($l = 1, 2, 3$), are the components of the vector k. The coefficients $A, B, C, D_l, E_{il}, F_{il}, \alpha_{ijl}$, β_{ijl} and γ_{ijl} depend on temperature and pressure. The thermodynamic potential must be invariant under G_0, and this imposes restrictions upon the coefficients $E_{il}, F_{il}, D_l, \alpha_{ijl}, \beta_{ijl}$ and γ_{ijl}. The following two restrictions can be taken into account explicitly, as they hold for any group G_0:

1) The structure described by the set of numbers $(a_j, b_j, \bar{k})$ is identical to that described by $(a_j, -b_j, -\bar{k})$. Taking Φ to be invariant under this transformation, one obtains:

$$C = D_l = \alpha_{ijl} = \beta_{ijl} = E_{il} = 0$$

2) The thermodynamic potential is invariant under a change of phase of the modulation $e^{i\bar{k}\cdot\bar{r}}$ imposed on the functions ϕ_j.[22] This means that the structure

$$\delta\rho = \sum_j (a_j \phi_j \cos \bar{k}\cdot\bar{r} + b_j \phi_j \sin \bar{k}\cdot\bar{r})$$

is degenerate with the structure

$$\delta\rho' = \sum_j [a_j \phi_j \cos(\bar{k}\cdot\bar{r} + \eta) + b_j \phi_j \sin(\bar{k}\cdot\bar{r} + \eta)]$$

for any phase η. Taking η to be $\pi/2$ it follows that the structure $(a_j, b_j, \bar{k})$ is degenerate with the structure $(b_j, -a_j, \bar{k})$. This fact imposes the following restrictions on Φ:

$$F_{il} = 0$$
$$A = B$$
$$\gamma_{ijl} = -\gamma_{jil}$$

The general form of Φ is therefore:

$$\Phi = A \sum_{j=1}^N (a_j^2 + b_j^2) + \sum_{ijl} \gamma_{ijl} a_i b_j k_l$$

where

$$\gamma_{ijl} = -\gamma_{jil}$$

The linear term in $\bar{k}$ does appear in the expansion if there exists an invariant under G_0 of the form $\Sigma \gamma_{ijl} a_i b_j k_l$ with $\gamma_{ijl} = -\gamma_{jil}$, or alternatively, if the antisymmetric part of $T^2_{(0,n)}$ has any representation in common with the vector representation. This is exactly Landau's condition for the case $\bar{f} = 0$.

The same result is obtained for $\bar{f} \neq 0$. The representation $(\bar{f}, n)$ contains functions which belong not only to $\bar{f}$ itself, but also to any vector which belongs to the star of $\bar{f}$. It is sufficient to expand the thermodynamic potential in terms of functions belonging only to the vector $\bar{f}$, with the restriction that this expansion be invariant under the small group of $\bar{f}$. Let $\psi_1 \ldots \psi_L$ be the basis functions of $(\bar{f}, n)$ which belong to the vector $\bar{f}$. Since we are considering real representations, the functions $\psi_1^* \ldots \psi_L^*$ which belong to the vector $-\bar{f}$ are also contained in this representation. From this set of $2L$ function one can construct another set of $2L$ real functions $\phi_1 \ldots \phi_{2L}$. From this point we can proceed as in the case where $\bar{f} = 0$. The result is that there exists a linear term in $\bar{k}$ in the expansion only if the antisymmetric part of $T^2_{\bar{f},n}$ has any representation in common with the vector representation of the small group of $\bar{f}$. This happens if and only if $\{T^2_{\bar{f}n}\}$ has a representation in common with the vector representation of the group G_0 itself.[15]

For a general point $\bar{f}$ of the Brillouin zone the small group if $\bar{f}$ is the trivial group and therefore $\{T^2_{\bar{f},n}\}$ and the vector representation of the small group of $\bar{f}$ are the same representation (they are both equal to the unit representation). In this case, there is a linear term in $\bar{k}$ in the expansion of the free energy and the representation $(\bar{f}, n)$ is not allowed by the second condition.

3 EXAMPLE: FERROMAGNETISM IN A NON-CENTROSYMMETRIC TETRAGONAL CRYSTAL

Let us apply the second condition of Landau's theory to investigate whether ferromagnetic transitions should be expected in a non-centrosymmetric tetragonal crystal. As an example let us consider a crystal whose symmetry group G_0 is P4mm. There are two ferromagnetic representations[23] of this group and they obviously belong to the vector $\bar{f} = 0$: The representation A_2 [or $(0, n = A_2)$ in our notation] which is one-dimensional, and describes ferromagnetic structure along the C_4 (=z) axis, and the representation E, which is two-dimensional and describes a ferromagnetic structure in the xy plane. The basis vector of A_2 is $\phi = S_z$ (a z-component of a magnetic moment), while those of E are $\phi_1 = S_x$, $\phi_2 = S_y$ (x- and y-components of a magnetic moment). In the case of A_2 the linear term $\Sigma_l \gamma_{11l} a_1 b_1 k_l$ vanishes since $\gamma_{11l} = -\gamma_{11l} = 0$. The band is therefore parabolic near $k = 0$. In the case of E there exists a linear term, $(a_1 b_2 - a_2 b_1) k_z$, which is invariant under P4mm. The representation $(k_z, n = E)$ is reducible and it splits into two irreducible representations whose bases are:

I. $S_x \cos kz + S_y \sin kz$
$S_x \sin kz - S_y \cos kz$

II. $S_x \cos kz - S_y \sin kz$
$-S_x \sin kz - S_y \cos kz$

They describe a right-handed spiral (set I) and a left-handed spiral (set II). The energy near $k = 0$ splits therefore into two bands as shown in Figure 2.

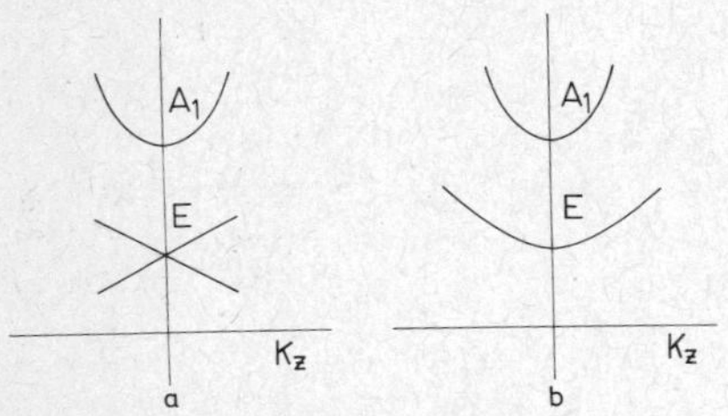

FIGURE 2 The form of the bands near the ferromagnetic representations ($k = 0$). (a) for a non-centrosymmetric tetragonal crystal, (b) for a centrosymmetric tetragonal crystal.

We conclude that ferromagnetism is allowed in non-centrosymmetric tetragonal crystals in the z-direction, but not in the xy plane (except accidentally).[24] For a centrosymmetric crystal the linear term of the expansion in $\bar{k}$ vanishes in both representations; thus, ferromagnetism can be expected in the xy plane as well as in the z direction.

ACKNOWLEDGEMENT

We are grateful to Profs. H. B. Callen, W. F. Brown, Jr., H. Thomas, S. Alexander and D. Treves for helpful discussions. Many thanks are due to Prof. M. Luban for his careful and critical reading of the manuscript. We wish to thank the referee for his comments.

REFERENCES

1. L. D. Landau and E. M. Lifshitz, *Statistical Physics* (Pergamon Press, London, 1968), 2nd edition.
2. This condition is due to Lifshitz (Ref. 3).
3. E. M. Lifshitz, *J. Phys.* **6**, 61 (1942).
4. For a review of representations of space groups see, for example, G. F. Koster, *Solid State Phys.* **5**, 173 (1967).
5. J. M. Hastings and L. M. Corliss, *J. Phys. Chem. Solids* **29**, 9 (1968); R. Plumier, *J. Phys.* **27**, 213 (1966); O. P. Aleshko-Ozhevskii, R. A. Sizov, I. I. Yamzin and V. A. Lubimtsev, *Sov. Phys. JETP* **28**, 425 (1969); D. E. Cox, B. C. Frazer and R. Newnham, *J. Appl. Phys.* **40**, 1124 (1969); H. Watanable, N. Kazama, Y. Yamaguchi and M. Ohashi, *J. Appl. Phys.* **40**, 1128 (1969); H. Umebayashi, G. Shirane, B. C. Frazer and W. B. Daniels, *Phys. Rev.* **165**, 688 (1969); G. Shirane and S. J. Pickart, *J. Appl. Phys.* **37**, 1032 (1966); V. A. Sizov and I. I. Yamzin, *Sov. Phys. JETP* **26**, 736 (1968).
6. J. Villain, *J. Phys. Chem. Solids* **11**, 303 (1959).
7. T. A. Kaplan, *Phys. Rev.* **116**, 888 (1959); D. H. Lyons and T. A. Kaplan, *Phys. Rev.* **120**, 1580 (1960).
8. A. Yoshimori, *J. Phys. Soc. Jap.* **14**, 807 (1959).
9. J. O. Dimmock, *Phys. Rev.* **130**, 1337 (1963).
10. T. A. Kaplan, *Bull. Acad. Sci. USSR* **28**, 328 (1964).
11. I. E. Dzyaloshinskii, *Sov. Phys. JETP* **19**, 960 (1964).
12. This derivation employs techniques which were used in calculations of electronic bands.[13]
13. G. Dresselhaus, *Phys. Rev.* **100**, 580 (1955).
14. The choice of the functions ϕ_{li} is not unique. The functions within any representations may be replaced by linear combinations of themselves. An example of a procedure for finding a set $\{\phi_{li}\}$ is the following: Apply all symmetry operators of G_0 on $\delta\rho$. A set of functions $\{\delta\rho_n\}$ which transform into one another under G_0 is obtained. This set forms a basis of representation of G_0 which, in general, is reducible. Decomposing this into irreducible representations a set $\{\phi_{li}\}$ is obtained.
15. G. Ya. Lyubarskii, *The Application of Group Theory in Physics* (Pergamon Press, London, 1960).
16. The validity of Landau's condition depends on the possibility of expanding Φ in terms of C_{li} near $T = T_c$. Landau's condition therefore will be valid not only for second-order transitions but also for first-order transitions of this type. It is argued[17,18] that transition from liquids to liquid crystals are of this type.
17. G. W. Gray, *Molecular Structure and Properties of Liquid Crystals* (Academic Press, New York, 1962).
18. S. Goshen, D. Mukamel and S. Shtrikman, *Solid State Comm.* **9**, 649 (1971).
19. Expanding ϕ in orders higher than 2 in C_{li}, invariants of the form $T_{l_0}^2 \cdot T_{l_1}$ have to be taken into account.[10] Here l_1 may be any representation for which such an invariant exists. In this case the density $\delta\rho$ is described by a linear combination of functions belonging to l_0 and l_1 as well. It can be shown that in the vicinity of the transition point C_{l_1} is proportional to $C_{l_0}^2$, thus, to first order in C_{l_0} the representations do not mix.
20. C. Haas, *Phys. Rev.* **140**, A863 (1965).
21. By the representation $(\bar{k}, n)$ we mean the representation whose basis functions are the functions (1) plus the functions obtained by applying the symmetry operations of G_0 on them. This representation may, in general, be reducible.
22. This may be proved in the following way: The function $\phi_j e^{i\bar{k}\cdot\bar{r}}$ is degenerate with any function obtained by applying any symmetry element g of G_0. Let g be a translation by a lattice vector $\bar{a}$, then $g(\phi_j e^{i\bar{k}\cdot\bar{r}}) = \phi_j e^{i(\bar{k}\cdot\bar{r}+\bar{k}\cdot\bar{a})}$. Since $\bar{k}\cdot\bar{a} \ll 1$, one can verify by applying

the translation g, n times that the energy is invariant under a change of phase of $n \cdot \bar{k} \cdot \bar{a}$ for any integer n. This means the free energy is invariant under an arbitrary change of phase.

23. By "ferromagnetic representation" we mean one which is contained in the axial vector representation.
24. The bands which are associated with ferromagnetism in the xy plane will have a minimum at some $k \neq 0$, thus leading to a helicoidal magnetic structure. Physically this structure is due to the Dzyaloshinskii–Moriya interaction which tends to split neighbouring magnetic moments and which exists in non-centrosymmetric tetragonal crystals. Helicoidal structures in non-centrosymmetric cubic crystals due to the Dzyaloshinskii–Moriya interaction were discussed by J. Villain.[25]
25. J. Villain, Thesis, CNRS document No. 1233 (1967, unpublished).

Discussion

D. LITVIN The central relation in your derivation of the Lifshitz condition is the expansion of the thermodynamic potential. As this expansion is invariant under G_0, I do not see how a term

$$\sum_{l=1}^{3} D_l k_l$$

containing the components of a single vector **k**, and not a combination of the vectors of the star of **k** can appear in this expansion.

S. SHTRIKMAN As the derivation shows D_l is found to vanish.

R. ENGLMAN In magneto-electric theory, one normally starts with a spin-Hamiltonian, rather than the thermodynamic potentials of Landau [depending on the order parameter (S)]. The conditions for the form of the spin-Hamiltonian are an important matter (they are occasionally incorrectly stated) and should constitute a basis for the conditions of the Landau theory.

S. SHTRIKMAN Certainly. But in practice getting the free energy from the Hamiltonian is a major undertaking, as you very well know.

MAGNETOELECTRIC STUDIES OF MAGNETIC TRANSITIONS IN ANTIFERROMAGNETIC CRYSTALS†

L. M. HOLMES‡

Bell Laboratories, Murray Hill, New Jersey 07974

Recent experimental studies of anomalies in the magnetoelectric (ME) effect associated with the magnetic ordering transition and with spin-flop and metamagnetic behavior in antiferromagnets are reviewed. The materials discussed include Cr_2O_3, $DyPO_4$, $TbPO_4$, $DyAlO_3$, $TbAlO_3$, and $MnGeO_3$. Some new results are presented on $MnGeO_3$. The status of present theoretical understanding of the ME behavior is evaluated critically. Promising new directions for research are pointed out and the potentialities of the technique are assessed.

INTRODUCTION

Since the time of the discovery[1-3] of the magnetoelectric (ME) effect in antiferromagnetic Cr_2O_3, it has been recognized that ME measurements can give information about magnetic space groups.[4] In antiferromagnets, ferrimagnets, and more complicated systems, this information is difficult to obtain by other than neutron-diffraction studies, and sometimes the neutron experiments may be difficult to do[5] (e.g., the sample may absorb neutrons rather than diffracting them). This paper will be concerned with using the ME effect to study magnetic behavior, but we will describe experiments which attempt to go beyond the determination of structures to give detailed information about magnetic phase transitions. These experiments have been more or less exploratory in nature, but they show the ME effect to be a potentially powerful tool for these purposes.

Not all antiferromagnetic crystals exhibit the ME effect. The occurrence of magnetoelectricity is limited by symmetry requirements which have been described elsewhere.[4,6,7] For our purposes, a material will be said to be "magnetoelectric" if the thermodynamic potential[8] contains a term of the form

$$\Phi_{me} = -\vec{E}\vec{\vec{\alpha}}\vec{H},$$

where $\vec{E}$ and $\vec{H}$ are the electric and magnetic fields, respectively, and $\vec{\vec{\alpha}}$ is the ME tensor. In such a material, an applied electric field gives rise to an ME component of magnetization[9] given by

$$4\pi\vec{M} = \vec{E}\vec{\vec{\alpha}}$$

while an applied magnetic field gives rise to an ME component of electric polarization[9] given by

$$4\pi\vec{P} = \vec{\vec{\alpha}}\vec{H}.$$

An understanding of the tensor nature of $\vec{\vec{\alpha}}$ is crucial to an understanding of the ME effect. The ME tensor may contain up to nine independent components, but the number of nonzero components and the form of $\vec{\vec{\alpha}}$ are greatly restricted by symmetry in many cases of interest. We refer the reader elsewhere for a discussion of these matters.[6] For the rest of this introduction, we shall consider the simplest possible situation in which $\vec{H}$, $\vec{E}$, $\vec{P}$, and $\vec{M}$ are all parallel, so we may deal with a scalar ME "susceptibility" α. Typical values of α, which is dimensionless in Gaussian units, are $\sim 10^{-3}$ in antiferromagnetic crystals. It may be helpful to note that in a material where $\alpha = 10^{-3}$ Gaussian units, an applied electric field $|\vec{E}| = 10^4$ V/cm produces a magnatization $4\pi|\vec{M}| = 0.03$ G, and an applied magnetic field $|\vec{H}| = 10^4$ Oe produces an electric polarization $4\pi|\vec{P}| = 3 \times 10^{-9}$ C/cm^2. We shall use the notation $|\vec{E}| \equiv E_0$ and $|\vec{H}| \equiv H_0$ in this paper.

Studies of two general classes of magnetic transitions will be considered: (1) the magnetic ordering transition in $H_0 = 0$, and (2) transitions produced by applying a magnetic field to the antiferromagnetic crystal. In the former case one is interested in the "critical behavior"

---

† Presented at the Symposium on Magnetoelectric Interaction Phenomena in Crystals, Seattle, Washington, U.S.A., May 21–24, 1973.

‡ Present address: Laboratorium für Festkörperphysik, ETH, Zürich, Switzerland.

of $\vec{\alpha}$ (see Figure 1a) at temperatures T close to the Néel point T_N. Recent experimental and theoretical studies have suggested[10] that when certain conditions are satisfied, $\alpha(T)$ is proportional to the sublattice

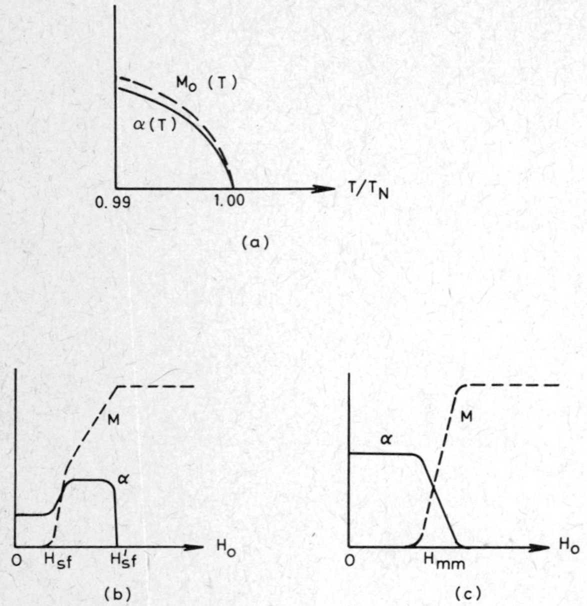

FIGURE 1 Idealized examples of magnetoelectric studies of phase transitions. (a) Critical behavior with $H_0 = 0$, comparing sublattice magnetization M_0 to the magnetoelectric susceptibility α. (b) Spin-flop in a low-anisotropy antiferromagnet at $T \ll T_N$, comparing net magnetization M and α. $H_{sf} \approx (2H_AH_E)^{1/2}$ and $H'_{sf} \approx 2H_E$. (c) Metamagnetic transition in a high-anistropy antiferromagnet at $T \ll T_N$. $H_{mm} \approx H_E$.

magnetization $M_0(T)$ in the critical region, which raises the exciting possibility of studying the critical behavior of M_0 in an antiferromagnet by measuring α for the bulk crystal. The second class of magnetic transitions to be considered comprises "spin-flop" (Figure 1b) and "metamagnetic" (Figure 1c) behavior. Spin-flop refers to an abrupt rotation of the sublattice magnetizations at a critical field $H_{sf} \approx (2H_AH_E)^{1/2}$, where H_A and H_E are the (internal) anisotropy and exchange fields, respectively.[11] Spin-flop is manifested in an abrupt increase in the net magnetization $M \equiv |\vec{M}|$, followed by a rather linear increase to saturation at a second critical field $H'_{sf} \approx 2H_E$. The spin-flop transition at H_{sf} may be first or second order depending on temperature and on the orientation of $\vec{H}$.[12,13] An abrupt change in α is expected at H_{sf}, since the rotation of the sublattice magnetizations generally involves a change in magnetic symmetry. At the higher-field transition, $\alpha \to 0$ (aside from possible higher-order ME effects[7,14] which we shall not consider here), since the material is paramagnetic in $H_0 \geqslant H'_{sf}$. Metamagnetic behavior (Figure 1c) is observed in high-anisotropy antiferromagnets where H_A is big enough to prevent the rotation of the sublattice magnetizations.[15] The transition to paramagnetism occurs in a single step (in the simplest case) at a critical field $H_{mm} \approx H_E$. In real crystals, the metamagnetic transitions (which are first order at $T \ll T_N$) are broadened by demagnetizing fields.

In the following we shall present examples of ME studies of critical behavior, and of spin-flop and metamagnetic transitions. Areas in which research is lacking or which look particularly promising for future work will be pointed out, and the potentialities of the technique will be evaluated critically.

EARLY STUDIES OF Cr_2O_3

The ME effect was first discovered[1–3] in a low-anisotropy antiferromagnet Cr_2O_3. This material crystallizes in a corundum-type crystal structure, with four Cr^{3+} ions positioned along the three-fold axis in the trigonal unit cell. The sublattice magnetizations are parallel to [111] and the magnetic point group is $\bar{3}'m'$ at $T < T_N = 308$ K. Taking the z axis along [111], the ME tensor has only three nonzero components, α_{zz} and $\alpha_{xx} = \alpha_{yy}$.[1] The former component has an unusual dependence on T, with a maximum at $T \approx 260$ K and a sign reversal at lower T, whereas the behavior of α_{xx} is more like that expected for M_0 in Cr_2O_3.[3,16,17] The data of $\vec{\alpha}$ *vs.* T have been fitted by various authors[18–20] to theoretical curves developed on the basis of assumed "single-ion" and "two-ion" ME mechanisms. There appears to be agreement that the single-ion effects dominate for α_{xx}. In an early study of the critical behavior of $\vec{\alpha}$, Astrov[16] found that α_{xx} could be fitted to a power law,

$$\alpha_{xx} \propto (T_N - T)^\beta$$

at temperatures $T \gtrsim 0.7\ T_N$, with $\beta = \frac{1}{2}$.

A spin-flop transition was observed in pulsed-field magnetization measurements[21] on Cr_2O_3 with $\vec{H}$ along [111]. The transition occurs at $H_{sf} \approx 60$ kOe at $T = 4.2$ K, but H_{sf} increases monotonically to over 100 kOe at $T = 280$ K. The sublattice magnetizations rotate from parallel to [111] at $H_0 < H_{sf}$ to some direction in the basal plane at $H_0 > H_{sf}$, but the orientations of the sublattice magnetizations have not been determined for H_0 between H_{sf} and H'_{sf}, and so the magnetic point group and the form of $\vec{\alpha}$ are not known. Nevertheless, an anomaly is to be expected in the ME signal if $\vec{\alpha}$ changes abruptly at H_{sf} as indicated schematically in Figure 1b.

An ME study of spin-flop in Cr_2O_3 was carried out by Foner and Hanabusa[22] using a Bitter solenoid to produce the magnetic field H_0 of up to 105 kOe, and using an a.c. modulation coil to give an axial component of magnetic field at a frequency f of 80–100 Hz. The α_{zz} component of the ME tensor was investigated by measuring essentially the open circuit voltage across the sample at the frequency f. In our notation, this quantity is proportional to

$$d(4\pi P_z)/dH_z = d(\alpha_{zz}H_z)/dH_z.$$

From Figure 1b, the measured quantity might be expected to show a sharp peak at the transition field and to be relatively constant for a range of fields somewhat below and above H_{sf}, and this behavior was observed at $T = 4.2$ K, as can be seen in Figure 2,

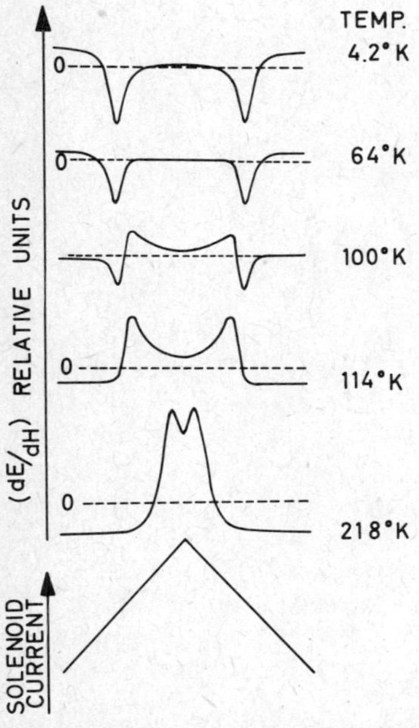

FIGURE 2 High-field magnetoelectric studies of spin-flop in Cr_2O_3 (after Ref. 22). The differential ME signal "dE/dH" is described in text. Bottom trace shows the common variation of magnetic current which corresponds to a field sweep from 20 to 105 kOe.

which has been reproduced from the original article. In the highest fields, α_{zz} was nearly zero at $T = 4.2$ K. It has been pointed out[23] that this result favors a model in which the sublattice magnetizations approach the two-fold axis perpendicular to [111] for $H_0 > H_{sf}$. This would correspond to magnetic point group $2'/m$, for which $\alpha_{zz} \equiv 0$ by symmetry. At higher T, however, the measured signal is nonzero at the maximum $H_0 = 105$ kOe, which suggests the magnetic symmetry may be other than $2'/m$.

The anomaly in Figure 2 moves to higher H_0 as T increases, in good agreement with the magnetization data, but at temperatures near $T_{rev} = 95$ K, double extrema appear in the ME curves, which makes the interpretation of the data somewhat uncertain. At about this same temperature, the curve of α_{zz} *vs.* T at $H_0 = 0$ passes through zero. The double extrema were ascribed tentatively to off-diagonal elements in $\vec{\alpha}$, but effects of crystal misorientation and depolarizing fields could not be excluded.[22] We shall return to this problem below when we present a model for calculating the form of the ME anomaly due to spin-flop.

A knowledge of the orientation of the sublattice magnetizations in the basal plane at $H_0 \gtrsim H_{sf}$ would be very helpful in understanding the spin-flop behavior in more detail. It was recently stated[24] that if H is misaligned by as little as 1° from the z axis, then it is the projection of $\vec{H}$ in the basal plane which determines the high-field symmetry. In this case it would be worthwhile carrying out ME experiments with $\vec{H}$ intentionally misaligned through a few degrees in known directions in order to study the effects of the different possible high-field symmetries.

CRITICAL BEHAVIOR

At the time of writing the critical behavior of $\vec{\alpha}$ has been studied in three ME materials: $DyPO_4$ ($T_N = 3.39$ K),[10], $DyAlO_3$ ($T_N = 3.52$ K),[25] and Cr_2O_3.[16,26] In this section we shall briefly summarize the results and point to some of the probelms and potentialities of this type of measurement.

The ME effect in $DyPO_4$ was first reported by Rado,[27] who showed that $\vec{\alpha}$ could be written in terms of two nonzero components, α_{xx} and $\alpha_{yy} = -\alpha_{xx}$, where x, y, and z are taken along the tetragonal axes a, a, and c of the crystal. The material forms in a zircon structure (space group $I4_1/amd$) in which each (magnetic) Dy^{3+} ion has four Dy^{3+} near neighbors arranged in a flattened tetrahedron. The antiferromagnetism may be described using a two-sublattice model, with the sublattice magnetizations parallel to [001]. The crystal-field splittings of the Dy^{3+} levels is such that the lowest state is in isolated Kramers doublet of extreme magnetic anisotropy (measured $g_{\parallel} = 19.4$ and $g_{\perp} = 0.51$).[28] The material is therefore Ising-like at $T \lesssim T_N$ with effective spin $\frac{1}{2}$.

Based on a model of localized moments in $DyPO_4$,

a quantum-mechanical mechanism was proposed to describe the ME coupling.[27] The mechanism might be described as an electric-field induced shift in the *g*-tensor of the Dy^{3+} ions. Later, Rado[10] was able to demonstrate using this single-ion mechanism that proportionality was to be expected between α_{xx} and the sublattice magnetization M_0 as a function of temperature in $DyPO_4$. An essential condition for this proportionality to hold was that the magnetic splitting (due to exchange and magnetic dipolar interactions) of the lowest Kramers doublet be small compared to the energy separations to the excited crystal-field levels; i.e., the material must be truly describable by the effective spin $\frac{1}{2}$ formalism, a condition which appears to be well satisfied in $DyPO_4$.

The theoretical proportionality between α_{xx} and M_0 could be tested by comparing the ME data with theoretical calculations of M_0. The latter were obtained from series expansions in the Ising diamond lattice, assuming only nearest-neighbor coupling.[29] Proportionality was found to hold over the complete range of validity of the series expansions. Agreement was obtained both at low temperatures (T/T_N between 0.5 and 0.7) and in the critical region at higher temperatures (T/T_N between 0.966 and 0.9999). In agreement with theory for M_0, it was found that $\alpha_{xx}(T)/\alpha_{xx}(0)$ could be fitted in the critical region to power law of the form $D(1 - T/T_N)^\beta$, with $D = 1.661$, $\beta = 0.314$, and $T_N = 3.39$ K.

The ME effect in $DyAlO_3$ was first reported by Mercier and Bauer,[30] who worked with polycrystalline material. The crystal structure of $DyAlO_3$ is orthorhombic (space group *Pbnm*) with four Dy^{3+} ions is the unit cell positioned in mirror planes which are perpendicular to the *c* axis. The antiferromagnetism may be described using a four sublattice model in which the sublattice magnetizations are confined to the mirror planes. The magnetic point group at $T < T_N$ is $m'm'm'$. The ME tensor therefore has three nonzero components, α_{xx}, α_{yy}, α_{zz}, where *x*, *y*, and *z* are taken along the unit cell axes *a*, *b*, and *c*, respectively. Structurally, $DyAlO_3$ is more complex than $DyPO_4$, but the magnetic behavior is similar in certain important ways. As in $DyPO_4$, the magnetic ordering is Ising-like and may be described in the effective spin $\frac{1}{2}$ formalism. Neutron-diffraction,[31] optical,[32] and magnetic[33] studies show that the sublattice magnetizations are closely confined to given directions in the mirror planes at $T \leqslant T_N$. The measured three principal values of the *g* tensor for the lowest crystal-field doublet are (18 ± 2) and (3 ± 1) in the mirror planes and (2 ± 1) along the *c* axis.[32] For $\vec{E}$ and $\vec{H}$ applied along the *c* axis, a single-ion *g*-shift mechanism for α_{zz} may be written down which is completely analogous to that proposed by Rado in $DyPO_4$.

A detailed study of α_{zz} was carried out in $DyAlO_3$ by Holmes, Van Uitert and Hull.[25] Some of the data are shown in Figure 3 in comparison with smoothed

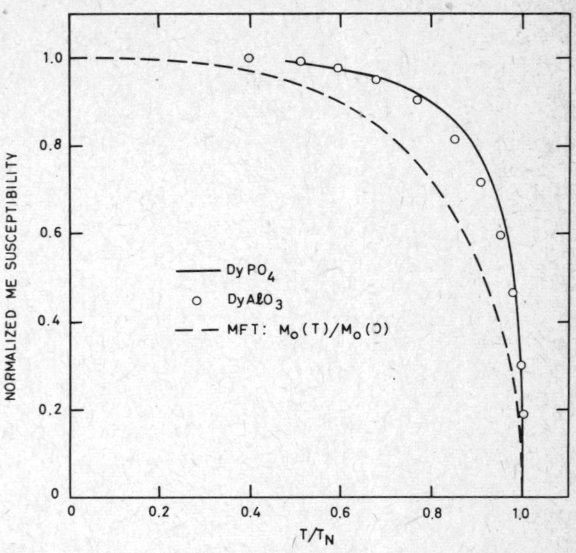

FIGURE 3 Comparison of magnetoelectric data on two Ising-like materials, $DyPO_4$ ($T_N = 3.39$ K) smoothed data from Ref. 10, and $DyAlO_3$ ($T_N = 3.52$ K) data from Ref. 25. Dashed line shows sublattice magnetization from molecular field theory, effective spin $\frac{1}{2}$.

data from Ref. 10 on $DyPO_4$. A molecular field calculation of $M_0(T)$ for spin $\frac{1}{2}$ is also shown. It may be seen that the results are quite similar for α_{xx} in $DyPO_4$ and α_{zz} in $DyAlO_3$. The later quantity was also compared[25] to a curve of $M_0(T)$ derived from a neutron-diffraction study[31] on polycrystalline material. Proportionality between $\alpha_{zz}(T)$ and $M_0(T)$ was found to hold reasonably well in view of estimated experimental errors (the discrepancy in the normalized curves was 5% at $T/T_N = 0.8$). A study of the critical behavior showed that for T/T_N between 0.96 and 0.995, $\alpha_{zz}(T)/\alpha_{zz}(1.4\text{ K})$ could be fitted to a power law $D(1 - T/T_N)^\beta$, with $D = 1.51 \pm 0.03$, $\beta = 0.311 \pm 0.005$, and $T_N = 3.525$ K. Thus, β is nearly the same as in $DyPO_4$. The derived exponent and coefficient agree closely with theoretical predictions[34] ($\beta \approx \frac{5}{16}$ and $D = 1.49$ to 1.66, depending on coordination) concerning $M_0(T)$ in an Ising-like magnet.

In contrast to the materials above, Cr_2O_3 is characterized by relatively low anisotropy. A new study of the ME effect in this material by Fischer, Gorodetsky, and Shtrikman[26] at temperatures close to T_N produced a power-law exponent $\beta = 0.35 \pm 0.01$. The measurements were done on a nonoriented crystal which showed the ME effect "spontaneously" (i.e., ME annealing was not applied—see below), and the data were said to show the same dependence on T as the sublattice magnetization.

In order to obtain meaningful and reproducible results, the crystals studied should be, insofar as is possible, in a single-domain antiferromagnetic state. This usually means applying biasing d.c. magnetic ($H_{d.c.}$) and electric ($E_{d.c.}$) fields as the sample is cooled through T_N, a procedure known as ME annealing.[35] The biasing fields cause one type of domain to be energetically more favorable than others, due to the ME coupling in the crystal. This procedure can create problems when studying critical behavior, because the magnetic ordering temperature varies with $H_{d.c.}$. This is illustrated in Figure 4 with data obtained while studying the critical behavior of $\overleftrightarrow{\alpha}$ in $DyAlO_3$. The ME signal was independent of $H_{d.c.}$ only for fields below a critical value $H'_{d.c.}$, which approached zero as $T \to T_N = 3.525$ K. Any misalignment of $H_{d.c.}$ from the c axis would tend to exaggerate the problem, for

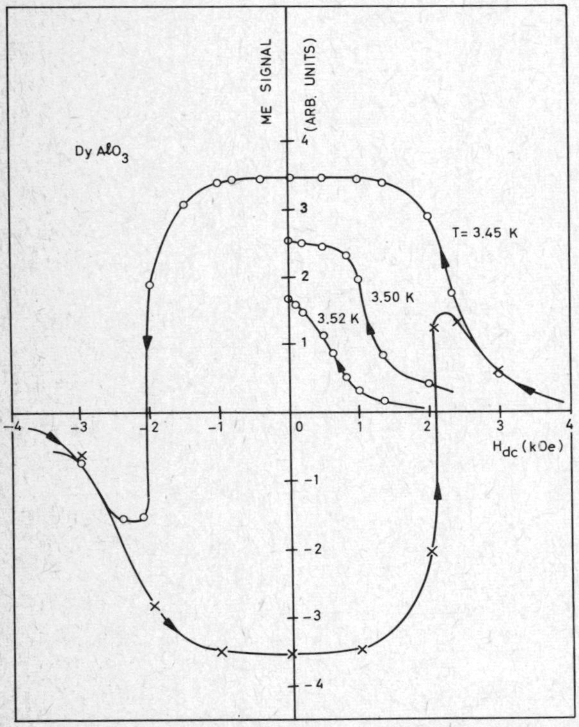

FIGURE 4 Magnetoelectric signal as a function of d.c. magnetic bias field at temperatures close to $T_N = 3.525$ K in $DyAlO_3$. The plotted signal is proportional to $d(4\pi P_z)/dH_z$. A constant electric bias field, $E_{d.c.} \approx 10$ statvolt/cm, was applied parallel to [001] during the measurements.

the g factors are largest for directions perpendicular to the c axis. A curve similar to that in Figure 4 was reported by Mercier and Bauer[30] at temperatures close to T_N in $TbAlO_3$. The effect is related to the metamagnetic behavior described below. Through repeated measurements of α_{zz} at $T = 1.4$ K in $DyAlO_3$, a value of $H_{d.c.} = 0.2$ kOe was judged sufficient to produce essentially a single domain in the crystal, but with $H_{d.c.} = 0.2$ kOe it was possible to obtain reliable α_{zz} data only at temperatures below 3.51 K, or $0.995T_N$.

The studies of critical behavior are an exciting development in ME measurements and should be extended to more systems. Experiments (and also theoretical studies) to delineate the limits of validity of the proportionality between the components of $\overleftrightarrow{\alpha}$ and M_0 would be particularly useful. It would be useful to know whether the proportionality holds in a non-Kramers system (such as, e.g., $TbAlO_3$), and to what extent the "two-ion" contributions to $\overleftrightarrow{\alpha}$ affect the behavior in the critical region. From molecular field theory, the two-ion contributions would be expected to vary as χM_0, where χ is the magnetic susceptibility (which itself may vary rapidly in the critical region).[19,25,36]

METAMAGNETIC TRANSITIONS

Metamagnetic transitions have been investigated magnetoelectrically in a number of insulating rare earth compounds, including $TbAlO_3$,[30] $DyPO_4$,[27] $DyAlO_3$,[37] and $TbPO_4$.[38] The behavior is qualitatively that shown in Figure 1(c), with $\overleftrightarrow{\alpha}$ decreasing abruptly to zero as H_0 is increased beyond the critical field H_{mm} for the transition to paramagnetism. An especially beautiful example is provided by the data of Rado[27] on $DyPO_4$, which have been reproduced in Figure 5. Two interesting complications of the behavior illustrated in Figure 1(c) can occur and have been investigated. The first complication might be called "switching effects." Under certain controlled conditions the sign of $\overleftrightarrow{\alpha}$ may be reversed by cycling H_0 through H_{mm}. The second complication occurs in some materials where the transition to paramagnetism takes place in two or more discrete steps, rather than at a single critical field H_{mm} as illustrated in Figure 1(c). In this case the ME effect can give useful information about the magnetic symmetries and about the order of the magnetic transitions.

Switching effects in a metamagnetic crystal were apparently first observed by Mercier and Bauer[30] in $TbAlO_3$. The phenomenon is exemplified in Figures 4 and 5. With an electric bias field applied to the crystal,

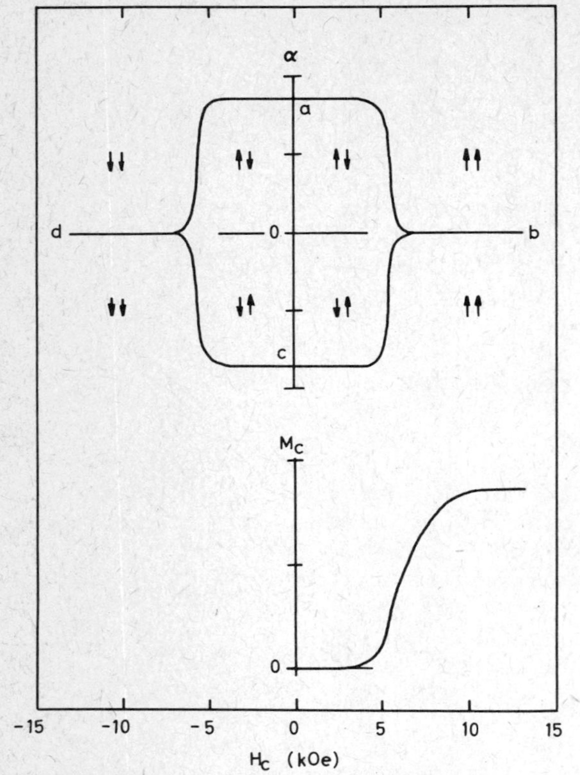

FIGURE 5 Metamagnetic behavior in $DyPO_4$ after Ref. 27. Upper curve shows α_{xx} component of the ME tensor and the lower curve shows the measured magnetization in a magnetic field H_c along the c axis at $T = 1.5$ to 1.6 K. As H_c was varied along the path 0–b–0–d–0– etc., the ME signal varied along the path a–b–c–d–a– etc.

the behavior may be understood as related to ME annealing, for cycling H_0 through H_{mm} has the effect of cycling the sample through the antiferromagnetic to paramagnetic transition.[37] It should be mentioned that switching effects of a different kind[39] were reported in Cr_2O_3 at temperatures close to T_N. In that material it is possible to rearrange the antiferromagnetic domain structure so as to reverse $\vec{\alpha}$ in the antiferromagnetic state (i.e., it is not necessary to cycle through the antiferromagnetic–paramagnetic transition).

The metamagnetic behavior of $DyAlO_3$ is such that both one-step and two-step transitions are possible, depending on the orientation of $\vec{H}$ in the a–b plane. Both types of transitions were investigated magnetoelectrically by Holmes and Van Uitert,[37] and the authors also made the first attempt at fitting the resulting ME data to analytical expressions. The attempted curve fitting was only qualitatively successful for the first-order, single-step transition. The discrepancies were tentatively attributed to uncontrollable changes in the ME domain structure as H_0 was increased through H_{mm}. The fit was quite good, however, for the data obtained on the first step of a two-step transition (Figure 6). For this transition, which was apparently second order, the ME data were quite reproducible on cycling H_0 through the transition field. The transition involved a reversal in the magnetization on one of the four sublattices, resulting in a change in the point symmetry from $m'm'm'$ at $H_0 = 0$ to m' for H_0 above the transition field.

The experiment was performed as shown in Figure 6 by rotating the crystal in a d.c. field of 10 kOe while measuring an a.c. signal proportional to $d(4\pi P_z)/dH_0$. If no transition occurred, then the signal would be proportional to $\cos\theta$. A transition was produced, however, when the component of $\vec{H}$ perpendicular to [001] reached a critical value. In Figure 6, this occurred for angular settings of between 50 and 60°.

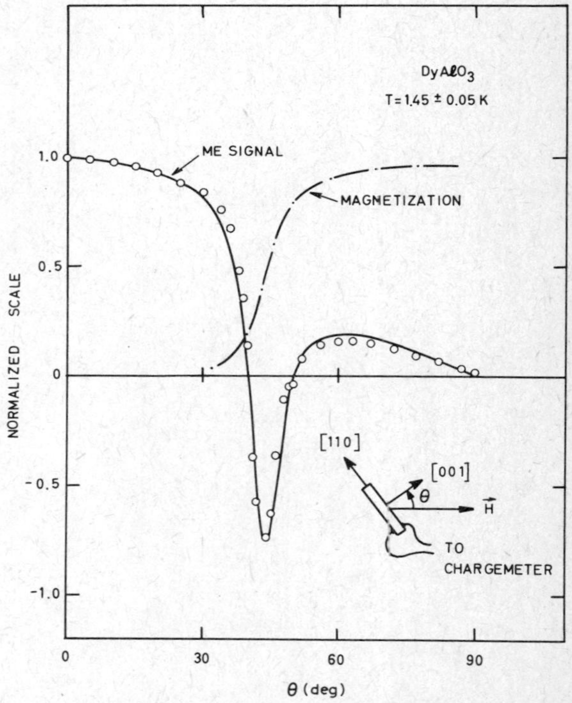

FIGURE 6 Comparison of normalized magnetoelectric signal (open circles) and magnetization (dot–dash curve) for metamagnetic transition in $DyAlO_3$ (after Ref. 37). Sample was rotated in a fixed magnetic field $\vec{H}$ of magnitude $H_0 = 10$ kOe, on which was superimposed a small a.c. component at a frequency of 150 Hz. The a.c. ME signal on the charge-meter was proportional to $d(4\pi P_z)/dH_0$. Solid curve: see text.

An analytical expression to be compared with the ME data was derived by assuming that the change in α_{zz} (from the initial value, called a_0, at $H_0 = 0$ to the new value, called a_1, above the transition field) could be expanded as a power series in ν, the normalized magnetization in the crystal. The latter quantity was measured independently on the same crystal. Retaining the leading terms in this expansion gave

$$4\pi P_z = [(1 - \nu)a_0 + \nu a_1] H_0 \cos\theta .$$

This expression must then be differentiated with respect to H_0 for comparison to the ME data, and it must be recalled that ν is itself a function of H_0. If the differentiated expression is normalized to the value at $\theta = 0$, one obtains

$$\frac{\alpha(\theta)}{\alpha(0)} = (\cos\theta)\left\{1 - \left(1 - \frac{a_1}{a_0}\right)\left[\nu(H') + H'\frac{d\nu}{dH'}\right]\right\}$$

where $H' = H_0 \sin\theta$. This formula contains a single adjustable parameter, a_1/a_0, and it was argued on physical grounds in Ref. 37 that a_1/a_0 should be expected to be of order of magnitude $\frac{1}{2}$. Indeed, choosing $(a_1/a_0) = 0.4$ gives excellent agreement with the data, as shown by the solid curve in Figure 6.

The work described above on $DyPO_4$ and $DyAlO_3$ was carried out at $T \ll T_N$. Only sketchy reports of the behavior at higher temperatures have been published, and this is clearly an area for more experimental work. It would appear, for example, that the ME effect might be a sensitive way of investigating the so-called "tricritical" region, where the character of the metamagnetic transition changes with temperature from first order to second order.

SPIN-FLOP

In addition to the measurements[22] described earlier on Cr_2O_3, spin-flop has been studied magnetoelectrically in the orthorhombic pyroxene material $MnGeO_3$.[40] In this section the work on $MnGeO_3$ will be reviewed and some new data will be presented. A model from Ref. 40 of the ME anomaly due to spin-flop will be described and curves will be calculated for comparison with the data on Cr_2O_3.

The ME effect was first reported in crystals of $MnGeO_3$ by Holmes and Van Uitert,[40] and, independently, in studies of polycrystalline material by Gorodetsky, Hornreich and Sharon.[41] In the single-crystal study two tensor components α_{yz} and α_{zy} were observed. The coordinates x, y, and z are taken along the crystal axes a, b, and c, respectively. The data are plotted as a function of temperature in Figure 7. The

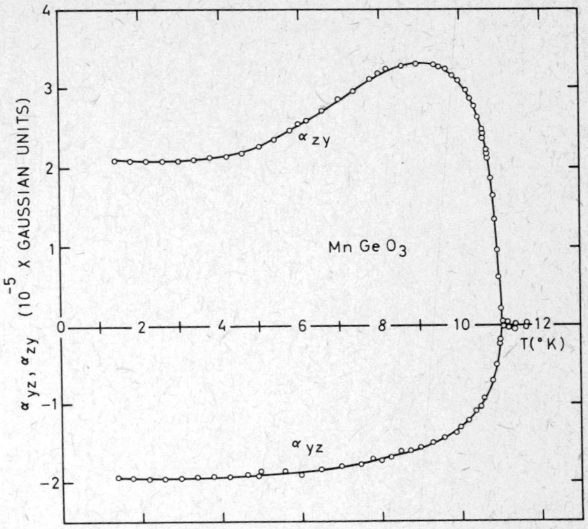

FIGURE 7 Measured components of the magnetoelectric tensor in crystals of $MnGeO_3$.

relative sign of α_{yz} and α_{zy} was not determined directly, but inferred[42] from the results in Ref. 41. The form of α is in accord with a neutron-diffraction study[43,44] in which the magnetic space group was shown to be $Pb'ca$. However, the neutron results (on polycrystalline samples) gave an anomalously high value for the ordering temperature, $T_N = 16$ K. Both ME data[40,41] and specific heat measurements[41] gave $T_N = 11.0 \pm 0.3$ K. Magnetic ordering in $MnGeO_3$ was first reported by Sawaoka, Miyahara, and Akimoto[45] based on magnetic susceptibility data which peaked at $T = 10$ K.

The neutron study revealed a spin-flop transition with $H_{sf} = 9$ kOe at $T = 1.1$ K, and this was confirmed in magnetization measurements on single crystals.[40] During spin-flop the sublattice magnetizations move from the $\pm b$ axes (at $H_0 = 0$) to the $\pm c$ axes (at $H_0 > H_{sf}$). The space group in $H_0 > H_{sf}$ is $Pb'c'a'$, for which $\vec{\vec{\alpha}}$ is a diagonal tensor. The crystal structure is quite complex, with 16 Mn^{2+} ions in the unit cell, but the spin-flop behavior at low temperatures may be described adequately using a conventional two-sublattice, Néel-type model.

In Figure 8 data are shown on a crystal of $MnGeO_3$ which was cut to measure the electric polarization along the c axis with $\vec{H}$ applied along the b axis. The anomaly at spin-flop shows up clearly for H_0 slightly above 9 kOe, in agreement with the magnetization and neutron data, and the curves are qualitatively similar to those obtained by Foner and Hanabusa[22] at $T = 4.2$ K on Cr_2O_3. The figure illustrates the pronounced sharpening of the transition as the misalignment

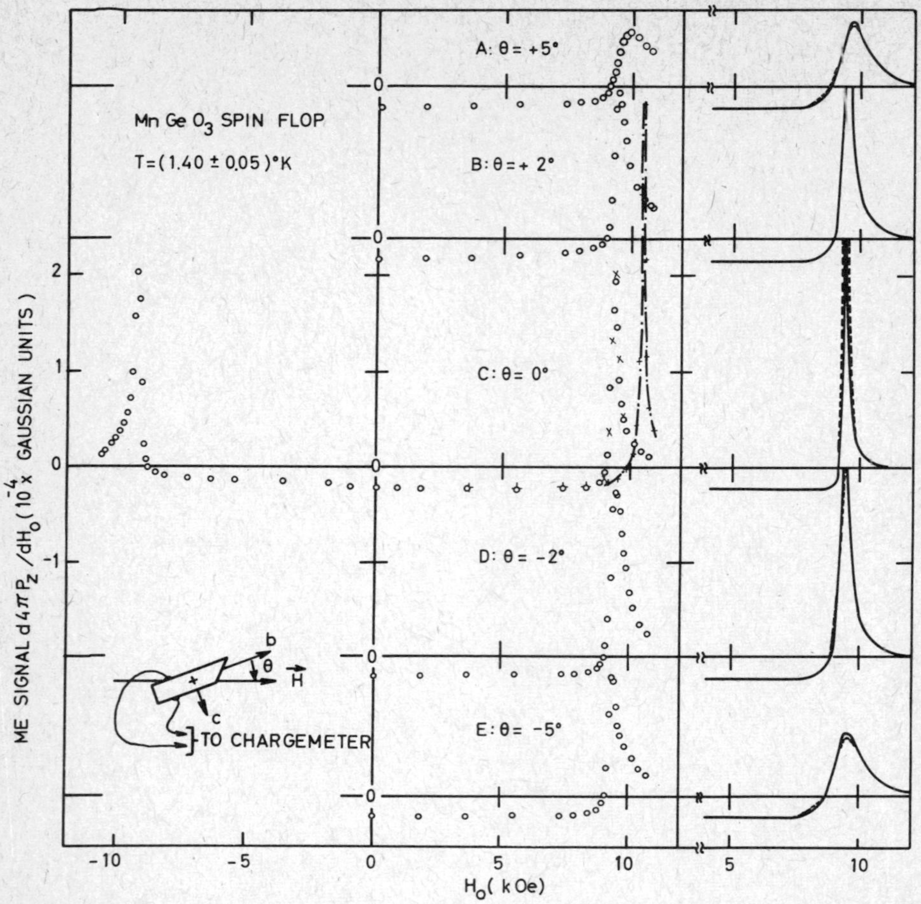

FIGURE 8 Magnetoelectric signal $d(4\pi P_z)/dH_0$ in $MnGeO_3$ as a function of applied magnetic field. Inset shows experimental arrangement. Open circles and x-shaped points were taken in decreasing and increasing magnetic fields, respectively. Plus-shaped points show data for $\theta = 0$ at $T = 4.2$ K.

between $\vec{H}$ and the b axis is diminished.[46] In this experiment α_{zy} changes from -2×10^{-5} Gaussian units at $H_0 = 0$ to zero at $H_0 > H_{sf}$. The theoretical curves in the right were calculated using the model described in Ref. 40, which is a simple extension of that described below.[47] The plus-shaped points were obtained at $T = 4.2$ K. The spin-flop anomaly in the ME signal is sharper and more intense at this higher temperature, and this is an effect which is not yet understood.

Measurements on a sample of $MnGeO_3$ cut to measure the electric polarization along the b axis, with the magnetic field applied close to that same direction, were reported in Ref. 40. Since α_{yy} increases from zero at $H_0 = 0$ to a finite value at $H_0 > H_{sf}$, the data in this experiment were sensitive to the symmetry in the high-field state. The spin-flop anomaly in the ME data was found to reverse in sign on changing the angle θ (as defined in Figure 8) from $-2°$ to $+2°$. This was shown to be in agreement with theory for the second-order transition. The direction of rotation of the sublattice magnetizations is dependent on the sign of θ, leading to essentially time-reversed spin-flop states for positive and negative θ. The ME signal became small at $\theta = 0$, which was attributed to a first-order transition at this angular setting.

The model used[40] in calculating ME spin-flop curves in $MnGeO_3$ is similar to that described above for metamagnetic transitions, in that it is based on the assumption that $\vec{\alpha}$ varies linearly with some parameters which change smoothly through the magnetic transition. We shall write down the equations for the situation described in Figure 9, in which an uniaxial antiferromagnet (e.g., Cr_2O_3) is subjected to an applied magnetic field $\vec{H}$ at an angle θ from the c axis. A signal propor-

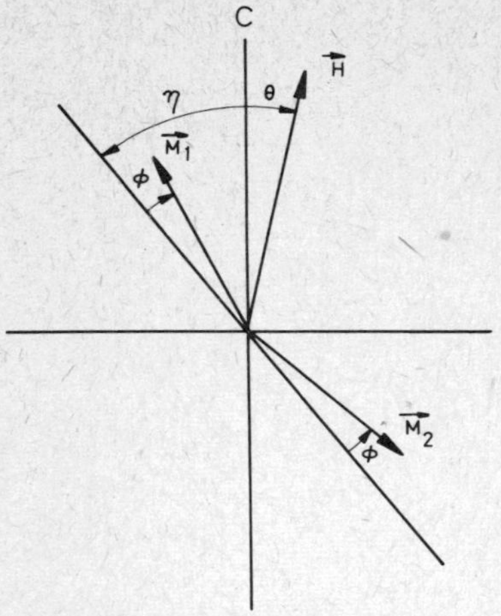

FIGURE 9 Two-sublattice model of an antiferromagnet at $T = 0$ K.

tional to $d(4\pi P_z)/dH_0$ is measured as H_0 is increased through H_{sf}. Let η be the angle between the vector $\mathbf{L} = \vec{M}_1 - \vec{M}_2$ and the c axis, where $\vec{M}_1$ and $\vec{M}_2$ are the sublattice magnetizations. This angle changes from zero at $H_0 = 0$ to $\approx\pi/2$ at $H_0 > H_{sf}$. Only the α_{zz} component of the ME tensor will be considered, and it will be assumed that

$$\alpha_{zz} = a_0 \cos\eta + a_1 \sin\eta,$$

where a_0 and a_1 are the values of α_{zz} in the low- and high-field states, respectively. Substituting into the definition

$$4\pi P_x = \alpha_{zz} H_z$$

one finds, for the measured quantity,

$$d(4\pi P_z)/dH_0 = \cos\theta\{a_0[\cos\eta - H_0(d\eta/dH_0)\sin\eta] + a_1[\sin\eta + H_0(d\eta/dH_0)\cos\eta]\}.$$

The angle η varies with H_0 in a manner which may be calculated from molecular field theory.[48] In an uniaxial material with isotropic exchange,

$$\eta = \tfrac{1}{2}\tan^{-1}\left(\frac{H_0{}^2 \sin 2\theta}{H_{sf}^2 - H_0{}^2\cos 2\theta}\right).$$

Curves calculated from these equations are shown in Figure 10 for various choices of a_0 and a_1, and with $\theta = 3°$. Choosing smaller values of θ tends to sharpen and intensify the anomaly at $H_0 \approx H_{sf}$. The constants have been chosen in Figure 10 to mimic the different types of curves obtained in Cr_2O_3 (Figure 2). The agreement is surprisingly good, considering the simplicity of the model. However, the curve labelled (b) in Figure 10 would have to be shifted vertically upwards and centered on the "zero" line in order to agree in detail with the 100 K curve in Figure 2.

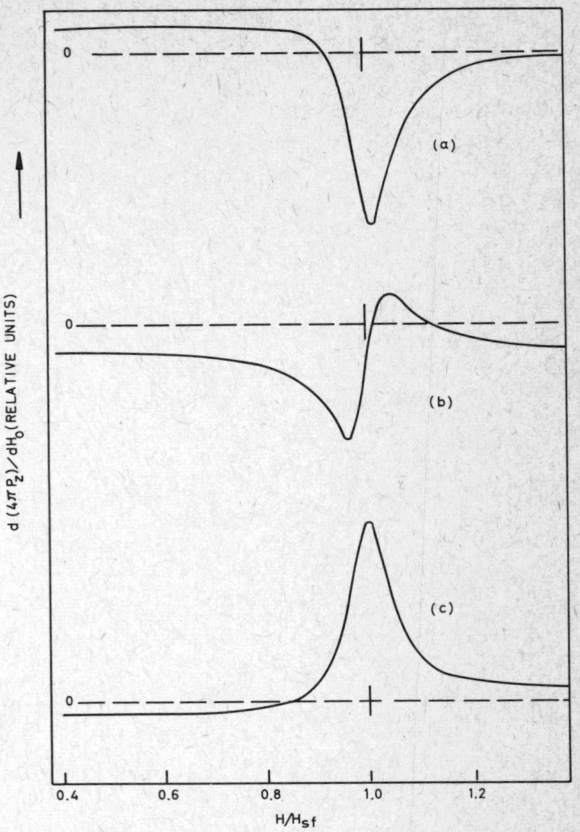

FIGURE 10 Calculated ME curves for spin-flop transition (see text). Curve (a): $a_0 = 1$, $a_1 = 0$. Curve (b): $a_0 = a_1 = -1$. Curve (c): $a_1 = -0.5$, $a_1 = +0.5$.

This model may easily be generalized to describe more complicated symmetries, as was done[40] for $MnGeO_3$, and to incorporate more sophisticated calculations of η. It is capable of explaining quantitatively many of the salient features of ME anomalies at spin-flop, but it is nevertheless heuristic and applicable only at very low temperatures, $T \ll T_N$. Theoretical work to place these matters on a firmer basis is certainly needed.

From the experimental side, it would be useful to study in more detail the difference in ME behavior between the first-order and second-order transitions, as described in Ref. 40. This could best be done on a material (e.g., $GdAlO_3$) in which the first-order transition occurs over a wider range of θ values[49] than in $MnGeO_3$. One of the potential uses of this technique is in mapping out magnetic phase diagrams in low-anisotropy antiferromagnets. For this purpose, studies at higher temperatures approaching T_N would be required.

CONCLUSIONS

The experimental studies to date have shown that ME measurements can provide a useful adjunct to the more classical techniques for studying magnetic phase transitions. More theoretical work is clearly needed, particularly in understanding the detailed form of the ME anomalies. Much can be learned, however, from general considerations of the change in symmetry involved, and from thermodynamic principles. The restriction that the material show the ME effect is a rather severe limitation on the number of substances which can be studied in this way, but by now the list of known ME materials[50] is sufficiently long that many useful and interesting experiments can be foreseen.

Note added in proof: ME studies of critical behavior have been carried out in $GdVO_4$ (G. Gorodetsky, R. M. Hornreich and B. M. Wanklyn, *Phys. Rev. B* **8**, 2263 (1973)) and $HoPO_4$ (A. H. Cooke, S. J. Swithenby and M. R. Wells, *Int. J. Magnetism*–in press). The ME effect has been used to study the pressure-dependence of T_N in Cr_2O_3 (G. Gorodetsky, R. M. Hornreich and S. Shtrikman, *Phys. Rev. Letters* **31**, 938 (1973)).

ACKNOWLEDGEMENT

The results in Figures 7 and 8 were obtained in collaboration with L. G. Van Uitert of this laboratory.

REFERENCES AND NOTES

1. I. E. Dzyaloshinskii, *Sov. Phys. JETP* **10**, 628 (1960).
2. D. N. Astrov, *Soviet Phys. JETP* **11**, 708 (1960).
3. V. J. Folen, G. T. Rado and E. W. Stalder, *Phys. Rev. Letters* **6**, 607 (1961).
4. V. L. Indenbom, *Sov. Phys.–Crystallography* **5**, 493 (1960).
5. An example is the ME determination of the magnetic symmetry in $GdAlO_3$. M. Mercier and G. Velleaud, *J. Phys. (Paris)* **32**, Colloque No. 1, Cl-499 (1971).
6. R. R. Birss, *Symmetry and Magnetism* (North Holland, Amsterdam, 1964), Chap. 4.
7. T. H. O'Dell, *The Electrodynamics of Magneto-Electric Media* (North Holland, Amsterdam, 1970).
8. We follow the approach of Ref. 1. For an elegant treatment based on special relativistic effects, see Ref. 7.
9. Demagnetizing and depolarizing fields may influence the measured quantity. G. T. Rado and V. J. Folen, *J. Appl. Phys.* **33** Supplement, 1126 (1962).
10. G. T. Rado, *Solid State Commun.* **8**, 1349 (1970).
11. L. Néel, *Ann. Phys.* **5**, 232 (1936).
12. C. K. Chepurnykh, *Soviet Phys.–Solid State* **10**, 1517 (1968).
13. H. Rohrer and H. Thomas, *J. Appl. Phys.* **40**, 1025 (1969)
14. E. Ascher, *Phil. Mag.* **17**, 149 (1968).
15. For a description of metamagnetic behavior see I. S. Jacobs and P. E. Lawrence, *Phys. Rev.* **164**, 866 (1967).
16. D. N. Astrov, *Soviet Phys. JETP* **13**, 729 (1961).
17. G. T. Rado and V. J. Folen, *Phys. Rev. Letters* **7**, 310 (1961).
18. G. T. Rado, *Phys. Rev.* **128**, 2546 (1962).
19. R. Hornreich and S. Shtrikman, *Phys. Rev.* **161**, 506 (1967); **166**, 598 (E) (1968).
20. R. Englman and H. Yatom, *Phys. Rev.* **188**, 803 (1969).
21. S. Foner and Shou-Ling Hou, *J. Appl. Phys.* **33**, Supplement 1289 (1962).
22. S. Foner and M. Hanabusa, *J. Appl. Phys.* **34**, 1246 (1963).
23. See Ref. 7, pp. 134–6.
24. K. L. Dudko, V. V. Eremenko and L. M. Semenenko, *Phys. Stat. Sol.* (b) **43**, 471 (1971).
25. L. M. Holmes, L. G. Van Uitert and G. W. Hull, *Solid State Commun.* **9**, 1373 (1971).
26. E. Fischer, G. Gorodetsky and S. Shtrikman, *J. Phys. (Paris)* **32**, Colloque No. 1, Cl-650 (1971).
27. G. T. Rado, *Phys. Rev. Letters* **23**, 644 (1969); **23**, 946 (E) (1969).
28. J. C. Wright and H. W. Moos, *Phys. Letters* **29A**, 495 (1969).
29. J. W. Essam and M. F. Sykes, *Physica* **29**, 378 (1963).
30. M. Mercier and P. Bauer, *Les Élements des Terres Rares* (Colloques Internationaux du Centre National de la Recherche Scientifique, No. 180, Paris, 1970), Vol. II, p. 377.
31. R. Bidaux and P. Mériel, *J. Phys. Rad.* **29**, 220 (1968).
32. H. Schuchert, S. Hüfner and R. Faulhaber, *Z. Phys.* **222**, 105 (1969).
33. L. M. Holmes, L. G. Van Uitert and R. R. Hecker, *Phys. Rev. B* **5**, 138 (1972).
34. M. E. Fisher, *Rep. Prog. Phys.* **30**, 615 (1967).
35. S. Shtrikman and D. Treves, *Phys. Rev.* **130**, 986 (1963).
36. M. Date, J. Kanamori and M. Tachiki, *J. Phys. Soc. Japan* **16**, 2589 (1961).
37. L. M. Holmes and L. G. Van Uitert, *Phys. Rev. B* **5**, 147 (1972).
38. G. T. Rado and J. M. Ferrari (Conference on Magnetism and Magnetic Materials, Denver, 1972), paper 6E-9.
39. T. J. Martin and J. C. Anderson, *IEEE Trans. on Magnetics* **MAG-2**, 446 (1966).
40. L. M. Holmes and L. G. Van Uitert, *Solid State Commun.* **10**, 853 (1972).
41. G. Gorodetsky, R. M. Hornreich and B. Sharon, *Phys. Letters* **39A**, 155 (1972).
42. In Ref. 41 it was inferred from the powder data that $\vec{\vec{\alpha}}$ is antisymmetric in $MnGeO_3$. The present results do not support this conclusion.
43. P. Herpin, A. Whuler, B. Boucher and M. Sougi, *Phys. Status Solidi* (b) **44**, 71 (1971).
44. B. Boucher, M. Sougi and A. Whuler, *J. Phys. (Paris)* **32**, Colloque No. 1, Cl-853 (1971).
45. S. Sawaoka, S. Miyahara and S. Akimoto, *J. Phys. Soc. Japan* **25**, 1253 (1968).
46. Owing to an accidental cleavage in the crystal the sample was not regular in shape (see drawing of cross-section in Figure 8). This may explain the differences in the curves for positive and negative θ.
47. The details will not be repeated here.
48. T. Nagamiya, K. Yosida and R. Kubo, *Adv. Phys.* **4**, 1 (1955).
49. K. W. Blazey, H. Rohrer and R. Webster, *Phys. Rev. B* **4**, 2287 (1971).
50. See, e.g., R. M. Hornreich, *IEEE Trans. on Magnetics*, **MAG 8**, 584 (1972).

THE MAGNETOELECTRIC EFFECT IN POLYCRYSTALLINE POWDERS—ANNEALING AT GENERAL ANGLES†‡

R. M. HORNREICH

Department of Electronics, The Weizmann Institute of Science, Rehovot, Israel

The relationships between the magnetoelectric (ME) susceptibilities in polycrystalline powders and single crystals following an annealing treatment in arbitrarily oriented electric and magnetic fields are derived. The case considered is that in which the single crystal ME susceptibility tensor is diagonal with two independent elements. It is shown that the elements of the single crystal ME tensor can be obtained from the powder measurements. The possible advantages of annealing at a general angle are discussed.

A magnetoelectric (ME) material is one in which there exists a linear relationship between an electric field and the medium's magnetic polarization and between a magnetic field and the medium's electric polarization.[1]

It is well known that a nonzero ME effect can exist in an antiferromagnetic randomly oriented polycrystalline powder only if an antiferromagnetic remanent state is first induced in the specimen by cooling it through its Néel temperature in the presence of simultaneous electric and magnetic annealing fields.[2–5] Then the symmetry of the remanent state will be at least that of the tensor E_iH_j where E_i and H_j are the fields used in the annealing treatment.

In previous work, calculations[2,5] and experiments have been restricted to the so-called parallel and perpendicular cases, wherein the annealing fields are parallel or perpendicular to each other respectively. For these cases the nonzero ME tensor elements will be[2] $\alpha_{xx} = \alpha_{yy}$, α_{zz} (annealing fields E_z, H_z) and α_{xy}, α_{yx} (annealing fields E_x, H_y).

It is also possible, however, to induce the required antiferromagnetic remanent state by carrying out the annealing process in electric and magnetic fields arbitrarily oriented with respect to each other. Indeed, such a procedure can have an advantage over the usual parallel/perpendicular annealing procedure. To see this, consider an anneal carried out in fields $\mathbf{E} = (0, 0, E)$ and $\mathbf{H} = (H \sin\beta_0, 0, H\cos\beta_0)$. The resulting ME powder tensor will then have the form

$$\begin{pmatrix} \alpha'_{xx} & 0 & \alpha'_{xz} \\ 0 & \alpha'_{yy} & 0 \\ \alpha'_{zx} & 0 & \alpha'_{zz} \end{pmatrix} \quad (1)$$

and the elements α'_{zz}, α'_{zx} can be measured following the annealing treatment.[6] Further, *both* of these elements can be determined following a *single* annealing treatment. On the other hand, the usual α_{zz}, α_{xy} measurements require two separate annealing operations, operations which do not necessarily yield the same antiferromagnetic remanent state. This can then lead to errors in the calculation of the single crystal susceptibilities from the powder results.

We are thus interested in deriving expressions for the ME susceptibilities expected in polycrystalline powders following an annealing treatment at a general angle in terms of the single crystal susceptibility tensor elements. We shall here consider only the simplest non-trivial case, one in which there are two independent elements in the single crystal ME susceptibility tensor. This tensor, which is appropriate to the magnetic crystal classes[7] 422, $\bar{4}'2m'$, $4/m'm'm'$, 32, $\bar{3}'m'$, 622, $\bar{6}'m'2$, and $6/m'm'm'$, has the form

$$\begin{pmatrix} \alpha_{\xi\xi} & 0 & 0 \\ 0 & \alpha_{\xi\xi} & 0 \\ 0 & 0 & \alpha_{\zeta\zeta} \end{pmatrix} . \quad (2)$$

The polycrystalline powder is composed of an aggregate of randomly oriented crystallites, each of

---

† This work was supported in part by the Joint Research Committee, Basic Research Branch, of the Israel Academy of Sciences and Humanities.

‡ This paper was presented at the Symposium on Magneto-electric Interaction Phenomena in Crystals, May 21–24, 1973, Seattle, Washington.

which is characterized, in its principal axis system, by the ME tensor given in (2). We assume that all interactions between the crystallites are negligible and describe the orientation of the body coordinates of a given crystallite with respect to the space axes by the Euler angles ϕ, ψ, and θ.[8] The contribution of this crystallite to α'_{zz} and α'_{zx} will then be

$$g_{zz} = \alpha_{\xi\xi} \sin^2 \theta + \alpha_{\zeta\zeta} \cos^2 \theta, \tag{3a}$$

$$g_{zx} = (\alpha_{\zeta\zeta} - \alpha_{\xi\xi}) \sin \theta \cos \theta \sin \phi. \tag{3b}$$

In integrating g_{zz} and g_{zx} over all crystallites, we assign to the ME tensor of each crystallite a plus or minus sign, corresponding to the remanent state induced during the annealing process. In general, the choice of sign will depend upon the ratios of the ME susceptibilities in the neighborhood of the Néel point[2] as well as the orientation of the crystallite. The basic criterion is that the ME contribution to the free energy of the system, $-\alpha_{ij}E_iH_j$, be negative in the neighborhood of the Néel point as a result of the anneal.

The free energy of a given crystallite, during the annealing process, is given by

$$F = -EH[(\alpha_{\xi\xi} \sin^2 \theta + \sigma_{\zeta\zeta} \cos^2 \theta) \cos \beta_0 + (\alpha_{\zeta\zeta} - \alpha_{\xi\xi}) \sin \theta \cos \theta \sin \phi \sin \beta_0]. \tag{4}$$

Note that F is independent of ψ.

Letting $p = (\alpha_{\zeta\zeta}/\alpha_{\xi\xi})_0$, the ratio of the single crystal susceptibilities in the neighborhood of the material's Néel temperature, we see that, for fixed θ, F reverses sign at

$$\sin \phi_0 = \frac{\cot \beta_0 (1 - q \cos^2 \theta)}{q \sin \theta \cos \theta}; \quad |\phi_0| \leqslant \pi/2, \tag{5}$$

where $q = 1 - p$. When integrating g_{zz} and g_{zx} over ϕ, we must therefore reverse the algebraic signs of $\alpha_{\xi\xi}$ and $\alpha_{\zeta\zeta}$ at the angles $\phi = \phi_0(\theta)$ and $\phi = \pi - \phi_0(\theta)$. Carrying out the integrals over ψ and ϕ then yields

$$\alpha'_{zz} = \frac{1}{\pi} \int d\theta \sin \theta \, (\alpha_{\xi\xi} \sin^2 \theta + \alpha_{\zeta\zeta} \cos^2 \theta) \phi_0(\theta), \tag{6a}$$

$$\alpha'_{zx} = -\frac{1}{\pi} \int d\theta \sin \theta \, [(\alpha_{\zeta\zeta} - \alpha_{\xi\xi}) \sin \theta \cos \theta](\cos \phi_0(\theta)). \tag{6b}$$

In the event that (5) has no solution, we set $\phi_0 = \pm\pi/2$ in (6), the choice of sign being such that F is always negative in the neighborhood of the Néel temperature.

To integrate (6) over θ we first determine for which value of θ $\phi_0(\theta)$ is given by (5) and for which it is equal to $\pm\pi/2$. We find that, for

$$\begin{gathered} 0 < \theta < \theta_2, \\ \theta_1 < \theta < \pi - \theta_1, \\ \pi - \theta_2 < \theta < \pi, \end{gathered} \tag{7}$$

we must set $\phi_0 = \pm\pi/2$, elsewhere it is given by (5). The angles θ_1, θ_2 in (7) are determined from

$$\cos(2\theta_1 - \beta_0) = \cos(2\theta_2 + \beta_0) = r \cos \beta_0, \quad r = (1 + p)/(1 - p). \tag{8}$$

Using (5), (7), and (8), the integration of (6) yields

$$\alpha'_{zz} = \tfrac{1}{3}(\alpha_{\zeta\zeta} + 2\alpha_{\xi\xi}) - (2/\pi q) \cos \beta_0 \times \left\{ \tfrac{1}{3}(\alpha_{\zeta\zeta} - \alpha_{\xi\xi})(2 - q)\chi_2 E(k) + \alpha_{\xi\xi} K(k)/\chi_2 - \frac{\pi}{6} p(\alpha_{\zeta\zeta} + 2\alpha_{\xi\xi})[(1 - \chi_1{}^2)(1 - \chi_2{}^2)]^{-1/2} \times \Lambda_0\left(\frac{\pi}{2} - \theta_2, k\right)\right\}, \tag{9a}$$

$$\alpha'_{zx} = \frac{2}{\pi}(\alpha_{\zeta\zeta} - \alpha_{\xi\xi})(\chi_2/3)[(\chi_1{}^2 + \chi_2{}^2)E(k) - 2\chi_1{}^2 K(k)]. \tag{9b}$$

Here

$$\chi_i = \cos \theta_i \quad (i = 1, 2), \qquad k = (1 - (\chi_1/\chi_2{}^2)^{1/2}. \tag{10}$$

K and E are the complete elliptic integrals of the first and second kind respectively and Λ_0 is Heuman's Lambda function.[9]

For the case $p \ll -1$ (e.g., Cr_2O_3[10]), (9) simplifies to

$$\alpha'_{zz} = \tfrac{1}{3}(\alpha_{\zeta\zeta} + 2\alpha_{\xi\xi})\left(1 - \frac{2}{\pi}\beta_0\right) + \frac{2}{3\pi}(\alpha_{\zeta\zeta} - \alpha_{\xi\xi}) \sin \beta_0 \cos \beta_0, \tag{11c}$$

$$\alpha'_{zx} = \frac{2}{3\pi}(\alpha_{\zeta\zeta} - \alpha_{\xi\xi}) \sin^3 \beta_0. \tag{11b}$$

Note that the single crystal susceptibilities $\alpha_{\zeta\zeta}$ and $\alpha_{\xi\xi}$ can be derived directly from the powder results. To do this, the ratio $p = (\alpha_{\zeta\zeta}/\alpha_{\xi\xi})_0$ is first determined from $(\alpha'_{zz}/\alpha'_{zx})_0$ using (9). Following this, $\alpha_{\zeta\zeta}$ and $\alpha_{\xi\xi}$ are simply the appropriate linear combination of α'_{zz} and α'_{zx}. Since α'_{zz} and α'_{zx} can be measured simultaneously following a single annealing treatment (by, for example, using two perpendicular pickup coils in a measurement of the magnetic polarization induced by an applied electric field), the ratio of the derived single crystal susceptibilities $\alpha_{\zeta\zeta}/\alpha_{\xi\xi}$ would be independent of the annealing treatment.

REFERENCES

1. L. D. Landau and E. M. Lifshitz, *Electrodynamics of Continuous Media* (Addison-Wesley Publishing Co., Inc., Reading, Massachusetts, 1960), p. 119; For a recent review, see R. M. Hornreich, *IEEE Trans. Magnetics* **8**, 584 (1972).
2. S. Shtrikman and D. Treves, *Phys. Rev.* **130**, 986 (1963).
3. T. J. Martin and J. C. Anderson, *Phys. Lett.* **11**, 109 (1964).
4. T. H. O'Dell, *Phil. Mag.* **13**, 921 (1966).
5. R. M. Hornreich, *J. Appl. Phys.* **41**, 950 (1970).
6. The usual method of applying an electric field during the annealing operation is to paint conducting electrodes on two opposite faces of the specimen. Thus, in order to measure α'_{xx}, α'_{xz}, or α'_{yy}, it would be necessary to transfer the electrodes to different faces of the specimen while maintaining it at a temperature below that at which magnetic ordering occurs.
7. R. R. Birss, *Rep. Progr. Phys.* **26**, 307 (1963).
8. H. Goldstein, *Classical Mechanics* (Addison-Wesley Publishing Co., Inc., Reading, Massachusetts, 1950), p. 107.
9. P. F. Byrd and M. D. Friedman, *Handbook of Elliptic Functions for Engineers and Physicists* (Springer-Verlag, Berlin, 1954).
10. V. J. Folen, G. T. Rado and E. W. Stalder, *Phys. Rev. Lett.* **6**, 607 (1961); S. Foner and M. Hanabusa, *J. Appl. Phys.* **34**, 1246 (1963).

CRYSTALLOGRAPHY OF THE FERROMAGNETOELECTRIC SWITCHING AND TWINNING

V. JANOVEC

Institute of Physics, Czech. Acad. Sci., 180 40 Prague 8, Na Slovance 2, Czechoslovakia

and

L. A. SHUVALOV

Institute of Crystallography, Acad. Sci. USSR, Moscow, V-333, Leninskii prospekt 59, USSR

Basic symmetry properties of ferromagnetoelectric domains are examined by a simple group theoretical procedure. Possible types of ferromagnetoelectric switching and the crystallographical coupling between ferroelectric and ferromagnetic switching are discussed. It is shown that the orientation and the charge of coherent stress-free domain walls can be determined by symmetry considerations. Domain structure and switching properties of the orthorhombic ferromagnetoelectric boracites are analysed as an illustrative example.

INTRODUCTION

Any analysis of possible domain configurations and switching properties starts with a crystallographical consideration. For ferroelectrics or ferromagnetics such examination is usually performed intuitively or by means of simple geometrical reasoning. For ferromagnetoelectric domains, characterized by magnetization M and polarization P, these approaches may become less effective due to different transformation properties of M and P. An algebraic method not relying on geometrical imagination can then be more reliable.

The basic idea of the procedure we are going to pursue is to examine domains, their symmetry relations and coexistence conditions in terms of symmetry operations of the group G describing the high symmetry phase. In this approach, domains are not characterized by distinct physical properties (e.g. M and P for ferromagnetoelectric domains) but by sets (complexes) of the symmetry operations from G. More specifically, to each domain there corresponds one left coset in the resolution of G into left cosets of the maximal subgroup of G that leaves one domain invariant.[1–4] Further, left cosets themselves describe the symmetry relations between domains[1–5] and can be utilized for an elementary discussion of switching. Finally, the presence of operations of the order two in a left coset provides sufficient conditions for the coherent stress-free coexistence of domains and for the formation of twins.[5–9]

Since the coset resolutions are the main tools of our analysis we first consider a few relevant concepts and theorems of abstract group theory (for more details see Ref. 10 or any other respectable book on group theory).

Let H be a subgroup of G. The complex gH, where $g \in G$, is called the *left coset of H in G* and g its representative. The group G can be partitioned into disjoint left cosets

$$G = H + g_2 H + \ldots + g_q H \tag{1}$$

Resolution (1) will be denoted by G/H. The number of distinct left cosets in G/H is called the *index of H in G* and is denoted by $[G:H]$.

If F is a subgroup of H, then each left coset of (1) consists of $[H:F]$ left cosets of F in G. It holds

$$[G:F] = [G:H][H:F] \tag{2}$$

The complex HgH, where $g \in G$, is called a *double coset of H in G*. The group G can be decomposed into disjoint double cosets

$$G = H + Hg_2 H + \ldots + Hg_p H \tag{3}$$

The resolution (2) will be denoted by $H \backslash G/H$.

Resolutions (1) and (2) are meaningful both for finite and infinite groups. In the former case, the index $[G:H]$ is equal to the ratio of the orders of G and H, respectively.

FERROMAGNETOELECTRIC DOMAINS, DEGENERACIES

We assume that a high symmetry para-phase (symmetry group G) changes at a ferromagnetoelectric transition (or during several successive transitions) into a ferromagnetoelectric phase of lower symmetry described by a group $F^{(i)}$. The ith ferromagnetoelectric (FME) domain is characterized physically by the spontaneous magnetization $M^{(i)}$ and the spontaneous polarization $P^{(i)}$; symmetrically, it is specified by a group $F_{M^{(i)}P^{(i)}}$ defined as the maximal subgroup of G that leaves both $M^{(i)}$ and $P^{(i)}$ invariant. To any FME domain there corresponds a distinct left coset in the resolution $G/F_{M^{(i)}P^{(i)}}$. Therefore, the number q_{MP} of FME domains equals the index of $F_{M^{(i)}P^{(i)}}$ in G,

$$q_{MP} = [G : F_{M^{(i)}P^{(i)}}] \tag{4}$$

Degenerate Ferromagnetoelectric Domains
If $F_{M^{(i)}P^{(i)}} = F^{(i)}$, then the appearance of M and P accounts fully for the symmetry reduction $G \to F^{(i)}$ and the domain structure consists of FME domains only. If $F^{(i)} \subset F_{M^{(i)}P^{(i)}}$, each FME domain can further split into d_{MP} "subdomains" distinct in some tensor property or in translation properties (antiphase domains.[5,11,12] The degeneracy d_{MP} of a FME domain, i.e. the number of subdomains that can appear within one FME domain, is

$$d_{MP} = [F_{M^{(i)}P^{(i)}} : F^{(i)}] \tag{5}$$

The total number of subdomains equals

$$q = [G : F^{(i)}] = q_{MP}\, d_{MP} \tag{6}$$

Hereafter, unless stated otherwise, we consider non-degenerate FME domains ($F_{M^{(i)}P^{(i)}} = F^{(i)}$).

Crystallographically Equivalent Low Symmetries
Besides the ith domain there are, in general, other domains going with the symmetry $F^{(i)}$. The total number d_F of domains compatible with $F^{(i)}$ is

$$d_F = [G_{F^{(i)}} : F^{(i)}] \tag{7}$$

where $G_{F^{(i)}}$ is the normalizer of $F^{(i)}$ in G.[3,4] All these domains can be constructed by applying to the ith domain representatives of left cosets of the resolution $G_{F^{(i)}}/F^{(i)}$. Every group conjugate (within G) with $F^{(i)}$ has the same chance to appear at the transition. There exist m such groups,

$$m = [G : G_{F^{(i)}}] \tag{8}$$

They can be written in the form $h\, F^{(i)}\, h^{-1}$, where h are the representatives of the left cosets of the resolution $G/G_{F^{(i)}}$.[2–5]

Magnetically Degenerate Ferroelectric Domains
Neglecting magnetic properties a FME crystal can be looked upon as a ferroelectric one. To investigate ferroelectric (FE) domains, differing in the spontaneous polarization P only, we introduce group $F_{P^{(i)}}$ that is a maximal subgroup of G leaving $P^{(i)}$ invariant. The number q_P of FE domains is

$$q_P = [G : F_{P^{(i)}}] \tag{9}$$

Within each of these domains d_P ferromagnetic domains may exist, where

$$d_P = [F_{P^{(i)}} : F_{M^{(i)}P^{(i)}}] \tag{10}$$

The total number q_{MP} of FME domains can be expressed as

$$q_{MP} = q_P\, d_P \tag{11}$$

If G is a grey Shubnikov group (which seems to be a physically reasonable assumption) then also $F_{P^{(i)}}$ is grey; since $F_{M^{(i)}P^{(i)}}$ is an ordinary or a black and white Shubnikov group, it must be a proper subgroup of $F_{P^{(i)}}$ and any FE domain is, therefore, magnetically degenerate.[2]

Electrically Degenerate Magnetic Domains
By analogy, using group $F_{M^{(i)}}$ defined as a maximal subgroup of G which does not change $M^{(i)}$, we get for the number q_M of distinct ferromagnetic (FM) domains

$$q_M = [G : F_{M^{(i)}}] \tag{12}$$

Any of these domains may contain d_M different FE domains, where

$$d_M = [F_{M^{(i)}} : F_{M^{(i)}P^{(i)}}] \tag{13}$$

It holds

$$q_{MP} = q_M\, d_M \tag{14}$$

Specifically, it may happen that $F_{M^{(i)}} = F_{M^{(i)}P^{(i)}}$. Then $d_M = 1$ and FME domains are unequivocally described by the spontaneous magnetization.

Obviously, group $F_{M^{(i)}P^{(i)}}$ is the intersection of $F_{M^{(i)}}$ and $F_{P^{(i)}}$,

$$F_{M^{(i)}P^{(i)}} = F_{M^{(i)}} \cap F_{P^{(i)}} \tag{15}$$

When we are interested only in tensor properties of domains all the above considerations can be performed within Shubnikov point groups. When sublattice formation and antiphase domains are to be examined, Shubnikov space groups must be used. Groups $F_{P^{(i)}}$ for all possible directions of $P^{(i)}$ can be obtained as grey isomorphs of groups tabulated in Ref. 13 (point groups) and in Ref. 14 (space groups). Magnetic point groups $F_{M^{(i)}}$ corresponding to all conceivable orientations of $M^{(i)}$ are available[15] and $F_{M^{(i)}P^{(i)}}$ has also been derived (within point groups) for all ferromagnetoelectric transitions.[16]

DOMAIN PAIRS, SWITCHING

First, we introduce a few auxiliary concepts which will be useful in our discussion on switching and twinning. The *ordered domain pair* (i,j) consists of the first, the ith, and the second, the jth, domain; both domains are derived from a common high symmetry phase and are considered irrespective of their coexistence conditions.[4] Any operation $s_{ij} \in G$ that brings the first domain of the ordered pair (i,j) into its second domain will be called an *F-operation*[2] of this pair. The set of all F-operations of (i,j), referred to as an *F-complex* S_{ij} of (i,j), is equal to the left coset $s_{ij}F^{(i)}$.

An ordered pair (i,j), or its F-complex, can be associated with the idealized switching $i \rightarrow j$ in which the ith domain is changed into the jth domain.† The one-to-one correspondence between ordered pairs, F-complexes and switching processes enables us to examine the basic crystallographical features of FME switching.

Enforced Switching

When the polarization $P^{(i)}$ of the ith FME domain is switched to another direction, say $P^{(k)}$, the question arises, what possible ferromagnetic (FM) switchings may accompany this process? The F-complex associated with the ferroelectric (FE) switching $P^{(i)} \rightarrow P^{(k)}$ is equal to $r_{ik}F_{P^{(i)}}$; it consists of d_P left cosets of $F_{M^{(i)}P^{(i)}}$. Each of these cosets represents a possible FM switching that goes with $P^{(i)} \rightarrow P^{(k)}$.

If $r_{ik}F_{P^{(i)}}$ contains an operation (and, consequently, the whole left coset of $F_{M^{(i)}P^{(i)}}$) belonging to $F_{M^{(i)}}$ then, and only then, the FE switching $P^{(i)} \rightarrow P^{(k)}$ may not be accompanied by a FM switching. If, on the other hand,

$$r_{ik}F_{P^{(i)}} \cap F_{M^{(i)}} = \phi \tag{16}$$

then the FE switching necessarily enforces a FM switching.

Considering specifically the trivial FE switching $P^{(i)} \rightarrow P^{(i)}$ with the F-complex $F_{P^{(i)}}$ we find that left cosets of $F_{P^{(i)}}/F_{M^{(i)}P^{(i)}}$ represent d_P FME domains with common $P = P^{(i)}$ but different M, and describe all possible FM switchings that preserve $P^{(i)}$. Hence a FME switching $i \rightarrow j$ characterized by the F-complex $s_{ij}F_{M^{(i)}P^{(i)}}$ reduces to a FM switching if, and only if,

$$s_{ij}F_{M^{(i)}P^{(i)}} \subset F_{P^{(i)}} \tag{17}$$

The FE switchings accompanying a FM switching $M^{(i)} \rightarrow M^{(l)}$ can be discussed in a similar way. Generally, there are always d_M *FE* switchings that go with $M^{(i)} \rightarrow M^{(l)}$. The condition for enforced FE switching reads

$$t_{il}F_{M^{(i)}} \cap F_{P^{(i)}} = \phi \tag{18}$$

where $t_{il}F_{M^{(i)}}$ is the switching complex of $M^{(i)} \rightarrow M^{(l)}$. A necessary and sufficient condition for the reduction of the FME switching $i \rightarrow j$ to a FE switching is

$$s_{ij}F_{M^{(i)}P^{(i)}} \subset F_{M^{(i)}} \tag{19}$$

If neither (17) nor (19) is fulfilled the FME switching $i \rightarrow j$ consists of simultaneous FM and FE switchings.

The crystallographical coupling between the FME switching and the ferroelastic[2] switching can be analysed by an analogous procedure. The relation between FE and ferroelastic switchings has already been discussed in detail elsewhere.[18]

Crystallographically Non-equivalent Switchings

Between q domains q^2 different ordered pairs and, therefore, q^2 different switchings are conceivable. Crystallographically significant are only crystallographically non-equivalent switchings. They correspond to crystallographically non-equivalent ordered pairs, the number of which equals the number of double cosets in the resolution $F^{(i)}\backslash G/F^{(i)}$.[4] A representative set of non-equivalent switchings can be found as follows: The initial state is repre-

† Strictly speaking, an F-operation s_{ij} relates the initial domain with the final domain of the switching process $i \rightarrow j$ only if the final state has been reached by a single domain process, i.e. via the high symmetry phase. For more realistic switching processes passing through multidomain intermediate states the relation between the initial and the final single domain states may, generally, differ from s_{ij} and might be treated more properly in terms of twinning. For our discussion, however, this difference is irrelevant.

sented by the ith domain. Domains of final states are obtained by applying on the ith domain representatives of all double cosets of the resolution $F^{(i)}\backslash G/F^{(i)}$.

Finally, we note that the left and double coset resolutions needed in the analysis of particular examples can be constructed from tabulated resolutions of ordinary point groups.[17]

COMPATIBILITY OF DOMAINS, TWINS

Coherent Stress-Free Walls, Ferromagnetoelectric Twins

Two domains join along a domain wall. Crystallographically, the most significant are planar walls along which both domains meet coherently without mechanical stresses. In the following we confine ourselves to this type of wall only.

Two FME domains joining along a wall form a FME *twin*. An operation that transforms the structure of the first domain into the structure of the second domain is called a *twinning operation*.†

In order to construct a twin from a given ordered pair (i,j) one has first to make sure that a wall between the ith and the jth domains really exists, then find its possible orientation(s) and, finally, rotate domains to bring them into mutual contact along the found wall.

More detailed examination can be preformed using the difference tensor $D = u^{(j)} - u^{(i)}$ (where $u^{(i)}$, $u^{(j)}$ are matrices of the spontaneous deformation in the ith and jth domain, respectively, expressed in the coordinate system of the para-phase) and its scalar invariants

$$\Lambda_3 = \det D$$

$$\Lambda_2 = \begin{vmatrix} D_{11} & D_{12} \\ D_{12} & D_{22} \end{vmatrix} + \begin{vmatrix} D_{11} & D_{13} \\ D_{13} & D_{33} \end{vmatrix} + \begin{vmatrix} D_{22} & D_{23} \\ D_{23} & D_{33} \end{vmatrix}$$

Three cases can occur:[9, 19]

i) $\Lambda_3 \neq 0$. No wall exists. It is impossible to form a twin from the domain pair.

ii) $\Lambda_3 = 0$, $\Lambda_2 = 0$. The ith and jth domains have identical spontaneous deformations and can, therefore, meet along a wall W_∞ of arbitrary orientation. Twins differing in the wall orientation can be formed with twinning operations identical with the F-operations of (i,j). A sufficient condition is the presence of inversion ($\bar{1}$ or $\bar{1}'$) or of three diads of different orientations in the F-complex S_{ij}. (The term "diad" signifies a two-fold rotatory axis or a two-fold rotatory-inversion axis, i.e. any of the operations 2, 2′, m, m').

iii) $\Lambda_3 = 0$, $\Lambda_2 \neq 0$. Two mutually perpendicular walls exist (in the coordinate system of the paraphase). Two twins can be formed from the ith and jth domain but additional rotations are always necessary so that the twinning operations may not coincide with the F-operations of (i,j). The walls can have either a fixed crystallographical orientation with respect to the para-phase (W_f walls) or a non-crystallographical orientation (S walls). The actual orientation can be determined by direct calculation[19] or by the "0-lattice" theory.[20] It turns out, however, that the presence of a diad in S_{ij} provides a sufficient (but not necessary) condition for the existence of two perpendicular walls, one of which is a crystallographical wall W_f with its normal parallel to that diad. The combination of two perpendicular S walls is possible only if S_{ij} does not contain any operation of the order two.

Wall Charge

The surface magnetic and electric charges of a wall are $\omega_m = (M^{(j)} - M^{(i)})\nu$ and $\omega_{el} = (P^{(j)} - P^{(i)})\nu$, respectively, where ν is the wall normal (oriented from the jth to the ith domain component) and $M^{(i)}$, $P^{(i)}$ and $M^{(j)}, P^{(j)}$ denote magnetization and polarization in the ith and jth domain of the FME twin, respectively.

The W_∞ wall can be either charged or neutral depending on its orientation. In the case of two perpendicular walls generated by a diad the charge of both walls is determined unequivocally (see Table I). We see that if one wall is (magnetically, electrically) charged then the other, perpendicular, wall is (magnetically, electrically) neutral.[21] Usually, the uncharged walls are energetically more favourable than the charged ones.

When looking for all *crystallographically non-equivalent twins* one has first to find the set of crystallographically non-equivalent ordered pairs and then for each pair determine all possible twins.

We emphasize that the energy balance, which determines the real domain structure, can modify the crystallographical conclusions we have obtained. Thus, for example, the energy of the W_∞ wall can depend on the wall orientation; then some orientations of W_∞ wall appear more often than the others. Further, W_f and S walls may differ for various reasons from the stress-free orientation; these

† The term twinning operation is often used also for an F-operation.

TABLE I

Surface magnetic (ω_m) and electric (ω_{el}) charge of walls generated by a diad

Generating diad	W_f wall perpendicular to the diad		S or W_f wall parallel to the diad	
	ω_m	ω_{el}	ω_m	ω_{el}
2	0	0	(a)	(b)
2′	$-2M^{(i)}\nu$	0	0	(b)
m	0	$-2P^{(i)}\nu$	(a)	0
m'	$-2M^{(i)}\nu$	$-2P^{(i)}\nu$	0	0

(a) Magnetically charged (for the S wall), $-2M^{(i)}\mu$ for the W_f wall), (b) electrically charged (for the S wall), $-2P^{(i)}\mu$ (for the W_f wall). $M^{(i)}$ and $P^{(i)}$ is the magnetization and polarization of the first domain of the twin, ν and μ are unit wall normals parallel and perpendicular to the diad, respectively.

deviations can be more pronounced for walls separating FM domains, since in this case the spontaneous deformation is considerably smaller than for domains differing in polarization.

EXAMPLE

Ferromagnetoelectric Cubic–Orthorhombic Transition in Ni–I boracite[3,12,22–24]

The symmetry of the cubic para-phase is $G=\bar{4}3m1'$ (order 48). In the orthorhombic ferromagnetoelectric phase of symmetry $F=mm'2'$ the magnetization $M^{(1)}(M_x, M_x, 0)$ and the polarization $P^{(1)}(0,0,P_z)$ appear simultaneously. In Refs. 13 and 15 we find $F_{M^{(1)}}=m_{xy}m'_{\bar{x}z}2'_z$ (order 4) and $F_{P^{(1)}}=m_{xy}m_{\bar{x}y}2_z1'$ (order 8), respectively (see caption to Table II for notation). From (15) it follows that $F_{M^{(1)}P^{(1)}}=m_{xy}m'_{\bar{x}y}2'_z=F_{M^{(1)}}$. Further, within the point groups, $F_{M^{(1)}P^{(1)}}=F^{(1)}$, and $G_{F^{(1)}}=m_{xy}m_{\bar{x}y}2_z1'$ (see below).

The following results can immediately be obtained: There are $q_p=48:8=6$ FE domains each of which can contain two FM domains ($d_P=8:4=2$), and $q_M=48:4=12$ non-degenerate ($d_M=4:4=1$) FM domains which represent faithfully $q_{MP}=48:4=12$ FME domains that are (within the point groups) non-degenerate ($d_{MP}=1$). The number of crystallographically equivalent orthorhombic groups is $m=48:8=6$; each of these groups is compatible with $d_F=8:4=2$ FME domains. From (16) it follows that any FE switching enforces a FM switching.

Resolutions $G/F_{M^{(1)}P^{(1)}}=G/F_{M^{(1)}}$, $F_{M^{(1)}P^{(1)}}\backslash G/F_{M^{(1)}P^{(1)}}$ and $G/F_{P^{(1)}}$ can easily be deduced[17] and are given in Tables II and III. As the union of all double cosets consisting of one left coset constitute the normalizer[4] we see from Table II that $G_{F^{(1)}}=m_{xy}m_{\bar{x}y}2_z1'$, Table II provide us, therefore, also with the resolution $G/G_{F^{(1)}}$.

The jth row of Table II contains the F-complex S_{ij} of the FME (and also FM) domain pair $(1,j)$. Similarly, the kth row of Table III represents the F-complex of the FE domain pair $(1,k)$. Moreover, any of the eight operations, say h, of the kth left coset produces from $m_{xy}m'_{\bar{x}y}2'_z$ the same group $F^{(k)}=hm_{xy}m'_{\bar{x}y}2'_zh^{-1}$ crystallographically equivalent with $m_{xy}m'_{\bar{x}y}2'_z$.

From Table II it follows that 122 non-trivial domain pairs can be divided into four classes of crystallographically equivalent pairs. The representative non-equivalent domain pairs consist of the first domain and domains with $j=2, 3, 4, 5$. The analysis of switching and twinning can, therefore, be confined to these pairs only.

Pair (1, 2). The switching reduces to a 180° FM switching. FM domains can meet along a wall of an arbitrary orientation. The wall is neutral only if it is parallel to $M^{(1)}$; in all other cases it is magnetically charged.

TABLE II

Resolutions of $G=\bar{4}3m1'$ into left and double cosets of $F_{M^{(1)}P^{(1)}}=m_{xy}m'_{\bar{x}y}2'_z$

j	left cosets $s_{ij}F_{M^{(1)}P^{(1)}}$				$\frac{M^{(j)}}{\lvert M^{(j)}\rvert}$	$\frac{P^{(j)}}{\lvert P^{(j)}\rvert}$
1	1	m_{xy}	$2'_z$	$m'_{\bar{x}y}$	[110]	[001]
2	1′	m'_{xy}	2_z	$m_{\bar{x}z}$	[$\bar{1}\bar{1}$0]	[001]
3	2_x	$\bar{4}^3_z$	$2'_y$	$4'_z$	[1$\bar{1}$0]	[00$\bar{1}$]
6	$2'_x$	$\bar{4}^{3\prime}_z$	2_y	$\bar{4}_z$	[$\bar{1}$10]	[00$\bar{1}$]
4	m_{xz}	3_x	$\bar{4}'_y$	$3'_z$	[0$\bar{1}$1]	[$\bar{1}$00]
7	$m_{\bar{x}z}$	3_y	$\bar{4}^{3\prime}_y$	$3'$	[0$\bar{1}\bar{1}$]	[100]
8	m_{yz}	3^2_y	$\bar{4}^{3\prime}_x$	$3^{2\prime}_z$	[$\bar{1}$01]	[0$\bar{1}$0]
9	$m'_{\bar{y}z}$	3^2_x	$\bar{4}'_x$	$3^{2\prime}$	[$\bar{1}$0$\bar{1}$]	[010]
5	m'_{xz}	$3'_x$	$\bar{4}_y$	3_z	[01$\bar{1}$]	[$\bar{1}$00]
10	$m'_{\bar{x}z}$	$3'_y$	$\bar{4}^3_y$	3	[011]	[100]
11	m'_{yz}	$3^{2\prime}_y$	$\bar{4}^3_x$	3^2_z	[10$\bar{1}$]	[0$\bar{1}$0]
12	$m'_{\bar{y}z}$	$3^{2\prime}_x$	$\bar{4}_x$	3^2	[101]	[010]

Notation: Subscripts indicate orientation of axes in the standard coordinate system of the cubic syngony; at diads zero components are omitted, at triads only positive components are given (e.g. 3_x, 3 mean rotations along the 3-fold axes in the directions [1$\bar{1}\bar{1}$] and [111], respectively). $M^{(j)}/|M^{(j)}|$ and $P^{(j)}/|P^{(j)}|$ denote directions of the spontaneous magnetization and polarization, respectively, in the jth ferromagnetoelectric domain. Each solid frame assembles all left cosets constituting one double coset.

TABLE III

Resolutions of $G = \bar{4}3m1'$ into left cosets of $F_{P(1)} = m_{xy}m_{\bar{x}y}2_z1'$

k	Left cosets $r_{1k}F_{P(1)}$								$\frac{P^{(k)}}{\lvert P^{(k)}\rvert}$
1	1	2_z	m_{xy}	$m_{\bar{x}y}$	$1'$	$2'_z$	m'_{xy}	$m'_{\bar{x}y}$	$[001]$
2	2_x	2_y	$\bar{4}^3_z$	$\bar{4}_z$	$2'_x$	$2'_y$	$\bar{4}^{3\prime}_z$	$\bar{4}'_z$	$[00\bar{1}]$
3	m_{xz}	$\bar{4}_y$	3_x	3_z	m'_{xz}	$\bar{4}'_y$	$3'_x$	$3'_z$	$[\bar{1}00]$
4	$m_{\bar{x}z}$	$\bar{4}^3_y$	3_y	3	$m'_{\bar{x}z}$	$\bar{4}^{3\prime}_y$	$3'_y$	$3'$	$[100]$
5	m_{yz}	$\bar{4}^3_x$	3^2_y	3^2_z	m'_{yz}	$\bar{4}^{3\prime}_x$	$3^{2\prime}_y$	$3^{2\prime}_z$	$[0\bar{1}0]$
6	$m_{\bar{y}z}$	$\bar{4}_x$	3^2_x	3^2	$m'_{\bar{y}z}$	$\bar{4}'_x$	$3^{2\prime}_x$	$3^{2\prime}$	$[010]$

The same notation as in Table II. $P^{(k)}/|P^{(k)}|$ denotes direction of the spontaneous polarization in the kth ferroelectric domain generated from the first domain by the left coset $r_{1k}F_{P(1)}$.

Pair (*1*, *3*). Simultaneous 180° FE and 90° FM switching (verified experimentally[22]). Two diads in the left coset generate two W_f walls. The wall (100) is neutral whereas the wall (010) is magnetically charged. For non-magnetic boracites these results conform to those obtained previously;[21, 24] walls were observed.[22, 23]

Pair (*1*, *4*). Simultaneous 90° FE and 120° FM switching. The only diad m_{xz} of the left coset generates electrically charged domain wall (101) and a perpendicular non-crystallographical wall ($hk\bar{h}$) which is charged magnetically. These orientations are in agreement with earlier calculations[19, 21] for non-magnetic boracites; the "0-lattice" theory gave the orientations (101) and (010). The wall (101) was detected experimentally;[22, 23] the occurrence of the perpendicular wall is not clear.[19]

Pair (*1*, *5*). Simultaneous 90° FE and 60° FM switching. The wall (101) is both magnetically and electrically charged and the S wall ($hk\bar{h}$) is neutral.

We may notice that the actual orientations of $M^{(j)}$ and $P^{(j)}$ in different FME domains, given in Tables II and III, were not needed in the analysis.

The above discussion has been performed within the point groups. The magnetic space group of the ferromagnetoelectric phase has not been determined yet but non-magnetic space groups are known:[25] $G = F\bar{4}3c$, $F = Pca2_1$. We find[14] $F_{P(1)} = Iba2$. Groups $F_{P(1)}$ and F differ in the Bravais lattices only, the first being orthorhombic body-centred and the second orthorhombic primitive. Hence $[F_{P(1)}:F] = 2$, and each ferroelectric domain can further split into two distinct antiphase domains.[12]

ACKNOWLEDGEMENTS

The authors would like to thank Drs. J. Fousek and M. Polcarová for valuable discussions.

REFERENCES

1. A. Cimino and G. S. Parry, *Il Nuovo Cimento* **19**, 971 (1961).
2. K. Aizu, *Phys. Rev.* **B2**, 754 (1970).
3. E. Ascher, "Symetries et changements de phase", in *Les Transitions de Phase*, 13e Cours de Perfectionnement de l'Association Vaudoise des Chercheurs en Physique, 1971, p. 133.
4. V. Janovec, *Czech. J. Phys.* **B22**, 974 (1972).
5. V. L. Indenbom, in *Fizika kristallov s defektami* (Akademiya nauk SSSR, Telavi, 1966), Vol. 1, pp. 55–64.
6. J. K. Mackenzie and J. S. Bowles, *Acta Met.* **2**, 138 (1954).
7. N. D. H. Ross and A. G. Crocker, *Scripta Met.* **3**, 37 (1969).
8. A. L. Rojtburd, in *Nesovershenstva kristallicheskogo stroeniya i martensitnye prevrashcheniya* (Nauka, Moskva, 1972), pp. 7–33 and references cited therein.
9. F. Kroupa and V. Janovec, *Proc. of the 3rd Conference on High Strength Martensitic Steels* (Havířov, 1972), p. 133.
10. M. Hall, *The Theory of Groups* (The Macmillan Co., New York, 1959), Ch. 1.
11. V. L. Indenbom, *Kristallografiya* **5**, 115 (1960).
12. E. Ascher, *J. Phys. Soc. Japan* **28**, Supplement, 7 (1970); *The Interaction between Magnetization and Polarization. Phenomenological Symmetry Considerations in Boracites* (Report of the Battelle Institute, Geneva, 1969).
13. I. S. Zheludev and L. A. Shuvalov, *Kristallografiya* **1**, 681 (1956).
14. A. S. Sonin and I. S. Zheludev, *Kristallografiya* **4**, 487 (1959).
15. L. A. Shuvalov, *Kristallografiya* **4**, 399 (1959).
16. L. A. Shuvalov, *Kristallograficheskie metody v fizike segnetoelektricheskikh yavlenii*, Thesis, Moskva (1971).
17. V. Janovec and E. Dvořáková, *Resolutions of Crystallographical Point Groups into Left and Double Cosets* (Report of the Institute of Physics, Prague, 1973).
18. K. Aizu, *J. Phys. Soc. Japan* **32**, 1287 (1972).
19. J. Fousek and V. Janovec, *J. Appl. Phys.* **40**, 135 (1969).
20. W. Bollmann, *Crystal Defects and Crystalline Interfaces* (Springer-Verlag, Berlin–Heidelberg, 1970).
21. J. Fousek, *Czech. J. Phys.* **B21**, 955 (1971).
22. H. Schmid, in *Rost Kristallov* (Nauka, Moskva, 1967), Vol. 7, pp. 32–65.
23. A. Zimmermann, W. Bollmann and H. Schmid, *Phys. Stat. Sol.* (*a*) **3**, 707 (1970).
24. O. V. Kovalev, *Fiz. Tverd. Tela* **14**, 307 (1972).
25. T. Ito, N. Morimoto and R. Sadanaga, *Acta Cryst.* **4**, 310 (1951).

SYMPOSIUM ON MAGNETOELECTRIC
INTERACTION PHENOMENA IN CRYSTALS

Battelle Seattle Research Center

May 21–24, 1973

Bottom Row (L–R)
Dr. D. B. Litvin
Dr. L. M. Holmes
Dr. R. M. Hornreich
Prof. E. J. Post
Dr. D. E. Cox
Prof. R. L. White
Dr. V. E. Wood

Middle Row (L–R)
Prof. W. Opechowski
Dr. G. T. Rado
Dr. P. Chandrasekhar
Dr. T. P. Srinivasan
Prof. T. Ogawa
Dr. J. M. Trooster
Dr. E. Ascher
Dr. P. E. Bierstedt
Dr. S. L. Hou
Prof. M. Mercier

Top Row (L–R)
Dr. V. J. Folen
Dr. K. G. Srivastava
Dr. R. M. White
Prof. S. Shtrikman
Dr. T. H. O'Dell
Dr. A. Kiel
Dr. H. Schmid
Prof. R. Englman
Dr. J. P. Rivera
Prof. S. Iida

Subject Index

Annealing, magnetoelectric 10, 13, 211–13
Antiferroelastics, magnetoelectric 127
Antiferroelectric
antiferromagnetic 122, 124, 143
diamagnetic 122, 124, 143
domain switching 127
ferromagnetic 122, 141
orthoelectric 122, 124, 143
paramagnetic 122, 124, 143
Antiferromagnetic
antiferroelectric 122, 124, 143
crystals, magnetic transitions 201–10
domains, magnetoelectric dynamic switching 151–3
structure in critical region 98
ferroelectric 88, 122, 124, 136–7
orthoelectric 142
resonance, electric field controlled 188
Antiferromagnetics, magnetic structural data 116
Antiphase domains 220
Applications for magnetoelectric materials 87, 181–93
Axio-polar vector 69

Bias, internal 131
Birefringence, non-reciprocal 194
Bismuth orthoferrites 184
Boracites 39, 130–3, 135, 138, 155–60, 169–78, 219–20
diamagnetic 138
ferromagnetoelectric 39, 135
$Co_3B_7O_{13}Cl$, $(ME)_H$-effect 131–3, 135
$Fe_3B_7O_{13}Br$, Mössbauer effect 155–60
$Ni_3B_7O_{13}Cl$, $(ME)_H$-effect 169–78
$Ni_3B_7O_{13}I$, switching and twinning 219–20
growth sectors 130
paramagnetic 138
Brillouin scattering 123
Bronzes 138
Bubble domains 85
"Butterfly" loop, $Ni_3B_7O_{13}Cl$ 176

"Callen decoupling" approximation 9, 23
Chalcogenides, antiferromagnetic rare earth 119
Chambersite (Mn–Cl boracite) 128
Chromium oxide
annealing 82
critical behaviour 201–3
domain wall width 86
electric field induced Faraday effect 84
magnetoelectic susceptibility 8–9, 17–29
in polycrystalline powder 212
memory effect 82–3
spin flop 202
Classification of magnetoelectric materials 121–45
Coercive field
asymmetries 131
modulation 121, 125–6
elastic 125
electric 125
magnetic 125
Ni–Cl/boracite 176
Congolite (Fe–Cl boracite) 128
Constitutive relations 47, 54
Compensation temperature, $Tb_3Fe_5O_{12}$ 100
Corundem type compounds, magnetic structural data 111, 114
Coupling constants 27–8
Critical behaviour
determination 3
magnetoelectric studies of phase transitions 202
magnetoelectric susceptibility 98, 201–5
Cr_2O_3 87, 95
$DyAlO_3$ 203
$DyPO_4$ 11–12, 203
$GdVO_4$ 87, 95
sublattice magnetization 98
Crystalline electric field 6
Crystalline potential energy 10

Deflector, electric field modulated stripe domain 189
Damagnetizing field
corrections 106
YIG 107
Density of stored free enthalpy 69, 71, 122–4, 126
Detectors
pyroelectric 183
IR optical rectification 188
Devices
classification of magnetoelectric 183
critical parameters for applications 184
non-linear optical 190
Dipolar interactions 10
Display 185
Domain switching 121, 127, 217
Cr_2O_3 83, 151–3
Domain walls
electrically charged 220
"magnetically charged" 220
motion in garnets 100
width in Cr_2O_3 86
Domains
antiferroelectric 127
antiferromagnetic 4, 127
antiferromagnetic bubbles 85
antiphase 220
ferromagnetoelectric 215–220
pinning in Cr_2O_3, $Ni_3B_7O_{13}I$ 131
D-term mechanism 6
Dynamic measurement techniques, magnetoelectric 102–3
Dzyaloshinskii interaction
term 17, 20, 24
electric field induced 87
Dzyaloshinskii-Moriya interaction 200

Elastic
compliance 123
constants of third order 123
Elasticity, density of stored free enthalpy 123
Electret effect, $Ni_3B_7O_{13}Cl$ 173
Electric field control, Precise 187
Electric susceptibility 123
magnetic field dependence, YIG 85
Electro-magneto-striction, Mn–Zn ferrite 161–7
Electron paramagnetic resonance (EPR) 41–2
Electro-optic
device, magnetically switched 189
effect 123
modulators 123
Electrostriction, $BaTiO_3$ 161
Energy absorption at variable electric field 66
Enthalpy
density of stored free, for domain switching 126
permitted terms of density of stored free 122–4
Ericaite (Fe–Cl boracite) 128
Exchange
antisymmetric 9
axial 10
biquadratic 9
interaction contributions to paramagnetoelectric effect 38
interaction, isotropic 20
mechanism 8

Formula Index

$BaAl_2O_4$ 130
$BaCoF_4$ 117, 119, 137
$BaFeF_4$ 137
$BaMnF_4$ 137
$BaNiF_4$ 137
$BaNd_2Fe_2Nb_8O_{30}$ 138
$BaTiO_3$ 161
$BaTiO_3$–$Sr_{0.3}La_{0.7}MnO_3$ 57
$Ba_2Bi_4Ti_3O_{18}$ 133
$Ba_2Sm_4Fe_3Nb_7O_{30}$ 138
$BiCuO_3$ 141
$BiFeO_3$ 112, 116, 130, 133–4, 136, 140, 184, 187
$Bi_{0.9}Nd_{0.1}FeO_3$ 184–5
$BiMnO_3$ 141
$BiMn_2O_5$ 137
$BiNb(O_5F)$ 139
$Bi_xPr_{1-x}FeO_3$ 136
$Bi_2Ta(O_5F)$ 139
Bi_2WO_3(Russelite) 139
$Bi_3(TiNb)O_9$ 139
$Bi_4Ti_3O_{12}$ 140
$Bi_5(Ti_3Fe)O_{15}$ 140
$Bi_6(Ti_3Fe_2)O_{18}$ 140
$Bi_9(Ti_3Fe_5)O_{27}$ 133, 140

$Ca(Fe_{1-x}Cr_x)_2O_4$ 116–17
$CaMn_2O_4$ 137
$Cd(Fe_{1/2}Nb_{1/2})O_3$ 143
$Cd_2Nb_2O_6S$ 130
$Cd_2Nb_2O_7$ 130
$CoAl_2O_4$ 113
$Co_3B_7O_{13}Br$ 135
$Co_3B_7O_{13}Cl$ 115, 130–2, 135, 159
$Co_3B_7O_{13}I$ 135
$CoCo_2O_4$ 113
$CoCs_3Cl_5$ 116
CoF_2 43, 127
$CoGeO_3$ 87–8, 91–3, 116
$CoMn_2O_4$ 114
$CrMn_2O_4$ 114
Cr_2O_3 4–14, 17–29, 82–3, 87, 96, 114, 127, 131, 142, 151–3, 201–3, 212
Cr_2TeO_6 112
$CrTiNdO_5$ 87–8, 93–4, 116
$CrUO_4$ 116
Cr_2WO_6 112
$Cu_3B_7O_{13}Br$ 135
$Cu_3B_7O_{13}Cl$ 135
$CuCl_2 \cdot 2H_2O$ 19
$CuCr_2O_4$ 114
$CuFeS_2$ 116
$Cu(HCOO)_2 \cdot 4H_2O$ 141
$Cu_{0.5}In_{0.5}Cr_2S_4$ 119

$DyAlO_3$ 16, 112, 142, 149, 201, 203–7
$DyCoO_3$ 112
$DyFeO_3$ 82, 86, 119
DyIG 87–98
DyOOH 87–8, 94, 116
$DyPO_4$ 9–12, 16, 27, 115, 142, 147, 149–50, 201, 203–7
$Dy_3Fe_5O_{12}$ 141

$ErMnO_3$ 137
ErOOH 87–8, 94, 116

$FeBO_3$ 186
$Fe_3B_7O_{13}Br$ 135
$Fe_3B_7O_{13}Cl$ 135
$Fe_3B_7O_{13}I$ 135
$FeCl_2 \cdot 2H_2O$ 28
$(Fe_{0.5}Cr_{0.5})_2WO_6$ 112
FeF_2 19
FeF_3 186
$FeGaO_3$ 115
$FeGeN_2$ 115
$FeGeO_3$ 87–8, 91–3, 116
α-FeOOH 116
FeS 116
$FeSb_2O_4$ 87–8, 116, 136
α-Fe_2O_3 (hematite) 42–3, 85, 146
Fe_2SiO_4 114
Fe_2TeO_6 87–90, 112, 142
$Fe_3B_7O_{13}Br$ 155–160
$Fe_3B_7O_{13}Cl$ 128, 159

$GaFeO_3$ 43, 5, 13
$Ga_{0.85}Fe_{0.15}O_3$ 188
$Ga_{0.92}Fe_{1.08}O_3$ 5–6
$Ga_{2-x}Fe_xO_3$ 42–3, 135
$GdAlO_3$ 112, 142
$GdAsO_4$ 147
$GdCrO_3$ 119
$GdFeO_3$ 112, 119
$Gd_3Fe_5O_{12}$ 99–110, 141
GdIG 99–110
$Gd_2(MoO_4)_3$ 125, 127, 136
$GdVO_4$ 26, 87–8, 94–5, 98, 115, 142, 147
$GeMnO_3$ 142

$HoCoO_3$ 112
$HoMnO_3$ 115, 137
$HoPO_4$ 115, 147–50

La(Pr)$Bi_4(Ti_3Fe)O_{15}$ 140
$LiCoPO_4$ 16, 115, 142
$LiCuCl_3 \cdot 2H_2O$ 116
$Li_{0.5}Fe_{2.5}O_4$ 113, 116
$LiFePO_4$ 115, 142
$Li(Fe_{1/2}Ta_{1/2})O_2F$ 130, 139
$LiMnPO_4$ 9, 115, 142
$LiNiPO_4$ 16, 115, 142
$Li_4B_7O_{12}X$ 138
$LuMnO_3$ 137

$Mg_3B_7O_{13}Cl$ 130, 143
$MnAl_2O_4$ 113
$Mn_3B_7O_{13}Br$ 135
$Mn_3B_7O_{13}Cl$ 128, 135
$Mn_3B_7O_{13}I$ 135
MnF_2 19
$MnGa_2O_4$ 113
$MnGeN_2$ 115
$MnGeO_3$ 87–8, 91–93, 116, 201, 207–10
$MnNb_2O_6$ 116
$MnNb_3S_6$ 116
$MnTa_2O_6$ 119
$MnTiO_3$ 114
$Mn_{0.6}Zn_{0.3}Fe^{2+}_{0.1}Fe^{3+}_2O_4$ 162

β-$NaFeO_2$ 115, 135
Na_2NiFeF_7 115
$Nb_2Co_4O_9$ 87–8, 91, 114, 142
$Nb_2Mn_4O_9$ 87, 88, 90, 142
$Nd(BrO_3)_3 \cdot 9H_2O$ 39, 142
$Ni_3B_7O_{13}Br$ 135
$Ni_3B_7O_{13}Cl$ 115, 135, 169–78
$Ni_3B_7O_{13}I$ 115, 127, 130–1, 135, 177, 219–20
$NiCr_2O_4$ 113–14
$NiSO_4 \cdot 6H_2O$ 31–40, 143
$NiSeO_4 \cdot 6H_2O$ 39, 143

$Pb(Cd_{1/3}Nb_{2/3})O_3$ 138
$Pb(Co_{1/3}Nb_{2/3})O_3$ 138
$Pb(Co_{1/3}Ta_{2/3})O_3$ 138
$Pb(Co_{1/2}W_{1/2})O_3$ 139, 143
$Pb(Fe_{1/2}Mn_{1/4}W_{1/4})O_3$ 136
$Pb(Fe_{1/2}Nb_{1/2})O_3$ 113, 128, 133–4, 136, 138–9
$Pb(Fe_{1/2}Ta_{1/2})O_3$ 136, 138, 189, 192
$Pb(Fe_{2/3}W_{1/3})O_3$ 139, 192
$(1-x)Pb(Fe_{2/3}W_{1/3})O_3$–$xPb(Mg_{1/2}W_{1/2})O_3$ 130, 136
$Pb(Lu_{1/2}Nb_{1/2})O_3$ 143
$Pb(Mg_{1/3}Ta_{2/3})O_3$ 138
$Pb(Mn_{1/2}Nb_{1/2})O_3$ 113, 133, 136, 143
$Pb(Mn^{2+}_{1/2}Re^{6+}_{1/2})O_3$ 141
$Pb(Mn^{3+}_{1/2}Re^{5+}_{1/2})O_3$ 141
$Pb(Mn_{2/3}W_{1/3})O_3$ 143
$Pb(Ni_{1/3}Nb_{2/3})O_3$ 138
$Pb(Ni_{1/3}Ta_{2/3})O_3$ 138
$PbTa_2O_6$ 138